Excel

YEAR 10

Mathematics Revision & Exam Workbook

ESSENTIAL skills

Get the Results You Want!

AS Kalra &
Lyn Baker

Updated in 2024 for the NSW Curriculum and Australian Curriculum Version 9.0 changes

ISBN 978 1 74125 765 6

Pascal Press
PO Box 250
Glebe NSW 2037
(02) 9198 1748
www.pascalpress.com.au

Publisher: Vivienne Joannou
Project editor: May McCool
Edited by Valerie McCool, May McCool and Rosemary Peers
Typeset by Typecellars Pty Ltd and lj Design (Julianne Billington)
Answers checked by Valerie McCool, Peter Little and Melinda Amaral
Cover by DiZign Pty Ltd
Printed by Vivar Printing/Green Giant Press

Dedication by AS Kalra

This book is dedicated to the new generation of young Australians in whose hands lies the future of our nation and who by their hard work, acquired knowledge and intelligence will take Australia successfully through the 21st century.

This book is also in the loving, living and lasting memory of my dear mum, dad and uncle, who will remain a great source of inspiration and encouragement to me for times to come.

Acknowledgements
I would especially like to express my thanks and appreciation to my dear wife and my dear son, who have helped me to find the time to write this book. Without their help and support, achievement of all this work would not have been possible.
AS Kalra

Contents

CHAPTER 5 – Right-angled triangles and trigonometry

CHAPTER 6 – Linear relationships

CHAPTER 7 – Non-linear relationships

CHAPTER 8 – Area, surface area and volume

CHAPTER 9 – Further surface area and volume

CHAPTER 10 – Properties of geometrical figures

CHAPTER 11 – Data analysis

CHAPTER 12 – Probability

EXAM PAPERS

ANSWERS

Introduction

There are two workbooks in this series for the Year 10 Australian Curriculum Mathematics course:

- ***Excel Essential Skills Year 10 Mathematics Revision & Exam Workbook*** (this book) and
- ***Excel Essential Skills Year 10 Advanced Mathematics Revision & Exam Workbook***.

This book should be completed before the Advanced book. It is the core book, written specifically for the Year 10 Australian Curriculum (Mathematics) and the fifth in a series of eight Revision & Exam Workbooks for Years 7 to 10. Each book in the series has been specifically designed to help students **revise** their work so that they can prepare for success in their **tests** during the school year and in their **half-yearly** and **yearly** exams.

The emphasis in this book is for students to master and consolidate the core skills and concepts of the course through extensive practice. This will ensure that students have a solid foundation on which to build towards both the Mathematics and Advanced Mathematics courses in senior years.

- This book is a **workbook**. Students write in the book, ensuring that they have all their questions and working in the same place. This is invaluable when revising for exams—no lost notes or missing pages!
- Each page is a **self-contained, carefully graded unit of work**; this means students can plan their revision effectively by completing set pages of work for each section.
- **Topics are covered in depth**, providing students with practice to enable them to focus on areas of weakness or areas for extension.
- A **Topic Test** is provided at the end of each chapter. These tests are designed to help students test their knowledge of each syllabus topic. Practising tests similar to those they will sit at school will build students' confidence and help them perform well in their actual tests.
- **Two Exam Papers** have been included to test students on the complete Year 10 Mathematics course, helping students prepare for their **half-yearly** and **yearly exams**.
- A **marking scheme** is included in both the Topic Tests and Exam Papers to give students an idea of their progress.
- A **Topic Test and Exam Paper Feedback Chart**, found on the inside back cover, enables students to record their scores in all tests and exams.
- **Answers to all questions** are provided at the back of the book.

A note from the author

Mathematics is best learned if you have pen and paper with you and do every question in writing. Do not just read through the book—work through it and answer the questions, writing down all working. If this approach is coupled with a menu of motivation, realistic goal-setting and a positive attitude, it will lead to better marks in the examinations.

My best wishes are with you; I believe this book will help you achieve the best possible results. Good luck in your studies!

AS Kalra, MA, MEd, BSc, BEd

Chapter 1
Financial mathematics

UNIT 1: Simple interest (1)

Question 1 Use the formula $I = Prn$ to find I if:

a $P = \$2000, r = 0.08, n = 3$

b $P = \$7000, r = 0.05, n = 6$

c $P = \$18\,000, r = 0.07, n = 4$

d $P = \$65\,000, r = 0.075, n = 5$

Question 2 Find the simple interest on an investment of:

a $5000 for 3 years at 6% p.a.

b $12 000 at 8% p.a. for 4 years

c $9000 for 7 years at 5% p.a.

d $30 000 for 10 years at 7% p.a.

e $6500 at 6.5% p.a. for 2 years

f $27 500 at 9% p.a. for 6 months

g $12 500 at 6% p.a. for 18 months

h $11 000 for 4 years at $7\frac{1}{2}\%$ p.a.

Question 3 $7500 is borrowed at 6% p.a. simple interest for 5 years. Find:

a the interest paid

b total amount to be repaid

Financial mathematics

UNIT 2: Simple interest (2)

QUESTION **1** Find the simple interest on:

a $15 000 at 6% p.a. for 4 years

b $8000 at 7.5% p.a. for 6 years

QUESTION **2** Find the amount that needs to be invested to earn an amount of simple interest of:

a $2000 if invested at 4% p.a. for 2 years

b $4375 invested at 7% p.a. for 5 years

QUESTION **3** Find the number of years that the amount must have been invested if:

a $7000 earned $560 interest at 8% p.a.

b $13 000 earned $3510 at 9% p.a.

QUESTION **4** Find the rate of simple interest if:

a $8000 earns $1200 interest in 3 years

b $15 000 earned $4800 interest in 4 years

QUESTION **5** $25 000 is invested and earns $12 000 simple interest. Find the:

a number of years if the interest rate is 8% p.a.

b interest rate if invested for 10 years

QUESTION **6** An amount of money was borrowed over 7 years at 5.5% p.a. simple interest. The total interest paid was $5390. Find the:

a amount of money borrowed

b total amount repaid on the loan

Financial mathematics

UNIT 3: Application of simple interest

Question 1 Maddie decides to buy a computer marked at $4500. She pays 20% deposit and the balance over 2 years with simple interest charged at 14% p.a. on the balance.

a Find the deposit paid.

b Calculate the balance owing.

c Calculate the interest paid.

d Find the total amount to be repaid.

e Find the monthly repayment.

Question 2 Suzy borrows $4500 and agrees to repay it in equal monthly instalments over 3 years. Simple interest at 7.2% p.a. is charged on the loan. Find the:

a total amount of interest paid

b amount of each instalment

Question 3 The cash price of a car is $32 000. Tyson buys the car on terms. He pays 15% deposit and agrees to pay $680 every month for 4 years. Find the:

a deposit

b amount borrowed

c total paid for the car

d total amount of interest paid

e yearly rate of simple interest

Question 4 Monique buys a house for $750 000, pays a deposit of $150 000 and then pays off the balance at $4100 per month for 25 years. Find the:

a total cost of the house

b yearly interest paid

Financial mathematics

UNIT 4: Interest rates

QUESTION 1 For an interest rate of 8% p.a. find the:

a monthly rate

b quarterly rate

c six-monthly rate

d four-monthly rate

QUESTION 2 Find the monthly interest rate if the annual rate is:

a 6.5%

b 10%

QUESTION 3 Find the quarterly interest rate if the annual rate is:

a 9%

b 6%

QUESTION 4 Find the number of:

a months in 6 years

b quarters in 4 years

c six-monthly periods in 8 years

d four-monthly periods in 2 years

QUESTION 5 Interest on an investment is to be paid quarterly. If the principal is invested for 5 years and the annual interest rate is 12%, find:

a the number of quarters

b the quarterly interest rate

QUESTION 6 Find the annual interest rate:

a 3.5% per quarter

b 0.8% per month

c 7.5% per six-monthly period

d 0.035% per day

Financial mathematics

UNIT 5: Compound interest by repeated use of simple interest

QUESTION **1** This table compares the interest earned on $1000 at 10% p.a. simple interest with the interest earned on $1000 at 10% p.a. compound interest compounded annually. (Amounts are given to the nearest dollar.)

Time (years)	1	2	3	4	5	6	7	8	9
Simple interest	$100	$200	$300	$400	$500	$600	$700	$800	$900
Compound interest	$100	$210	$331	$464	$611	$772	$949	$1144	$1358

a How much more interest is earned at the compound interest rate than at the simple interest rate over a period of:

i 2 years? ______ **ii** 5 years? ______ **iii** 7 years? ______

b The interest earned after 1 year by either simple interest or compound interest is the same. Why?

______

______

QUESTION **2** Find the total compound interest earned in each case by repeated use of the simple interest formula. (Interest is compounded yearly.)

a $7200 is invested for 2 years at 8% p.a.

b $4500 is invested for 2 years at 7% p.a.

c $14 000 is invested for 3 years at 6% p.a.

d $6800 is invested for 3 years at 6.5% p.a.

e $9300 is invested for 4 years at 10% p.a.

Financial mathematics

UNIT 6: Compound interest

QUESTION **1** Use the future value formula $FV = PV(1 + r)^n$ to find the total amount returned when:

a $9000 is invested for 5 years at 12% p.a. compounded yearly

b $25 600 is invested for 4 years at 9% p.a. compounded six-monthly

c $120 000 is invested for 6 years at 8.5% p.a., compounded monthly

d $48 000 is invested for 3 years at 12% p.a., compounded quarterly

e $72 500 is invested for 5 years at 18% p.a., compounded monthly

QUESTION **2** Find the compound interest earned on the following investments.

a $18 000 at 6% for 3 years, compounded annually

b $12 000 at 12% p.a. for 2 years, compounded six-monthly

c $45 000 at 8% p.a. for 2 years, compounded quarterly

d $64 000 for 3 years at 18% p.a. compounded monthly

e $85 000 for 10 years at 4% p.a. interest compounded monthly

f $8600 for 6 years at 8.5% p.a. compounded daily

Financial mathematics

Unit 7: Applying the compound interest formula

Question **1** What sum of money would need to be invested to be worth \$10 000 at the end of 5 years at the given interest rate?

a 7% p.a. compounded yearly

b 6% p.a. compounded quarterly

c 5% p.a. compounded six-monthly

d 12% p.a. compounded monthly

e 13% p.a. compounded weekly (1 year = 52 weeks)

f 18% p.a. compounded annually

Question **2** Find the amount of compound interest earned on \$75 000 at the end of the stated period if invested at the given rate.

a At 7% p.a. compounded annually for 4 years

b At 10% p.a. compounded quarterly for 3 years

c At 8% p.a. compounded monthly for 2 years

d At 12% p.a. compounded six-monthly for 3 years

e At 9% p.a. compounded four-monthly for 5 years

Financial mathematics

UNIT 8: Compound interest tables

QUESTION 1 The table shows the total amount $1 increases to if invested at the given interest rate for the given number of periods, where interest is compounded per period.
Use the table to find the total amount returned in each situation.

Periods	Interest rate per period							
	1%	2%	2.5%	4%	5%	6%	10%	12%
1	1.0100	1.0200	1.0250	1.0400	1.0500	1.0600	1.1000	1.1200
2	1.0201	1.0404	1.0506	1.0816	1.1025	1.1236	1.2100	1.2544
3	1.0303	1.0612	1.0769	1.1249	1.1576	1.1910	1.3310	1.4049
4	1.0406	1.0824	1.1038	1.1699	1.2155	1.2625	1.4641	1.5735
5	1.0510	1.1041	1.1314	1.2167	1.2763	1.3382	1.6105	1.7623
6	1.0615	1.1262	1.1597	1.2653	1.3401	1.4185	1.7716	1.9738
7	1.0721	1.1487	1.1887	1.3159	1.4071	1.5036	1.9587	2.2107
8	1.0829	1.1717	1.2184	1.3686	1.4775	1.5938	2.1436	2.4760

a $8000 invested for 8 years at 6% p.a. compounded annually

b $20 000 invested for 1 year at 10% p.a. compounded quarterly

c $15 000 invested for 4 years at 8% p.a. compounded six-monthly

QUESTION 2 Use the above table to find the amount of money which could be invested now to give $50 000 at the end of 5 years at 10% p.a. compounded annually.

Financial mathematics

UNIT 9: Depreciation

Question 1 Use the depreciation formula $S = V_0(1 - r)^n$ to find the value of the following items after the given time.

a If a car bought for $16000 depreciates at 8% p.a., find its value after 3 years.

b If a coffee machine is worth $5000 and depreciates at 7% p.a., find its value after 5 years.

c If a computer costs $3500 and depreciates at 25% p.a., find its value after 4 years.

d If a photocopier costs $20000 and depreciates at 15% p.a., find its value after 3 years.

Question 2 A business buys new computers for $90000. They depreciate at the rate of 20% p.a. Calculate:

a their value after 3 years

b the amount of depreciation

Question 3

a Each year the population of a town decreases by 7%. If its population is now 20000 people, what will it be in 4 years?

b A library depreciates by 10% p.a. If it is now worth $50000, what will its value be after 5 years?

Financial mathematics

UNIT 10: Solving problems involving interest

QUESTION 1 Find the amount of interest earned if $12 000 is invested for 5 years at 7% p.a. if the interest is:

a simple interest

b compounded yearly

QUESTION 2 An amount of $20 000 is invested for 8 years at 6% p.a. interest, compounded monthly. Find the annual rate of simple interest (as a percentage to one decimal place) that will give the same result.

QUESTION 3 An amount of $15 000 is to be invested for 4 years. Find the interest earned if it is:

a simple interest at 9% p.a.

b compounded yearly at 8% p.a.

c Which gives the best result and by how much?

QUESTION 4 What sum of money (to the nearest $100) could be invested at 7% p.a., compounded yearly, to give $35 000 at the end of 5 years?

QUESTION 5 Use a 'guess and check' method to find the number of years $8000 needs to be invested at 6% p.a., compounded yearly, to produce $5515 interest.

Financial mathematics

TOPIC TEST

PART A

Time allowed: 15 minutes **Total marks: 15**

	Marks
1 The simple interest on \$5400 at 7% pa for 8 years is: (A) \$302.40 (B) \$387.80 (C) \$3024 (D) \$3878	1
2 \$800 invested for 3 years at 12% p.a. compound interest becomes: (A) \$545.18 (B) \$944 (C) \$1088 (D) \$1123.94	1
3 \$4000 invested for 5 years at 8% p.a. compound interest becomes: (A) \$5877.31 (B) \$5600 (C) \$5400 (D) \$5870	1
4 Which amount of money will give \$2400 simple interest when invested at 8% p.a. for 5 years? (A) \$600 (B) \$960 (C) \$6000 (D) \$9600	1
5 Find the simple interest on \$4500 at 7% p.a. for 5 years. (A) \$6075 (B) \$1575 (C) \$6311.48 (D) \$1811.48	1
6 \$500 invested for 3 years at 15% p.a. simple interest becomes: (A) \$225 (B) \$725 (C) \$760.44 (D) \$26.44	1
7 \$2000 invested for 4 years at 10% p.a. interest, compounded annually, becomes: (A) \$800 (B) \$2800 (C) \$928.20 (D) \$2928.20	1
8 Find the simple interest on \$1200 at 12% p.a. for 5 years. (A) \$720 (B) \$1920 (C) \$914.81 (D) \$2114.81	1
9 A sum of \$8500 amounted to \$8925 after being invested for 6 months at simple interest. What was the interest rate earned? (A) 8% p.a. (B) 9% p.a. (C) 10% p.a. (D) 11% p.a.	1
10 Calculate the compound interest earned on \$10 000 at 9% p.a. for 3 years compounded monthly (correct to the nearest dollar). (A) \$13 086 (B) \$3086 (C) \$2700 (D) \$12 700	1
11 If \$15 000 is invested for 5 years at 10% p.a. interest, compounded quarterly, it becomes: (A) \$22 500 (B) \$7500 (C) \$9579.25 (D) \$24 579.25	1
12 A computer costs \$2800 and depreciates at 20% p.a. Find its value after 3 years. (A) \$1433.60 (B) \$1366.40 (C) \$156.80 (D) \$1665.20	1
13 A mobile phone is worth \$800 and depreciates at 20% p.a. Find its value after 5 years. (A) \$537.86 (B) \$262.14 (C) \$360.80 (D) \$315.34	1
14 A debt of \$30 000 is to be paid in equal instalments of \$625. How many instalments are needed? (A) 36 (B) 48 (C) 60 (D) 72	1
15 After how many years will a sum of money double if invested at 5% p.a. simple interest? (A) 25 (B) 20 (C) 10 (D) 5	1

Total marks achieved for PART A ___ / 15

Financial mathematics

TOPIC TEST — PART B

Instructions
- This part consists of 8 questions.
- Write only the answer in the answer column.
- For any working use the question column.

Time allowed: 20 minutes — **Total marks: 15**

Questions	Answers	Marks
1 \$8000 is invested for 4 years at 10% p.a. interest compounded annually. Find:		
a the total amount at the end of 4 years		1
b the compound interest earned		1
c the rate of simple interest that would produce the same result		1
2 If \$8000 is invested for 8 years at 8% p.a., find the:		
a simple interest		1
b extra amount earned if interest is compounded annually		1
3 James decides to buy a car marked at \$13 500. He pays a 15% deposit and the balance over 4 years with interest charged at 7% p.a. on the balance. Find the:		
a deposit paid		1
b balance owing		1
c simple interest paid		1
d total to be repaid		1
e monthly repayment		1
4 Find the simple interest on \$9500 at 7.5% p.a. for 12 months.		1
5 Find the length of time for \$1200 to be the simple interest on \$4800 at 5% p.a.		1
6 Find the compound interest on \$24 000 at 7% p.a. for 2 years.		1
7 Calculate the total amount of interest earned when \$7200 is invested for 3 years at 8% p.a. compounded half-yearly.		1
8 Each year a property increases in value by $10\frac{1}{2}$%. What is the value of a \$600 000 property after 5 years? Answer to the nearest \$1000.		1

Total marks achieved for PART B

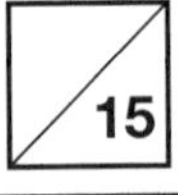

Chapter 2
Algebraic techniques

UNIT 1: Addition and subtraction of pronumerals

Question 1 Simplify the following expressions by collecting like terms.

a $2a + 5a =$ ______ **b** $7p - 3p =$ ______

c $4a + 8a =$ ______ **d** $9x - x =$ ______

e $3m + m =$ ______ **f** $6q - q =$ ______

g $-4a + 5a =$ ______ **h** $5ab + 6ab =$ ______

i $9t - 12t =$ ______ **j** $3x^2 + 5x^2 =$ ______

k $-7n + 5n =$ ______ **l** $-4k - 3k =$ ______

Question 2 Simplify the following.

a $3a + 4a + 5a =$ ______ **b** $12t - 7t + 4t =$ ______

c $8x - 3x - 2x =$ ______ **d** $-2k - 3k - 5k =$ ______

e $m - 4m + 2m =$ ______ **f** $6xy - 4xy - xy =$ ______

g $7a + 2a - 3a - 5a -$ ______ **h** $-8p + 3p - 2p =$ ______

i $15x^2 - 7x^2 - 6x^2 =$ ______ **j** $7q - 3q - 4q =$ ______

k $-5m + 2m - 3m - m =$ ______ **l** $-6y - 2y - 3y + y =$ ______

Question 3 Simplify by collecting like terms.

a $8a + 5b - 3a =$ ______ **b** $6x + 4y + 2x - 2y =$ ______

c $5a^2 + 2a - 3a^2 - 4a =$ ______ **d** $-3c - 2c - 3d + d =$ ______

e $7x - 3y - 4x - 3y =$ ______ **f** $9a - 4b - 3b + a =$ ______

g $6m + 7 - 3m - 1 =$ ______ **h** $12 - 4m - 3m =$ ______

i $-3a - 5b - 6a + 7b =$ ______ **j** $7xy - 3x - 4y + x =$ ______

k $-2 + 6x + 5 =$ ______ **l** $8t - 4u + 3t - 5u =$ ______

Question 4 Simplify the following.

a $3a + 7 - 9a =$ ______ **b** $5x^2 + 2x^2 - 7x^2 - x^2 =$ ______

c $7n + 8n - 3n - 5 =$ ______ **d** $4x - x + 3x - 6x =$ ______

e $8x + 3y - 5x - 3y =$ ______ **f** $-2y - 3y + 4y =$ ______

g $7a^2 - a + 4a^2 - 2 =$ ______ **h** $3m + 4m - 9m =$ ______

i $5x - 2y - 4y - 5x =$ ______ **j** $4k + 5k - 2n - 7n =$ ______

k $-5ab + 3a + b - a =$ ______ **l** $-3a - 2a + b - 5a =$ ______

m $15 - 3x - 2x - 1 =$ ______ **n** $3p + 5p - 9p =$ ______

o $9m + 2n - 5m + 4n =$ ______ **p** $5x^3 + 2x^2 + 3x - 4x^2 =$ ______

Algebraic techniques

UNIT 2: Index laws

Question 1 Simplify.

a $x^2 \times x^6 =$ ______
b $n^3 \times n^4 =$ ______
c $y^5 \times y =$ ______

d $3m^5 \times 2m^4 =$ ______
e $6a^6 \times 3a^3 =$ ______
f $a^4 \times a \times a^5 =$ ______

g $4x^3 \times x^7 =$ ______
h $8x^8 \times 2x^2 =$ ______
i $5m \times 3m^2 =$ ______

Question 2 Simplify.

a $x^8 \div x^2 =$ ______
b $y^{10} \div y^5 =$ ______
c $a^8 \div a^3 =$ ______

d $15m^{15} \div 5m^5 =$ ______
e $12n^{12} \div 2n^4 =$ ______
f $6a^9 \div a^3 =$ ______

g $12y^7 \div 2y^6 =$ ______
h $9a^7 \div 9a =$ ______
i $a^9b^5 \div a^3b^3 =$ ______

Question 3 Simplify.

a $(x^2)^3 =$ ______
b $(a^4)^5 =$ ______
c $(x^6)^2 =$ ______

d $(4m^3)^2 =$ ______
e $4(m^3)^2 =$ ______
f $(2a^4)^3 =$ ______

g $(a^3b^2)^4 =$ ______
h $(3ab^4)^3 =$ ______
i $5(x^2y)^7 =$ ______

Question 4 Simplify.

a $7^0 =$ ______
b $x^0 =$ ______
c $(3n)^0 =$ ______

d $5m^0 =$ ______
e $(ab)^0 =$ ______
f $a^0 + b^0 =$ ______

g $x^0 - y^0 =$ ______
h $8a^0 + (8a)^0 =$ ______
i $7m^0 + 4n^0 =$ ______

Question 5 Simplify the following.

a $a^8 \div a^2 =$ ______
b $3x^3 \times 5x^5 =$ ______

c $2(a^5)^5 =$ ______
d $(5x^2)^3 =$ ______

e $18x^4 \div 9x =$ ______
f $3a^2b^3 \times 2a^3b^2 =$ ______

g $m^4n^3 \times mn^2 =$ ______
h $10x^6y^4 \div 2x^3y^4 =$ ______

i $4x^2y^3 \times 5x^4y^6 =$ ______
j $5x^2y^2 \div 5xy =$ ______

k $5x^0 =$ ______
l $(6x^2)^2 =$ ______

Algebraic techniques

UNIT 3: Further products

QUESTION 1 Find the following products.

a $4 \times 3a =$ ____________ b $5x \times 3y =$ ____________

c $2m \times 3m =$ ____________ d $-6a \times 2b =$ ____________

e $-3x \times -4y =$ ____________ f $5ab \times 3 =$ ____________

g $11t \times -5t =$ ____________ h $4q^2 \times 3q =$ ____________

i $xy \times x^2y =$ ____________ j $2a \times 3b \times 4c =$ ____________

k $8x^8 \times 3x^3 =$ ____________ l $4x^2y \times 2xy =$ ____________

QUESTION 2 Find the following products.

a $x^3 \times x^7 =$ ____________ b $3x \times x^6 =$ ____________

c $a^5 \times 2a =$ ____________ d $-5q \times -q =$ ____________

e $4a^2 \times 3b =$ ____________ f $6x^2 \times 3y^2 =$ ____________

g $8t^2 \times 3t =$ ____________ h $4x^2y^2 \times -3 =$ ____________

i $-3p \times 2q \times 4r =$ ____________ j $5a \times -2b \times -6c =$ ____________

k $-a \times -a \times -a =$ ____________ l $ab \times ab \times ab =$ ____________

QUESTION 3 Simplify the following.

a $3a^2b \times 2ab =$ ____________ b $6ab^2 \times 4a =$ ____________

c $5p^3q^2 \times 7q =$ ____________ d $x^2y^3 \times x^4y^5 =$ ____________

e $a^7b^2 \times a^3b^5 =$ ____________ f $2m^5n^6 \times 3m^4n^2 =$ ____________

g $5p^3q^2 \times pq =$ ____________ h $4a^2b^4 \times 5b^3 =$ ____________

i $7xy^3 \times 2xy^2 =$ ____________ j $9ab^7 \times 3a^2b =$ ____________

k $10a^5b^2c^3 \times 2a^3b^4c^7 =$ ____________ l $3x^4y^2z \times 5xy^3z^5 =$ ____________

QUESTION 4 Simplify the following.

a $2a^{-3} \times 3a^5 =$ ____________ b $7a^2 \times 4a^{-3} =$ ____________

c $6m^4 \times 3m^{-4} =$ ____________ d $-2x^5 \times -4x^{-4} =$ ____________

e $5t^{-2} \times 7t^{-3} =$ ____________ f $8k^{-3} \times 3k^2 =$ ____________

g $9n^{-2} \times 4n =$ ____________ h $2a^{-2} \times 3a^{-3} \times 4a^{-4} =$ ____________

i $-e \times e^{-3} =$ ____________ j $4q^8 \times 2q^{-2} \times 6q =$ ____________

Algebraic techniques

UNIT 4: Further quotients

QUESTION 1 Divide the following.

a $10a \div 2 =$ ____________

b $9b \div 3b =$ ____________

c $8c \div c =$ ____________

d $6ab \div ab =$ ____________

e $12k \div -3 =$ ____________

f $-15m \div -5 =$ ____________

g $4k \div 4k =$ ____________

h $-32mn \div -8n =$ ____________

i $6x^2 \div 2x =$ ____________

j $-8a^2 \div 4a^2 =$ ____________

k $12abc \div -2b =$ ____________

l $xyz \div xz =$ ____________

m $a^{12} \div a^4 =$ ____________

n $20b^{20} \div 5b^5 =$ ____________

o $x^6y^4 \div x^2y^3 =$ ____________

p $a^5b^8 \div ab^3 =$ ____________

q $27p^7q^8 \div 3p^2q^7 =$ ____________

r $15a^2b^3c^5 \div 5abc^2 =$ ____________

QUESTION 2 Simplify.

a $\frac{2x^3}{3x} =$ ____________

b $\frac{5a^2}{7a^5} =$ ____________

c $\frac{9t^4}{10t} =$ ____________

d $\frac{5ab}{3a} =$ ____________

e $\frac{7xy}{8y} =$ ____________

f $\frac{3n^2}{5mn} =$ ____________

g $\frac{6a^3}{8a} =$ ____________

h $\frac{8x^4}{12x^7} =$ ____________

i $\frac{15a^2b^3}{10ab} =$ ____________

j $\frac{24e^9}{27e^{10}} =$ ____________

k $\frac{12m^4n^8}{9m^6n^2} =$ ____________

l $\frac{10a^5b^3}{15a^2b^4} =$ ____________

m $\frac{5a^4}{10a^2} =$ ____________

n $\frac{3t^7}{9t^5} =$ ____________

o $\frac{12x^3}{4x^8} =$ ____________

p $\frac{6a^2b^2}{3ab^4} =$ ____________

q $\frac{2m^5n^3}{8m^4n^2} =$ ____________

r $\frac{18xz}{9xyz} =$ ____________

s $\frac{3}{9x} =$ ____________

t $\frac{7x^2}{14x^8} =$ ____________

u $\frac{6a^2}{3a} =$ ____________

v $\frac{5n^3}{25n^4} =$ ____________

w $\frac{21a^3b^2}{7ab} =$ ____________

x $\frac{4x^2y^2}{20x^2y^3} =$ ____________

Algebraic techniques

UNIT 5: Mixed operations

QUESTION 1 Simplify, where possible.

a $9x + 2x =$ ______

b $5k - k =$ ______

c $2x \times x =$ ______

d $8p \div p =$ ______

e $8p \div 8 =$ ______

f $8p \div 8p =$ ______

g $12x^2 + 3x =$ ______

h $12a^{12} \div 2a^2 =$ ______

i $5x^2 \times 3xy =$ ______

j $x \times x^6 =$ ______

k $6ab \div -2b =$ ______

l $-5m - 3m =$ ______

m $-a \times -3ab =$ ______

n $-4n^2 + 4n^2 =$ ______

o $-8abc \div 4bc =$ ______

p $7x^8 - 6x^8 =$ ______

q $-3a \times -4b =$ ______

r $-x - x =$ ______

QUESTION 2 Simplify.

a $5x \times 3x + x^2$

______

______

b $9a \times 4 \div 12a$

______

______

c $16x^2yz \div 4xy \div 2xz$

______

______

d $2a^2 \times 4a^3 + 5a^5$

______

______

e $(2p^2)^3 \div 4p^4$

______

______

f $18a^2b^3 \div 6ab^2 \times 2a$

______

______

QUESTION 3 Simplify, where possible.

a $4x^2 \times 3x^3 - 5x^2 \times 2x^3$

______

______

b $(4a^5)^2 \div 8(a^3)^3$

______

______

c $9a^2b^3 \div 3ab^2 + 5a \times 2b$

______

______

d $12m^9 \div 3m^3 - 2m^2 \times 4m^4$

______

______

e $8p^2q \times 3p \div 4pq \div 2p$

______

______

f $12 - 3n^0 + 5n \times 6n \div 10n - (3n)^0$

______

______

Algebraic techniques

UNIT 6: Substitution

QUESTION 1 If $a = 3$, $b = 5$ and $c = 9$, find the value of:

a $a + 5$ __________ **b** bc __________ **c** $ab + c$ __________

d $5a$ __________ **e** $2c - a$ __________ **f** abc __________

g $4b - 2c$ __________ **h** $a + b - c$ __________ **i** $bc - 7a$ __________

j c^2 __________ **k** $4a^2$ __________ **l** ab^2 __________

QUESTION 2 If $x = -2$, and $y = -5$, find the value of:

a $6xy$ __________ **b** $3x^2$ __________

c $4x - 5y$ __________ **d** $x^2 + 4x$ __________

e $12 - 2y$ __________ **f** xy^2 __________

QUESTION 3 Given that $A = \frac{h}{2}(a + b)$, find A when:

a $a = 12, b = 22$ and $h = 15$

b $a = 9, b = 14$ and $h = 7$

QUESTION 4 Given that $m = \frac{y_2 - y_1}{x_2 - x_1}$, find m when:

a $x_1 = 2, y_1 = 7, x_2 = 4$ and $y_2 = -1$

b $x_1 = 5, y_1 - 6, x_2 = 2$ and $y_2 = 9$

QUESTION 5 Given that $c^2 = a^2 + b^2$ and that $c > 0$, find c when:

a $a = 112$ and $b = 441$

b $a = 40.8$ and $b = 14.5$

QUESTION 6 Given that $B = \frac{m}{h^2}$, find B when:

a $m = 81$ and $h = 1.8$

b $m = 64$ and $h = 1.65$

Algebraic techniques

UNIT 7: Expanding

Question 1 Expand the following expressions.

a $5(x + 2) =$ ______

b $7(x - 3) =$ ______

c $4(2x + 5) =$ ______

d $3(5x - 3y) =$ ______

e $6(2t - 1) =$ ______

f $x(x + 7) =$ ______

g $a(a - 1) =$ ______

h $3x(2x - 5) =$ ______

i $4n(3n + 2) =$ ______

j $8(2a + b - c) =$ ______

k $2a(5a + 4b + 3) =$ ______

l $-2(3x + 4) =$ ______

m $-5(2x - 3) =$ ______

n $-4x(1 - 2x) =$ ______

o $-7a(a + 4) =$ ______

p $-(x - y) =$ ______

q $-(m + n) =$ ______

r $-(3p - 1) =$ ______

s $2x(x^2 - 5) =$ ______

t $3a^2(ab + 5) =$ ______

Question 2 Expand and simplify.

a $5(2x + 3) + 4x$

______

______

b $4(a - 2) - 3a + 5$

______

______

c $12 - (x - 3)$

______

______

d $7x + 5y + 3(2x - 3y)$

______

______

e $7(x + 4) + 5(x + 2)$

______

______

f $9(a - 1) + 3(a - 2)$

______

______

g $6(2x + 5) - 4(3x + 2)$

______

______

h $5(4m - 2) - 3(m + 2)$

______

______

i $3(4a + 7) - 2(5a - 3)$

______

______

j $x(x + 5) - 3(x - 4)$

______

______

k $3a(2a - 1) - 2a(3a - 1)$

______

______

l $x(x + 3y) - y(x - 3y)$

______

______

Algebraic techniques

UNIT 8: Binomial products (1)

QUESTION **1** Expand and simplify.

a $x(x + 3) + 2(x + 3)$

b $x(x + 7) - 2(x + 7)$

c $x(x - 3) + 7(x - 3)$

d $x(x + 5) - 3(x + 5)$

e $2x(x + 3) + 5(x + 3)$

f $3x(x - 2) - 2(x - 2)$

g $x(2x + 3) - 5(2x + 3)$

h $x(3x - 5) + 2(3x - 5)$

i $3x(2x + 1) + 2(2x + 1)$

j $2x(3x - 2) - 1(3x - 2)$

QUESTION **2** By matching with an expansion in question 1, write down each binomial product.

a $(x + 2)(x + 3) =$ ____________

b $(x - 3)(x + 5) =$ ____________

c $(x - 5)(2x + 3) =$ ____________

d $(2x - 1)(3x - 2) =$ ____________

e $(2x + 5)(x + 3) =$ ____________

f $(x - 2)(x + 7) =$ ____________

g $(3x + 2)(2x + 1) =$ ____________

h $(x + 7)(x - 3) =$ ____________

i $(3x - 2)(x - 2) =$ ____________

j $(x + 2)(3x - 5) =$ ____________

QUESTION **3** Write the areas in each part of the rectangle and hence find the binomial product.

a

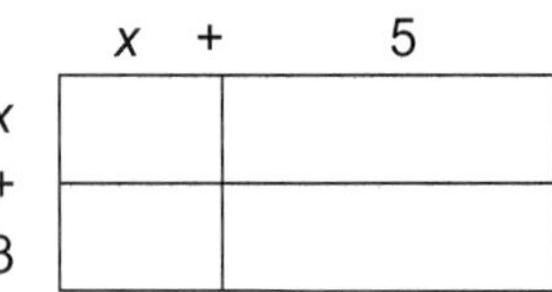

$(x + 5)(x + 3)$ = ____________

= ____________

b

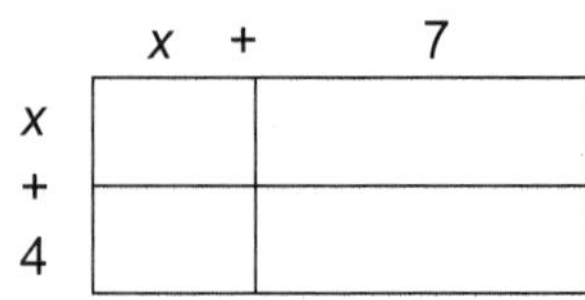

$(x + 7)(x + 4)$ = ____________

= ____________

Algebraic techniques

UNIT 9: Binomial products (2)

QUESTION 1 Expand and simplify the following.

a $(x + 1)(x + 2)$ **b** $(x + 2)(x + 3)$ **c** $(a + 3)(a + 5)$ **d** $(m + 6)(m + 1)$

e $(p + 8)(p + 2)$ **f** $(y + 3)(y + 7)$ **g** $(a + 4)(a + 7)$ **h** $(d + 3)(d + 9)$

i $(2a + 3)(a + 5)$ **j** $(3a + 1)(2a + 6)$ **k** $(4a + 6)(2a + 3)$ **l** $(2x + 5)(3x + 1)$

QUESTION 2 Expand and simplify.

a $(a + 3)(a - 2)$ **b** $(x - 3)(x + 2)$ **c** $(y - 4)(y + 6)$ **d** $(y + 5)(y - 3)$

e $(a + 7)(a - 3)$ **f** $(x + 6)(x - 2)$ **g** $(2y + 1)(y - 2)$ **h** $(3x + 2)(x - 3)$

i $(2x + 1)(3x - 1)$ **j** $(x + 7)(2x - 1)$ **k** $(x + 8)(3x - 5)$ **l** $(x - 3)(x - 4)$

QUESTION 3 Find the following products and simplify.

a $(a + 3)(4 + a)$ **b** $(a + 5)(6 + a)$ **c** $(2a + 1)(3 - a)$ **d** $(4 + x)(x + 9)$

e $(5 - n)(n + 7)$ **f** $(x - 6)(7 - x)$ **g** $(3x + 2)(2 + x)$ **h** $(3n + 1)(5 - n)$

i $(2a - 6)(a + 7)$ **j** $(x + y)(x - y)$ **k** $(2m + n)(2m - n)$ **l** $(a - b)(a + b)$

m $(2x - 3y)(2x + 3y)$ **n** $(a - b)(a - \text{b})$ **o** $(2x + 3)(2x - 3)$ **p** $(3x - 4)(2x + 5)$

Algebraic techniques

UNIT 10: Addition and subtraction of algebraic fractions

QUESTION 1 Find the sum of these algebraic fractions.

a $\frac{m}{5}+\frac{3m}{5}=$ __________

b $\frac{x}{3}+\frac{2x}{3}=$ __________

c $\frac{2t}{7}+\frac{3t}{7}=$ __________

d $\frac{5y}{8}+\frac{3y}{8}=$ __________

e $\frac{6a}{11}+\frac{3b}{11}=$ __________

f $\frac{19k}{8}+\frac{14k}{8}=$ __________

g $\frac{15x}{17}+\frac{2x}{17}=$ __________

h $\frac{10p}{7}+\frac{4p}{7}=$ __________

i $\frac{6m}{19}+\frac{3m}{19}=$ __________

QUESTION 2 Subtract the following algebraic expressions.

a $\frac{12y}{7}-\frac{9y}{7}=$ __________

b $\frac{5x}{11}-\frac{3x}{11}=$ __________

c $\frac{8a}{9}-\frac{5a}{9}=$ __________

d $\frac{12a}{17}-\frac{5a}{17}=$ __________

e $\frac{9m}{23}-\frac{7m}{23}=$ __________

f $\frac{5m}{12}-\frac{3m}{12}=$ __________

g $\frac{9a}{7}-\frac{5a}{7}=$ __________

h $\frac{12x}{10}-\frac{7x}{10}=$ __________

i $\frac{5a}{11}-\frac{2b}{11}=$ __________

QUESTION 3 Simplify the following.

a $\frac{x}{2}+\frac{x}{3}=$ __________

b $\frac{a}{4}-\frac{a}{5}=$ __________

c $\frac{m}{3}+\frac{m}{5}=$ __________

d $\frac{2x}{5}+\frac{x}{4}=$ __________

e $\frac{2a}{5}-\frac{a}{10}=$ __________

f $\frac{y}{8}-\frac{y}{32}=$ __________

g $\frac{8y}{3}-\frac{3y}{5}=$ __________

h $\frac{5p}{8}-\frac{p}{4}=$ __________

i $\frac{x}{4}-\frac{3x}{8}=$ __________

j $\frac{8y}{3}-\frac{3y}{4}=$ __________

k $\frac{4m}{7}-\frac{2m}{21}=$ __________

l $\frac{3a}{8}-\frac{2b}{4}=$ __________

QUESTION 4 Write in simplest form.

a $\frac{3x}{5}-\frac{x}{10}=$ __________

b $\frac{2x}{3}-\frac{x}{6}=$ __________

c $\frac{3t}{7}-\frac{2t}{21}=$ __________

Algebraic techniques

UNIT 11: Multiplication and division of algebraic fractions

Question 1 Find the products of these algebraic fractions.

a $\frac{a}{5} \times \frac{b}{7} =$ ______________

b $\frac{m}{2} \times \frac{n}{6} =$ ______________

c $\frac{x}{3} \times \frac{y}{8} =$ ______________

d $\frac{5x}{3} \times \frac{4y}{9} =$ ______________

e $\frac{a}{3} \times \frac{a}{11} =$ ______________

f $\frac{2x}{5} \times \frac{3x}{7} =$ ______________

g $\frac{a}{2} \times \frac{1}{4} =$ ______________

h $\frac{2}{3} \times \frac{n}{7} =$ ______________

i $\frac{5x}{7} \times \frac{5}{7} =$ ______________

Question 2 Find these products. Give the answer in simplest form.

a $\frac{2a}{3} \times \frac{b}{4} =$ ______________

b $\frac{5x}{2} \times \frac{x}{5} =$ ______________

c $\frac{4t}{3} \times \frac{3t}{4} =$ ______________

d $\frac{3c}{5} \times \frac{2d}{9} =$ ______________

e $\frac{3a}{4} \times \frac{8b}{9} =$ ______________

f $\frac{5x}{2} \times \frac{4y}{15} =$ ______________

g $\frac{6x}{5} \times \frac{2x}{3} =$ ______________

h $\frac{9m}{10} \times \frac{5n}{6} =$ ______________

i $\frac{9t}{8} \times \frac{4t}{3} =$ ______________

Question 3 Divide the following algebraic fractions.

a $\frac{a}{2} \div \frac{b}{3} =$ ______________

b $\frac{3x}{7} \div \frac{2y}{9} =$ ______________

c $\frac{m}{3} \div \frac{2n}{5} =$ ______________

d $\frac{a}{5} \div \frac{a}{15} =$ ______________

e $\frac{3n}{8} \div \frac{5n}{16} =$ ______________

f $\frac{20p}{11} \div \frac{10p}{22} =$ ______________

g $\frac{k}{18} \div \frac{km}{36} =$ ______________

h $\frac{xyz}{15} \div \frac{yz}{5} =$ ______________

i $\frac{pq}{20} \div \frac{q}{40} =$ ______________

j $\frac{4a^2b^3}{7} \div \frac{16a^3b^2}{21} =$ ______________

k $\frac{20ab}{27} \div \frac{4a}{36} =$ ______________

l $\frac{3xy}{4} \div \frac{5xy}{6} =$ ______________

Algebraic techniques

UNIT 12: Harder algebraic fractions

QUESTION 1 Find, giving the answer in simplest form.

a $\frac{8a}{5b}+\frac{9a}{5b}=$

b $\frac{5a}{7x}-\frac{4a}{7x}=$

c $\frac{16}{5a}+\frac{4}{5a}=$

d $\frac{14}{5t}-\frac{9}{5t}=$

e $\frac{16}{3x^2}-\frac{10}{3x^2}=$

f $\frac{2a}{5x}+\frac{8a}{5x}=$

g $\frac{9}{x}+\frac{3}{4x}=$

h $\frac{5a}{7b}-\frac{3a}{14b}=$

i $\frac{8m}{5n}-\frac{3m}{20n}=$

QUESTION 2 Simplify the following.

a $\frac{5}{y}\times\frac{3}{t}=$

b $\frac{4}{a}\times\frac{5}{b}=$

c $\frac{6}{5a}\times\frac{3}{2b}=$

d $\frac{8}{3t}\times\frac{4}{2t}=$

e $\frac{2x}{3y}\times\frac{3x}{2y}=$

f $\frac{4}{m}\times\frac{2a}{3n}=$

g $\frac{5b}{3c}\times\frac{2b}{9c}=$

h $\frac{3t}{20}\times\frac{10}{21t}=$

i $\frac{15x}{11y}\times\frac{33y}{60x}=$

j $\frac{8ab}{c}\times\frac{ac}{4b}=$

k $\frac{24a}{17b}\times\frac{34b}{16a}=$

l $\frac{x}{y}\times\frac{y}{z}\times\frac{z}{x}=$

m $\frac{x^2y}{m}\times\frac{m}{xy}=$

n $\frac{4ab}{5c}\times\frac{10c}{8a}=$

o $\frac{15x}{8y}\times\frac{32x^2y}{25x^3y^2}=$

QUESTION 3 Divide the following fractions.

a $\frac{9}{2n}\div\frac{3p}{8n}=$

b $\frac{8}{x}\div\frac{5}{x}=$

c $\frac{7}{2y}\div\frac{14}{10y}=$

d $\frac{a}{b}\div\frac{5a}{b}=$

e $\frac{9n}{5m}\div\frac{27n}{15m}=$

f $\frac{xy}{z}\div\frac{z}{xy}=$

g $\frac{15t}{m}\div\frac{5t}{7m}=$

h $\frac{18mn}{11p}\div\frac{48m}{33p}=$

i $\frac{35mn}{6p}\div\frac{7m^2}{12p}=$

Algebraic techniques

UNIT 13: Common factors

QUESTION 1 Factorise the following by taking out the common factor.

a $5x + 10 =$ ______ b $3x + 6 =$ ______ c $8y + 16 =$ ______

d $m^2 + m =$ ______ e $2x^2 + 4x =$ ______ f $3xy + 6x =$ ______

g $6a^2 - 3a =$ ______ h $3m + 15 =$ ______ i $9x + xy =$ ______

j $4x + 16 =$ ______ k $5b^2 + 10ab =$ ______ l $3m + 21 =$ ______

m $6m - 3mn =$ ______ n $5x + 15 =$ ______ o $ay - y =$ ______

p $7mn - 14mp =$ ______ q $x^2y^2 - xyz =$ ______ r $8m^2n^2 - 16m^2n =$ ______

QUESTION 2 Factorise the following by taking out the negative common factor.

a $-3x - 6 =$ ______ b $-4a - 8 =$ ______ c $-5y - 15 =$ ______

d $-m^2 + m =$ ______ e $-x^2 + 5x =$ ______ f $-l^2 + 2lm =$ ______

g $-x + 4x^2 =$ ______ h $-4m + m^2 =$ ______ i $-3x - 2x^2 =$ ______

j $-6a - 18a^2 =$ ______ k $-7y + 21 =$ ______ l $-8x + 16xy =$ ______

m $-3a - 9 =$ ______ n $-5xy + 15x^2y^2 =$ ______ o $-a^2y^2 + ay =$ ______

QUESTION 3 Factorise the following.

a $ab + ac + ad =$ ______ b $px + py + pz =$ ______

c $2a^2b + 3a^2b^2 - 5abc =$ ______ d $5m^3 + 10m^2 + 15m =$ ______

e $2a + 4b + 6c =$ ______ f $12x^2 + 15xy + 18xz =$ ______

g $x^2y^2 + xy^2 + x^2y =$ ______ h $9a^2b - 12a^2b^2 =$ ______

i $5a^2 - 5b^2 - 10c^2 =$ ______ j $6mp + 12m^2p - 18m^2p^2 =$ ______

k $3ab - 6ac - 9ad =$ ______ l $12x^2y^2 - 36x^3y^3 =$ ______

QUESTION 4 Factorise each of the following.

a $8a^2b^3 - 10a^3b^5 =$ ______ b $16xy + 6x^3 =$ ______

c $9p^3q^2 + 12pq^5 =$ ______ d $6a^2bc^3 - 9abc^2 =$ ______

e $12x^3y^4 - 15x^2y^6 =$ ______ f $2a^2b^3c - 8ab^2c =$ ______

g $10p^2q^5 - 25p^3q^5 =$ ______ h $28x^4y^7 + 42x^4y =$ ______

i $2a^4b^2c^6 - 12a^5b^3c^7 =$ ______ j $9t^2u^3 - 6tu^4 =$ ______

k $15x^2y^2 - 10x^3y + 20xy^3 =$ ______ l $24pq^4 + 16p^2q^3 + 8pq^2 =$ ______

Algebraic techniques

UNIT 14: Factorising trinomials

Question 1 Factorise the following.

a $x^2 + 7x + 12 =$

b $x^2 - 5x + 6 =$

c $x^2 + 3x + 2 =$

d $x^2 + 4x + 4 =$

e $y^2 - 7y + 12 =$

f $m^2 + 8m + 12 =$

g $a^2 + 6a + 9 =$

h $x^2 + 11x + 28 =$

i $n^2 + 2n - 3 =$

j $x^2 + 9x + 14 =$

Question 2 Factorise.

a $x^2 - 8x + 15 =$

b $y^2 - 4y - 12 =$

c $x^2 + 5x - 6 =$

d $x^2 + 19x + 90 =$

e $x^2 + 4x - 12 =$

f $m^2 - m - 56 =$

g $x^2 - 3x - 4 =$

h $y^2 - 6y - 7 =$

Question 3 Factorise.

a $x^2 - 8x =$

b $m^2 + 6m + 5 =$

c $t^2 - t - 6 =$

d $y^2 - 9y + 20 =$

e $a^2 - 7a - 18 =$

f $x^2 + 8x + 16 =$

g $x^2 - 12x =$

h $y^2 - 11y + 24 =$

Algebraic techniques

TOPIC TEST — PART A

Time allowed: 15 minutes — **Total marks: 15**

Marks

1 $x^4 + x^4 =$

(A) x^8 (B) x^{16} (C) $2x^4$ (D) $2x^8$ — 1

2 When $(6m - 2)$ is factorised, one of the factors is:

(A) m (B) 3 (C) $3m - 2$ (D) $3m - 1$ — 1

3 $6x^6 \times 3x^3 =$

(A) $9x^9$ (B) $18x^9$ (C) $9x^{18}$ (D) $18x^{18}$ — 1

4 $7p^2 - 9p - 4p^2 + 5p =$

(A) $3p^2 - 4p$ (B) $3p^2 - 14p$ (C) $3p^2 + 4p$ (D) $3p^2 + 14p$ — 1

5 If $x = -5$ then $2x^2 =$

(A) 50 (B) -50 (C) 100 (D) -100 — 1

6 $4x^0 + 4^0 =$

(A) 1 (B) 2 (C) 4 (D) 5 — 1

7 $(x - 4)(x - 3) =$

(A) $x^2 - 7x - 12$ (B) $x^2 - 7x + 12$ (C) $x^2 + 7x - 12$ (D) $x^2 + 7x + 12$ — 1

8 $\frac{x}{4} + \frac{x}{5} =$

(A) $\frac{2x}{9}$ (B) $\frac{x}{10}$ (C) $\frac{x^2}{20}$ (D) $\frac{9x}{20}$ — 1

9 $x^2 - 5x + 6$ expressed as a product of factors is:

(A) $(x + 3)(x + 2)$ (B) $(x + 3)(x - 2)$ (C) $(x - 3)(x + 2)$ (D) $(x - 3)(x - 2)$ — 1

10 $12x^{12} \div 3x^3 =$

(A) $4x^4$ (B) $4x^9$ (C) $9x^4$ (D) $9x^9$ — 1

11 $x^2y(2x^3 - y^2) =$

(A) $2x^6y - x^2y^2$ (B) $2x^5y - x^2y^2$ (C) $2x^6y - x^2y^3$ (D) $2x^5y - x^2y^3$ — 1

12 $\frac{a}{3} \times \frac{a}{5} =$

(A) $\frac{a}{4}$ (B) $\frac{a^2}{8}$ (C) $\frac{2a}{15}$ (D) $\frac{a^2}{15}$ — 1

13 $5 - 2(x - 4) =$

(A) $-2x - 3$ (B) $1 - 2x$ (C) $9 - 2x$ (D) $13 - 2x$ — 1

14 $\frac{2a^2b^3}{6ab^4}$

(A) $\frac{a}{3b}$ (B) $\frac{3a}{b}$ (C) $\frac{b}{3a}$ (D) $\frac{3b}{a}$ — 1

15 $(x + 2)(x - 5) =$

(A) $x^2 - 3x - 10$ (B) $x^2 + 3x - 10$ (C) $x^2 - 3x + 10$ (D) $x^2 - 7x - 10$ — 1

Total marks achieved for PART A — /15

Algebraic techniques

TOPIC TEST — PART B

Instructions
- This part consists of 5 questions.
- Write only the answer in the answer column.
- For any working use the question column.

Time allowed: 20 minutes — **Total marks: 15**

Questions	Answers	Marks
1 Simplify.		
a $(2x^3y^2)^2$		1
b $\frac{3xy}{6x^2y}$		1
c $12n^{12} \times 3n^2 \div 4n^6$		1
d $9x^2 \times 5x^3 + 3x^4 \times 6x$		1
2 Expand and simplify. $5(2x - 3) - 2(x + 1)$		1
3 Find these binomial products.		
a $(x - 6)(x + 4)$		1
b $(2x + 5)(3x + 7)$		1
4 Factorise fully.		
a $3a + 6b - 12$		1
b $6a^2b^3c - 12ab^4c^2$		1
c $x^2 + 7x + 12$		1
d $p^2 + 5p - 36$		1
5 Find in simplest form.		
a $\frac{4x}{5} - \frac{3x}{5}$		1
b $\frac{6x}{25} \times \frac{5y}{6}$		1
c $\frac{a}{3} \div \frac{3}{4}$		1
d $\frac{3x}{14} + \frac{x}{2}$		1

Total marks achieved for PART B ___ / 15

Chapter 3

Equations

UNIT 1: Simple equations

Question 1 Solve:

a $x + 5 = 9$ **b** $x - 4 = 7$ **c** $x + 6 = 3$ **d** $x - 8 = -2$

e $7 + x = 23$ **f** $a - 21 = 15$ **g** $m + 6 = -4$ **h** $14 - n = 8$

i $5x = 45$ **j** $3a = -21$ **k** $7p = 56$ **l** $-4t = -36$

m $\frac{x}{2} = 8$ **n** $\frac{a}{3} = 3$ **o** $\frac{x}{5} = -2$ **p** $\frac{m}{4} = 20$

Question 2 Solve the following equations.

a $2x + 8 = 18$ **b** $3a + 5 = 11$ **c** $6m - 1 = 41$

d $5n - 7 = 23$ **e** $4p + 9 = -3$ **f** $3k - 8 = -5$

g $9 - 2p = 1$ **h** $x - 7 = 18$ **i** $12 - 3a = 24$

j $\frac{x}{3} + 4 = 8$ **k** $\frac{a}{7} - 1 = 1$ **l** $\frac{t}{5} - 7 = -2$

m $\frac{2x}{3} = 6$ **n** $\frac{4a}{5} = 12$ **o** $\frac{3b}{7} = -6$

p $\frac{3x}{5} - 4 = 8$ **q** $\frac{5a}{6} + 3 = 13$ **r** $\frac{2n}{3} - 1 = -7$

Equations

UNIT 2: Equations with pronumerals on both sides

QUESTION 1 Solve the following equations.

a $8x + 5 = 7x + 12$

b $3x - 2 = 2x + 7$

c $5x + 8 = 4x + 3$

d $5x + 6 = 2x + 15$

e $7x - 3 = 5x + 9$

f $8x - 1 = 3x + 4$

g $9a + 5 = 4a + 5$

h $6p - 2 = p + 8$

i $5e - 6 = 2e - 3$

j $7k + 2 = 3k - 10$

k $4m + 3 = 9 - m$

l $11k - 2 = 5k - 8$

m $5x - 4 = 10 - 2x$

n $8n = 5n + 12$

o $7y + 8 = -3y + 28$

p $9n + 15 = 5n + 47$

q $8q + 7 = 31 - 4q$

r $6m - 16 = 2m + 52$

QUESTION 2 Solve, after first collecting like terms.

a $5x - 9 + 2x = 3x + 35$

b $7q + 5 - 2q - 8 = 3q - 7$

c $11a + 18 - 3a = 9a + 6 + a$

Equations

UNIT 3: Equations with grouping symbols

QUESTION **1** Solve the following equations.

a $4(x + 5) = 28$

b $3(x + 2) = 27$

c $5(x + 6) = 20$

d $7(x - 1) = 35$

e $2(3x - 2) = 8$

f $5(2x + 7) = -5$

g $3(x - 7) = 2x + 5$

h $2(4x + 5) = 7x - 3$

i $3(2x + 3) = 5x + 4$

j $7(x - 3) = 2x + 9$

k $4(3x + 1) = 7x - 11$

l $3 + 7x = 2(6x - 1)$

QUESTION **2** Solve the following equations.

a $5(a + 3) = 4(a + 4)$

b $3(x - 5) = 2(x + 4)$

c $7(m - 1) = 3(2m + 1)$

d $8(y + 2) = 3(y - 3)$

e $10(a - 1) = 4(2a + 3)$

f $3(5k - 1) = 7(k + 3)$

g $5(3m - 2) = 2(4m + 9)$

h $9(3a + 5) = 5(5a - 3)$

i $2(5m + 7) = 3(2m - 1)$

j $6(x + 5) + 5(x - 2) = 9$

k $5(2x + 3) - 2(3x - 4) = 31$

l $8k - 3(3k + 1) = 5$

Equations

UNIT 4: Equations with fractions

QUESTION **1** Solve the following equations.

a $\frac{x}{3}+2=7$

b $\frac{a}{2}-5=3$

c $\frac{n}{5}+4=7$

d $\frac{m}{6}-3=10$

e $\frac{x+3}{5}=2$

f $\frac{a-1}{4}=4$

g $\frac{x-5}{3}=-1$

h $\frac{t+6}{2}=\frac{1}{2}$

i $\frac{3x+2}{5}=4$

j $\frac{5x-3}{2}=1$

k $\frac{2k+7}{3}=5$

l $\frac{4p-1}{5}=-3$

m $\frac{4n}{5}-2=6$

n $\frac{7x}{3}+4=-3$

o $\frac{5k}{4}-\frac{1}{2}=2$

p $\frac{3e}{4}+6=0$

QUESTION **2** Solve.

a $\frac{x+4}{3}-2=2$

b $\frac{a-2}{5}+1=7$

c $\frac{m+5}{2}+3=-4$

d $\frac{4c+5}{3}-1=2$

e $\frac{3b-2}{5}+7=12$

f $\frac{7h+3}{5}+5=0$

Equations

UNIT 5: Solving problems

QUESTION 1 Write an equation for each of the following and then solve it to find the value of the unknown number.

a If 9 is added to the product of 5 and a number, the result is 29.

b If 12 is subtracted from 3 times a number, the result is 48.

c The product of a certain number and 7 is subtracted from 63 and the result is twice the number.

d If 4 times a certain number is subtracted from 25, the result is 85.

e If 15 is subtracted from a certain number we are left with $\frac{5}{6}$ of the number.

f Eight more than twice a number equals that number plus 20.

g When 24 is subtracted from 3 times a number, the result equals the number increased by 30. Find the number.

h Thirty-one more than 5 times a number equals 83 more than 3 times the number. Find the number.

i Six times a number is subtracted from 72. The result equals 27 less than 5 times the number. Find the number.

QUESTION 2 Write an equation and solve to find the unknown.

a The sum of 3 consecutive even numbers is 96. Find the numbers.

b If 12 years are added to a man's present age and this value is doubled, it is equal to 96. Find the man's present age.

c Sarah's age is 3 times Nick's age. If Sarah is 28 years older than Nick, find their ages.

Equations

UNIT 6: Equations arising from substitution in formulae

QUESTION 1 Given the formula $S = ut + \frac{1}{2}at^2$, find the value of S:

a when $u = 12$, $a = 10$ and $t = 8.5$

b when $u = 14$, $a = 9$ and $t = 7.6$

QUESTION 2 Given that $P = 2L + 2B$:

a find L when $P = 100$, $B = 8$

b find B when $P = 120$, $L = 15$

QUESTION 3 Given that $A = 84$ and $b = 12$, find the value of h in the following formulas:

a $A = \frac{1}{2}bh$

b $A = \frac{1}{2}h(a + b)$ and $a = 30$

QUESTION 4 If $v^2 = u^2 + 2as$, find:

a u if $v = 15$, $a = 7$ and $s = 16$

b a if $v = 40$, $u = 5$ and $s = 60$

QUESTION 5 Find the value of r ($r > 0$), correct to one decimal place if:

a $C = 2\pi r$ and $C = 360$

b $A = \pi r^2$ and $A = 240$

QUESTION 6 Given the formula $v = u + at$, find:

a u if $v = 48$, $a = 5$ and $t = 6$

b t if $v = 78$, $u = 30$ and $a = 8$

QUESTION 7 Given the formula $A = P(1 + \frac{r}{100})^n$, find the value of P correct to one decimal place.

a $A = 9750$, $r = 5$ and $n = 7$

b $A = 12\,500$, $r = 8$ and $n = 8$

Equations

UNIT 7: Simple inequalities

QUESTION 1 State the inequality that is graphed on each number line.

a

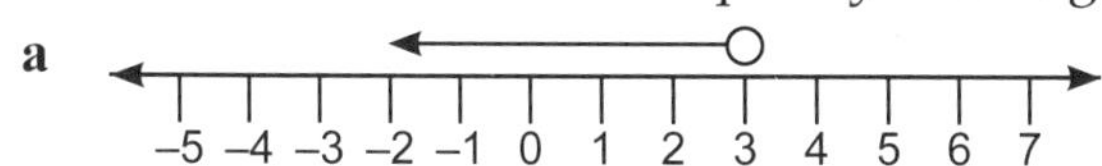

____

b

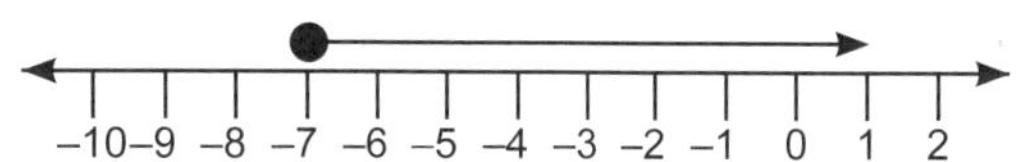

____

c

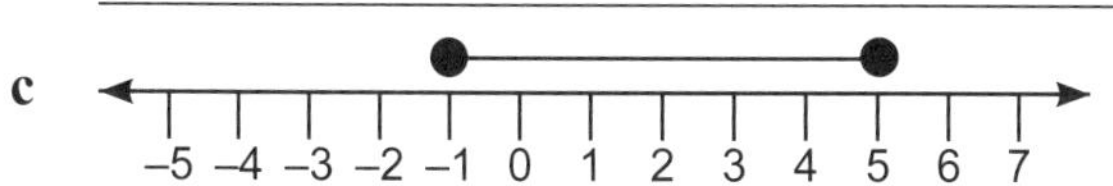

____

d

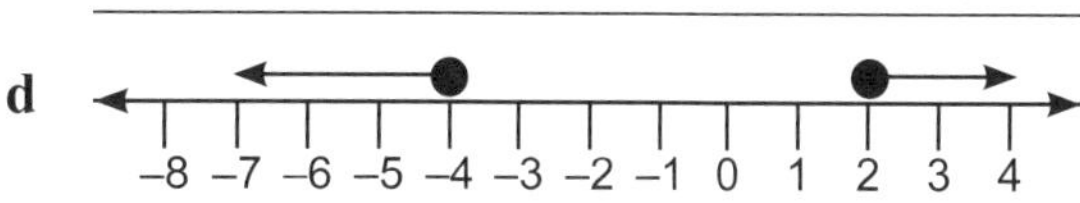

____

e

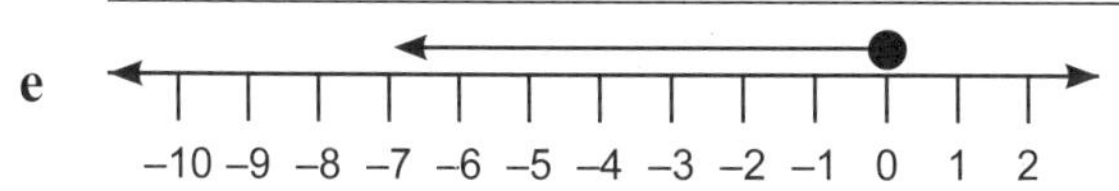

____

f

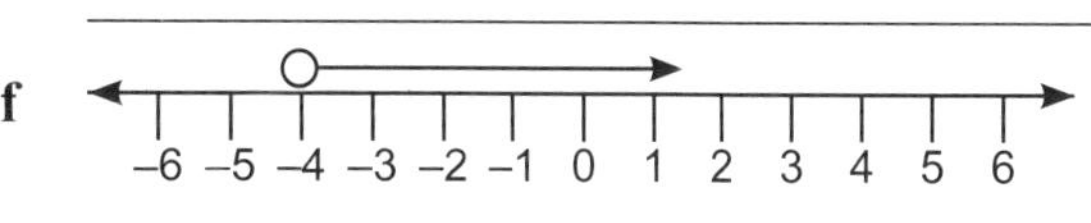

____

QUESTION 2 Graph each inequality on a number line.

a $x \geq 2$

b $x > 0$

c $x < 1$ or $x > 2$

d $x < -1$

e $-3 < x < 7$

f $x \geq 2\frac{1}{2}$

g $x \leq 3$ or $x \geq 5$

h $1 \leq x < 6$

i $-3 \leq x \leq 4$

QUESTION 3 Solve each inequality and graph the solution on a number line.

a $x + 3 < 7$

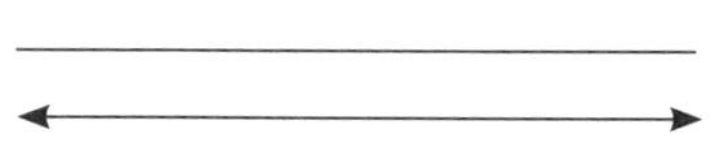

b $a - 3 > 5$

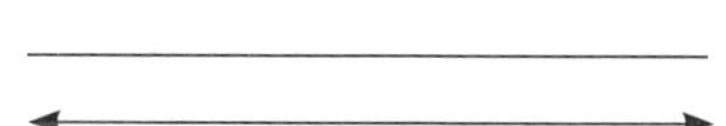

c $m + 4 > 8$

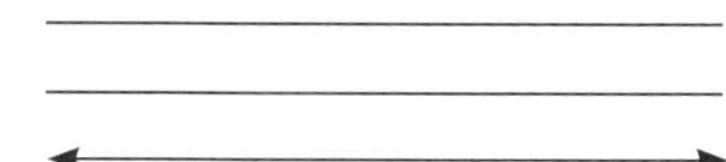

d $6 + y \leq 10$

e $15 > 8 + y$

f $x - 3 > 2$

g $m - 4 \leq 5$

h $9 \geq m - 2$

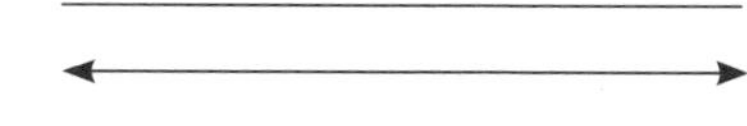

i $8 \geq x + 3$

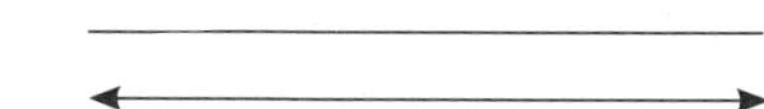

j $m - 4 \leq 10$

k $m - 2 \geq 8$

l $-6 > y + 3$

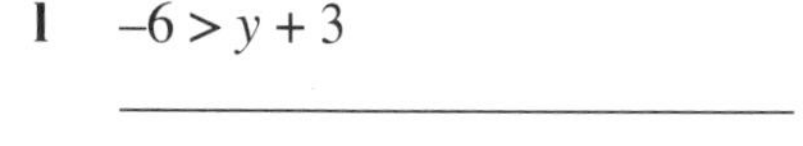

m $-3 \leq a + 1$

n $-8 \geq x - 4$

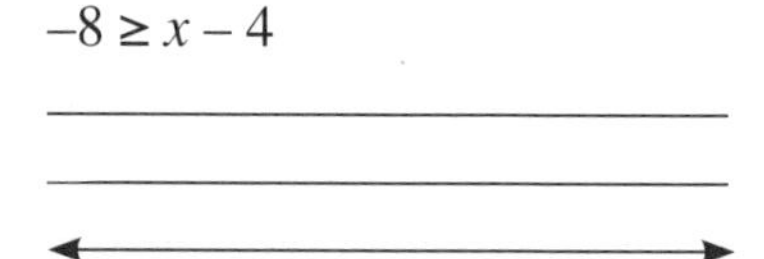

o $-5 < y - 7$

Equations

UNIT 8: Inequalities

QUESTION 1 Solve the following inequalities and graph the solutions on the number lines provided.

a $x + 1 < 3$

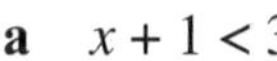

b $m + 5 > 7$

c $t - 3 < 1$

d $p + 5 \geq 2$

e $y - 2 \leq 1$

f $x + 4 \geq 6$

g $3x < 12$

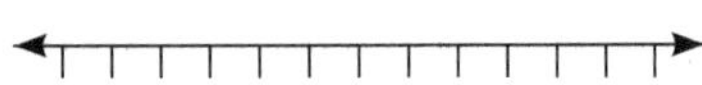

h $4p \geq 8$

i $6m < 12$

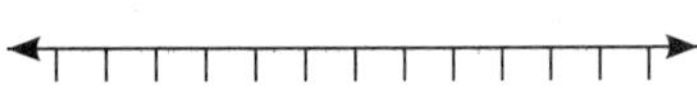

j $\frac{x}{3} \geq 1$

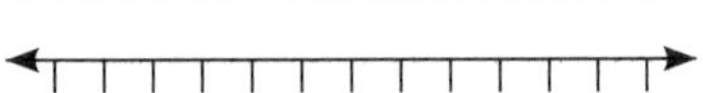

k $\frac{x}{2} < 3$

l $\frac{x}{4} \leq 2$

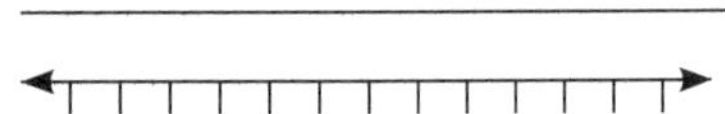

QUESTION 2 Solve the following inequalities.

a $6x < 24$

b $5x - 2 < 18$

c $3x - 1 \leq 5$

d $3x - 2 \geq -5$

e $4x - 7 \leq 9$

f $3t + 1 \geq 7$

g $6y - 7 < 5$

h $3(x + 2) \geq 12$

i $4(x - 3) \leq -16$

j $2(3x - 1) \geq 28$

k $8x < 5x + 15$

l $7x - 5 \leq 9$

m $\frac{x}{3} - 1 < 1$

n $\frac{y - 1}{2} > 2$

o $\frac{m}{3} + \frac{m}{2} \leq 1$

Equations

UNIT 9: Inequalities involving negatives

Question 1 Determine whether each inequality is True or False.

a $3 > 2$ **b** $-3 > -2$ **c** $5 < 8$ **d** $-5 < -8$

e $-2 < 8$ **f** $2 < -8$ **g** $3 > -4$ **h** $-3 > 4$

Question 2 Solve, remembering to reverse the inequality sign if multiplying or dividing by a negative number.

a $2x > 8$ **b** $-2x > 8$ **c** $-2x > -8$ **d** $2x > -8$

e $-3a \leq 9$ **f** $-5p > -10$ **g** $6k \geq 12$ **h** $-4q < -56$

i $-x < 7$ **j** $-m \geq -2$ **k** $-\frac{x}{3} \leq -6$ **l** $-\frac{x}{4} > 1$

Question 3 Solve.

a $9 - 2x \leq 7$ **b** $-3a + 5 > 11$ **c** $-4m - 1 < 23$

d $-5x + 8 \geq -2$ **e** $12 - 7x < 82$ **f** $9 + 2x \geq -5$

g $7 - \frac{x}{2} \geq 3$ **h** $\frac{7-x}{2} \geq 3$ **i** $5 - \frac{a}{10} < -2$

j $13 - 3x \geq -14$ **k** $-(\frac{x-2}{3}) \leq -5$ **l** $\frac{9-2x}{5} > 7$

UNIT 10: Solving simultaneous equations by substitution

Question 1 Solve the following pairs of simultaneous equations by substitution.

a $x + y = 10$
$y = x - 8$

b $2p - q = 12$
$p = 3 - q$

c $x + 3y = 15$
$y = x + 1$

d $x + 4y = 21$
$x = 12 - y$

e $y = 6 - x$
$2x - y = -6$

f $2x + y = 7$
$x = 4 - y$

Question 2 Use the method of substitution to solve the following pairs of simultaneous equations.

a $2m + 3n = 8$
$3m + 3n = 5$

b $2x + 3y = 12$
$4x - 3y = 6$

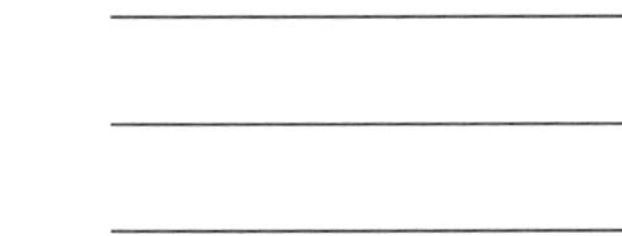

c $2x - 5y = 11$
$2x - 3y = 7$

d $y = 2x + 1$
$y = x + 4$

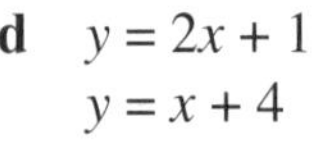

Equations

UNIT 11: Solving simultaneous equations by elimination

QUESTION **1** Solve simultaneously after adding the two equations together.

a $a + b = 11$
$a - b = 3$

b $2x + y = 17$
$x - y = 1$

c $3m - n = 22$
$2m + n = 23$

d $5p + 2q - 26 = 0$
$3p - 2q - 6 = 0$

e $7x + 4y = 113$
$5x - 4y = 19$

f $6k + 7d = 16$
$3k - 7d = 29$

QUESTION **2** Solve simultaneously after subtracting one equation from the other.

a $5x + 2y = 39$
$4x + 2y = 32$

b $9p + q = 79$
$5p + q = 47$

c $6a + 5b = 62$
$4a + 5b = 38$

d $3a - b = 17$
$a - b = 3$

e $8m - 3n = 102$
$5m - 3n = 57$

f $12x - 7y = 94$
$x - 7y = -5$

Equations

UNIT 12: Further simultaneous equations

Question 1 Multiply the second equation so that the coefficients of y are the same and then solve simultaneously.

a $3x + 2y = 21$
$x - y = 2$

b $5x + 8y = 64$
$3x + 2y = 30$

c $7x - 6y + 5 = 0$
$4x - 3y + 2 = 0$

Question 2 Solve simultaneously, using the elimination method after multiplying one or both equations.

a $5a - 4b = 9$
$3a + b = 2$

b $6x + y = 25$
$2x + 3y = 27$

c $3x + 2y = -1$
$x - 4y = -33$

d $3m + 2n = 10$
$5m + 3n = 17$

e $9a - 4b = 2$
$7a - 3b = 1$

f $a + 5b = 18$
$6a - 2b = 44$

g $2x + 7y - 1 = 0$
$5x - 3y - 64 = 0$

h $8y + 3z = 104$
$5y - 6z = 2$

i $m - 6n = 23$
$8m - 9n = 28$

Equations

UNIT 13: Equations involving powers

Question 1 Solve.

a $x^2 = 9$

b $x^2 = 16$

c $x^2 = 25$

d $x^2 = 1$

e $x^2 = 4$

f $x^2 = 64$

g $x^2 = 36$

h $x^2 = 49$

Question 2 Solve these cubic equations.

a $x^3 = 8$

b $x^3 = 1000$

c $x^3 = -1$

d $x^3 = -27$

Question 3 Solve the following equations.

a $x^2 - 100 = 0$

b $x^2 - 81 = 0$

c $x^2 - 169 = 0$

d $x^2 - 900 = 0$

e $2x^2 = 72$

f $3x^2 = 27$

g $5x^2 = 125$

h $7x^2 = 1008$

i $4x^2 - 25 = 0$

j $9x^2 - 16 = 0$

k $16x^2 - 25 = 0$

l $x^2 - 2\frac{1}{4} = 0$

m $9x^2 - 1 = 0$

n $3x^2 - 3 = 0$

o $9 - x^2 = 0$

p $2x^2 - 18 = 0$

q $4x^2 - 9 = 0$

r $25x^2 - 36 = 0$

s $5x^2 - 20 = 0$

t $(x + 5)^2 - 4 = 0$

Question 4 Solve, giving each answer to two decimal places.

a $x^2 = 23$

b $m^2 = 53$

c $5y^2 = 29$

d $k^2 - 19 = 0$

Equations

UNIT 14: Quadratic equations in factorised form

Question 1 Solve the following quadratic equations that are already expressed in factorised form.

a $(x-1)(x-2)=0$

b $(x-2)(x+3)=0$

c $(x-1)(x-3)=0$

d $x(x+5)=0$

e $2x(x-4)=0$

f $(x-3)(x-7)=0$

g $(x-3)(x-5)=0$

h $(x+1)(x-3)=0$

i $(x+2)(x-4)=0$

j $(x+3)(x-3)=0$

k $(x-2)(x+2)=0$

l $(x-5)(x+5)=0$

m $(x+1)(x-6)=0$

n $(x+2)(x+3)=0$

o $x(x+8)=0$

Question 2 Solve the following quadratic equations.

a $x(2x-1)=0$

b $(x+6)(2x-1)=0$

c $(3x-2)(x+1)=0$

d $(x-2)(3x-1)=0$

e $5x(2x-1)=0$

f $3x(x-2)=0$

g $(x+3)(3x-1)=0$

h $4x(2x-5)=0$

i $-2x(x-1)=0$

j $(3x+1)x=0$

k $(x-3)^2=0$

l $3x(x-3)=0$

Question 3 Solve the following equations.

a $(x-4)(x-5)=0$

b $(x-8)(x+8)=0$

c $x(x-3)=0$

d $2x(x-2)=0$

e $(x-7)(x-9)=0$

f $(x+1)(x-5)=0$

g $(2x-1)(x+4)=0$

h $(2x+3)(2x-3)=0$

i $(4x+5)(5x-4)=0$

Equations

UNIT 15: Equations involving a common factor

QUESTION **1** Solve the following quadratic equations.

a $x^2 - 5x = 0$

b $x^2 - 4x = 0$

c $x^2 - 2x = 0$

d $x^2 + 7x = 0$

e $x^2 + 5x = 0$

f $x^2 + 9x = 0$

g $x^2 = 4x$

h $x^2 = 9x$

i $x^2 = 12x$

j $6x^2 - 12x = 0$

k $x^2 + 8x = 0$

l $x^2 - 10x = 0$

m $3x^2 + 21x = 0$

n $5x^2 - x = 0$

o $4x^2 = -12x$

QUESTION **2** Solve the following equations.

a $6x^2 - 24x = 0$

b $5x^2 + 25x = 0$

c $9x^2 - 9x = 0$

d $8x^2 - 16x = 0$

e $3x^2 - 3x = 0$

f $6x^2 - 6x = 0$

g $6x^2 + 2x = 0$

h $3x^2 - 7x = 0$

i $5x^2 - 3x = 0$

j $7x^2 - 21x = 0$

k $9x^2 - 27x = 0$

l $8x^2 - 4x = 0$

Equations

UNIT 16: Solving quadratic equations by factorising

Question 1 Solve the following quadratic equations by factorising.

a $x^2 + 5x + 6 = 0$

b $x^2 - 2x - 35 = 0$

c $x^2 - 5x - 6 = 0$

d $x^2 + 7x + 12 = 0$

e $x^2 - 5x + 6 = 0$

f $x^2 + 2x - 48 = 0$

g $x^2 - 8x + 16 = 0$

h $x^2 + 2x - 15 = 0$

i $x^2 + 9x + 20 = 0$

j $x^2 - 8x + 15 = 0$

k $x^2 + 4x - 12 = 0$

l $x^2 - 3x - 10 = 0$

m $x^2 + 11x + 30 = 0$

n $x^2 - 9x + 14 = 0$

o $x^2 + 3x - 28 = 0$

p $x^2 - 2x - 99 = 0$

q $x^2 + 6x + 8 = 0$

r $x^2 + 6x - 7 = 0$

s $x^2 - 6x + 5 = 0$

t $x^2 + 8x + 16 = 0$

u $x^2 - 4x - 60 = 0$

Question 2 Factorise and solve the following quadratic equations.

a $x^2 = 3x + 18$

b $x^2 + 40 = 13x$

c $x^2 + 5x = 36$

d $x^2 = 15x - 54$

e $x^2 - 2x = 24$

f $x^2 = 24 - 5x$

Equations

TOPIC TEST — PART A

Instructions
- This part consists of 10 multiple-choice questions.
- Fill in only ONE CIRCLE for each question.
- Each question is worth 1 mark.

Time allowed: 15 minutes — **Total marks: 10**

Marks

1 If $2x + 5 = 19$, find the value of x.

(A) 7 (B) 9 (C) 11 (D) 12 — 1

2 If $\frac{x}{5} + 3 = 9$, then x is equal to:

(A) 1.2 (B) 2.5 (C) 30 (D) 60 — 1

3 If $2y - x = 5$, then x is equal to:

(A) $2y - 5$ (B) $5 - 2y$ (C) $-2y - 5$ (D) $2y + 5$ — 1

4 Find the value of x that satisfies the equation $5(x - 4) = 20$.

(A) 0 (B) 4 (C) 8 (D) none of these — 1

5 If $(x + 2)(x - 3) = 0$ then the values of x must be:

(A) 2 or −3 (B) −2 or −3 (C) 2 or 3 (D) −2 or 3 — 1

6 When $3(a + 7) = 42$, find the value of a.

(A) 5 (B) 6 (C) 7 (D) 8 — 1

7 If $x^2 - 9 = 0$ then the value(s) of x must be:

(A) 0 (B) 3 (C) ± 3 (D) 9 — 1

8 Find the solution of $2x - 1 > 3$:

(A) $x < 1$ (B) $x < 2$ (C) $x > 1$ (D) $x > 2$ — 1

9 Find the solution of $9 - x < 8$.

(A) $x < 1$ (B) $x > 1$ (C) $x > -1$ (D) $x < -1$ — 1

10 The solution to the simultaneous equations $2x - y = 2$ and $x + y = -5$ is:

(A) $x = -1, y = 4$ (B) $x = 1, y = 4$ (C) $x = -1, y = -4$ (D) $x = 1, y = -4$ — 1

Total marks achieved for PART A /10

Equations

TOPIC TEST — PART B

Instructions
- This part consists of 4 questions.
- Write only the answer in the answer column.
- For any working use the question column.

Time allowed: 20 minutes — **Total marks: 15**

Questions	Answers	Marks
1 Solve the following equations.		
a $9y - 15 = 30$		1
b $\frac{x+4}{5} + 2 = 3$		1
c $7x - 5 = 3x + 11$		1
d $5(x + 3) + 2(x - 1) = -8$		1
2 Solve the inequality and graph the solution on the number line provided.		
a $4x - 3 > 2x + 7$		2
b $12 - 3x \le 15$		2
3 Solve.		
a $x^2 = 144$		1
b $x^2 - 16 = 0$		1
c $x^2 - 12x + 27 = 0$		1
d $x^2 + 13x + 36 = 0$		1
e $x^2 + x - 90 = 0$		1
f $x^2 - 3x - 4 = 0$		1
4 Solve $7x + 2y = 8$ and $3x + 2y = -8$ simultaneously.		1

Total marks achieved for PART B /15

CHAPTER 4
Further algebra and indices

UNIT 1: Expanding and factorising trinomials

QUESTION 1 Write down the expansion of:

a $(x + 2)(x + 5)$

b $(x + 4)(x + 7)$

c $(x + 3)(x + 6)$

d $(x - 2)(x + 8)$

e $(x + 2)(x - 8)$

f $(x - 2)(x - 8)$

g $(x + 3)(x - 4)$

h $(x - 1)(x + 7)$

i $(x - 10)(x - 9)$

j $(x - 5)(x - 3)$

k $(x + 5)(x + 3)$

l $(x - 7)(x + 6)$

m $(x - 6)(x + 7)$

n $(x - 7)(x - 6)$

o $(x + 6)(x + 7)$

QUESTION 2 Factorise.

a $x^2 + 8x + 12$

b $x^2 + 9x + 14$

c $x^2 + 10x + 21$

d $x^2 + 11x + 30$

e $x^2 + 5x + 4$

f $x^2 + 15x + 56$

g $x^2 - 9x + 18$

h $x^2 - 3x - 28$

i $x^2 - 11x + 28$

j $x^2 + 3x - 28$

k $x^2 + 7x - 30$

l $x^2 - 2x - 35$

m $x^2 - 4x - 12$

n $x^2 + 5x - 14$

o $x^2 - x - 12$

p $x^2 + x - 72$

q $x^2 + 6x + 5$

r $x^2 - 5x + 6$

QUESTION 3 Expand and simplify.

a $(2x + 5)(x + 2)$

b $(4x + 1)(2x - 1)$

c $(7x - 5)(2x + 3)$

d $(5a - 4)(3a - 5)$

e $(3m - 7)(m + 6)$

f $(8p + 5)(7p - 9)$

Further algebra and indices

UNIT 2: Factorising by grouping

QUESTION 1 Factorise each of the following.

a $x(y + z) + 2(y + z) =$ ____________

b $a(b + 3) + 7(b + 3) =$ ____________

c $2x(m + 5) + 3(m + 5) =$ ____________

d $7(y^2 + 8) + x^2(y^2 + 8) =$ ____________

e $p(p - 3) - 2(p - 3) =$ ____________

f $t(a + 7) - 5(a + 7) =$ ____________

g $x(y - 2) - (y - 2) =$ ____________

h $a(m - n) - b(m - n) =$ ____________

i $6(x + y) + z(x + y) =$ ____________

j $x(m - n) - y(m - n) =$ ____________

k $3x(2a - 1) - 5(2a - 1) =$ ____________

l $3a(p - q) - 2(p - q) =$ ____________

QUESTION 2 Find the factors.

a $ax + ay + bx + by$

b $2a + 2b + ay + by$

c $ax + 7a + bx + 7b$

d $x^2 - x^2y + z^2 - z^2y$

e $x^3 + x^2 + x + 1$

f $ab + a + b + 1$

QUESTION 3 Factorise the following.

a $ab + ac + db + dc$

b $a^2 - ab + 7a - 7b$

c $a^3 - a^2 + 5a - 5$

d $am + an - bm - bn$

e $p^2q^2 - pq + apq - a$

f $3x^2 - 9yx + 8x - 24y$

QUESTION 4 Factorise.

a $x^3 - x^2 + 3x - 3$

b $y^3 + y^2 + y + 1$

c $9a - 9b + 4a^2 - 4ab$

d $pq^2 - p^2q + 7q - 7p$

e $am - 2m - 5a + 10$

f $3xy + 3xz + 2y + 2z$

Further algebra and indices

UNIT 3: Special products: Sum by difference

QUESTION 1 Expand and simplify the following.

a $(x + 2)(x - 2) =$ ______

b $(x + 3)(x - 3) =$ ______

c $(y + 1)(y - 1) =$ ______

d $(m + 5)(m - 5) =$ ______

e $(n + 7)(n - 7) =$ ______

f $(p + 4)(p - 4) =$ ______

g $(8 + x)(8 - x) =$ ______

h $(y + 6)(y - 6) =$ ______

i $(a + b)(a - b) =$ ______

j $(x + y)(x - y) =$ ______

k $(m + n)(m - n) =$ ______

l $(l + m)(l - m) =$ ______

QUESTION 2 Expand the following products and simplify.

a $(3a + 1)(3a - 1) =$ ______

b $(2x + 3)(2x - 3) =$ ______

c $(4a + 5)(4a - 5) =$ ______

d $(7m + n)(7m - n) =$ ______

e $(4q - 3)(4q + 3) =$ ______

f $(5x + 7)(5x - 7) =$ ______

g $(4a + 3b)(4a - 3b) =$ ______

h $(2x + y)(2x - y) =$ ______

i $(5x + 4y)(5x - 4y) =$ ______

j $(x + 9y)(x - 9y) =$ ______

k $(2a + 7b)(2a - 7b) =$ ______

l $(5m + n)(5m - n) =$ ______

m $(9a + 11b)(9a - 11b) =$ ______

n $(3a + 8b)(3a - 8b) =$ ______

QUESTION 3 Express the following as the difference of two squares.

a $(5x + 1)(5x - 1) =$ ______

b $(7a + 2)(7a - 2) =$ ______

c $(8x + 7)(8x - 7) =$ ______

d $(2x + 3y)(2x - 3y) =$ ______

e $(4x - 9y)(4x + 9y) =$ ______

f $(6x - 7y)(6x + 7y) =$ ______

g $(a - 12)(a + 12) =$ ______

h $(2x - 9)(2x + 9) =$ ______

i $(3x - 10)(3x + 10) =$ ______

j $(2m - n)(2m + n) =$ ______

k $(5 - 2q)(5 + 2q) =$ ______

l $(5x - 11)(5x + 11) =$ ______

m $(8a + 11b)(8a - 11b) =$ ______

n $(3a + 7b)(3a - 7b) =$ ______

Further algebra and indices

UNIT 4: Special products: Perfect squares

QUESTION 1 Expand and simplify the following.

a $(x + 3)^2 =$ ______

b $(y + 2)^2 =$ ______

c $(m + 7)^2 =$ ______

d $(x - 4)^2 =$ ______

e $(x - 9)^2 =$ ______

f $(x - 3)^2 =$ ______

g $(y + 11)^2 =$ ______

h $(x - 5)^2 =$ ______

i $(m - 2)^2 =$ ______

j $(x + y)^2 =$ ______

k $(a - b)^2 =$ ______

l $(m + n)^2 =$ ______

QUESTION 2 Expand and simplify.

a $(2x + 3)^2 =$ ______

b $(2m + 1)^2 =$ ______

c $(3y - 1)^2 =$ ______

d $(4a + 1)^2 =$ ______

e $(3x - 4)^2 =$ ______

f $(2x - 3y)^2 =$ ______

g $(2a + 1)^2 =$ ______

h $(5m - 1)^2 =$ ______

i $(6y + 1)^2 =$ ______

j $(3n + 2)^2 =$ ______

k $(2x + 5y)^2 =$ ______

l $(a + 3b)^2 =$ ______

m $(2x + y)^2 =$ ______

n $(x - 3y)^2 =$ ______

QUESTION 3 Expand and simplify the following.

a $(x + 3)^2 + 3(x - 1)$

b $(2a - 1)^2 - 4(a - 3)$

c $(y - 2)^2 + (y + 3)(y - 3)$

d $(a + b)^2 - (a + b)(a - b)$

e $(a + b)^2 + (a - b)^2$

f $(a + 3b)(a - 3b) + (a + b)^2$

g $(3x + 4y)(3x - 4y) + (x + y)^2$

h $(x + 1)^2 + (x + 2)^2$

Further algebra and indices

UNIT 5: Difference of two squares

Question 1 Factorise the following.

a $x^2 - 4 =$ ______ **b** $x^2 - 9 =$ ______ **c** $x^2 - 16 =$ ______

d $x^2 - 1 =$ ______ **e** $x^2 - 25 =$ ______ **f** $x^2 - 36 =$ ______

g $a^2 - b^2 =$ ______ **h** $x^2 - y^2 =$ ______ **i** $m^2 - n^2 =$ ______

j $a^2 - 49 =$ ______ **k** $y^2 - 64 =$ ______ **l** $t^2 - 81 =$ ______

m $p^2 - 4q^2 =$ ______ **n** $x^2 - 9y^2 =$ ______ **o** $m^2 - 25n^2 =$ ______

p $25a^2 - b^2 =$ ______ **q** $49x^2 - y^2 =$ ______ **r** $64p^2 - q^2 =$ ______

s $4x^2 - 9y^2 =$ ______ **t** $9m^2 - 16n^2 =$ ______ **u** $16x^2 - 25y^2 =$ ______

Question 2 Factorise each of the following.

a $x^2 - 121 =$ ______ **b** $25y^2 - 16 =$ ______ **c** $1 - 4y^2 =$ ______

d $100x^2 - 49y^2 =$ ______ **e** $y^? - 4z^2 =$ ______ **f** $1 - 25m^2 =$ ______

g $49m^2 - 100n^2 =$ ______ **h** $16a^2 - 49 =$ ______ **i** $9x^2 - 25y^2 =$ ______

j $9x^2 - 16y^2 =$ ______ **k** $a^2 - b^2c^2 =$ ______ **l** $a^2b^2 - c^2 =$ ______

m $36x^2 - 49y^2 =$ ______ **n** $p^2 - 64q^2 =$ ______ **o** $25 - 64a^2 =$ ______

Question 3 Find the factors of the following.

a $144 - 25a^2 =$ ______ **b** $a^2 - x^2 =$ ______ **c** $16x^2 - 9y^2 =$ ______

d $4x^2 - 25 =$ ______ **e** $81a^2 - 121b^2 =$ ______ **f** $4x^2 - 1 =$ ______

g $81 - z^2 =$ ______ **h** $4 - a^2b^2 =$ ______ **i** $9y^2 - 100 =$ ______

j $4a^2 - 49 =$ ______ **k** $36y^2 - x^2 =$ ______ **l** $16x^2 - 81y^2 =$ ______

m $1 - 100x^2 =$ ______ **n** $m^2 - 169 =$ ______ **o** $25x^2 - 121y^2 =$ ______

Question 4 Factorise fully.

a $x^4 - 16$ **b** $1 - x^4$ **c** $(x + 2)^2 - 9$

______ ______ ______

______ ______ ______

d $(y + 1)^2 - 25$ **e** $(x - 3)^2 - 16$ **f** $(x + 5)^2 - (x + 3)^2$

______ ______ ______

______ ______ ______

______ ______ ______

Further algebra and indices

UNIT 6: Further factorising of quadratic trinomials

QUESTION 1 Take out the highest common factor and then factorise the monic quadratic trinomial.

a $2a^2 + 10a + 12 =$ ______________________

____

b $3x^2 + 9x - 12 =$ ______________________

____

c $4x^2 + 36x + 80 =$ ______________________

____

d $2x^2 - 8x + 6 =$ ______________________

____

e $3m^2 - 27m + 60 =$ ______________________

____

f $3t^2 + 24t - 27 =$ ______________________

____

g $2x^2 + 22x + 36 =$ ______________________

____

h $4a^2 - 32a + 48 =$ ______________________

____

i $5y^2 - 15y + 10 =$ ______________________

____

j $6n^2 - 42n + 36 =$ ______________________

____

QUESTION 2 Factorise these trinomials.

a $3x^2 - 27x + 54 =$ ______________________

____

b $2y^2 - 20y + 48 =$ ______________________

____

c $4a^2 - 44a + 120 =$ ______________________

____

d $5m^2 + 25m - 70 =$ ______________________

____

e $3n^2 + 12n - 63 =$ ______________________

____

f $6p^2 + 18p - 168 =$ ______________________

____

g $4y^2 + 8y - 140 =$ ______________________

____

h $2n^2 - 2n - 84 =$ ______________________

____

QUESTION 3 Factorise.

a $am^2 + am - 20a =$ ______________________

____

b $2t^2 + 14t + 20 =$ ______________________

____

c $2y^2 - 18y + 36 =$ ______________________

____

d $3x^2 - 30x + 63 =$ ______________________

____

e $pn^2 - 12pn + 27p =$ ______________________

____

f $2x^2 - 26x + 60 =$ ______________________

____

g $a^2b + 6ab - 7b =$ ______________________

____

h $2y^2 + 16y + 14 =$ ______________________

____

Further algebra and indices

UNIT 7: Combining methods of factorising

QUESTION 1 Factorise.

a $3x^2 - 27$

b $5a^2 - 20$

c $3x^2 - 15x + 18$

d $14a - 42a^2$

e $a^4 - 16b^4$

f $16a^2 - 81b^2$

g $3x^2 - 21x + 36$

h $12t - 48t^2$

i $a^2 - 25b^2 + 4a + 20b$

j $(2m - 3n)^2 - 25p^2$

k $1 - 49t^2$

l $3a^2 - 4ab + 6a - 8b$

QUESTION 2 Factorise.

a $8y - 12y^2$

b $x^3 - x$

c $4a^2 - 8a$

d $4x^2 + 8x - 12$

e $9x - 9$

f $5t^2 + 35t + 50$

g $64 - a^2b^2c^2$

h $ab + ac + b + c$

i $a^2b^2 - c^2$

j $x^2 + 2x - 24$

k $3x^2 + 9x + 6$

l $x^2 - 16x + 39$

m $m^2n^2 - 1$

n $4a^2 - 4ax$

o $am + an - m - n$

Further algebra and indices

UNIT 8: Mixed factorisations

QUESTION 1 Factorise the following.

a $7x - 7 =$

b $x^2 - 9 =$

c $m^2 - 25 =$

d $x^2 - 2xy =$

e $-5m - 5n =$

f $ay + ab =$

g $4a^2 - 8a =$

h $2(x + y) + m(x + y) =$

i $x^2 - 121 =$

j $a^3 - 3a^2b =$

k $n^2 - 9n =$

l $9x^2 - 16y^2 =$

m $3x - 6 =$

n $-a^2 - 2a - ay =$

QUESTION 2 Factorise.

a $18y - 12y^2 =$

b $4a^2 - 4ax =$

c $ab + ac + b + c =$

d $m^2n^2 - 1 =$

e $a^2b^2 - c^2 =$

f $(x - y)^2 - z^2 =$

g $x^3 - x =$

h $m^3 + m^2 + m + 1 =$

i $xy + my - 7x - 7m =$

j $am + an - m - n =$

QUESTION 3 Factorise the following.

a $x^2 + 2x - 24 =$

b $x^2 - 6x - 27 =$

c $t^2 - 2t - 8 =$

d $x^2 - 10x + 21 =$

e $a^2 - 5a + 6 =$

f $x^2 - x - 2 =$

g $m^2 + 10m + 25 =$

h $y^2 - 9y + 20 =$

QUESTION 4 Find the factors.

a $4x^2 + 8x - 12 =$

b $2x^2 - 10x + 12 =$

c $3x^2 + 9x + 6 =$

d $2x^2 + 6x + 4 =$

e $9x^2 - 9x - 18 =$

f $3x^2 - 9x - 30 =$

Further algebra and indices

UNIT 9: Further algebraic fractions

Question 1 Add the following algebraic fractions.

a $\frac{5a}{3}+\frac{a}{3}=$ ____________

b $\frac{7x}{10}+\frac{3x}{10}=$ ____________

c $\frac{8m}{5}+\frac{2m}{5}=$ ____________

d $\frac{a}{2}+\frac{a}{3}=$ ____________

e $\frac{2a}{5}+\frac{a}{3}=$ ____________

f $\frac{3a}{7}+\frac{2a}{5}=$ ____________

g $\frac{3}{x}+\frac{5}{2x}=$ ____________

h $\frac{7}{9x}+\frac{4}{3x}=$ ____________

i $\frac{3m}{2n}+\frac{m}{6n}=$ ____________

Question 2 Simplify the following.

a $\frac{7a}{4}-\frac{3a}{4}=$ ____________

b $\frac{5x}{9}-\frac{2x}{9}=$ ____________

c $\frac{11m}{15}-\frac{6m}{15}=$ ____________

d $\frac{2x}{3}-\frac{x}{2}=$ ____________

e $\frac{3a}{10}-\frac{a}{5}=$ ____________

f $\frac{4q}{5}-\frac{q}{15}=$ ____________

g $\frac{3}{x}-\frac{5}{2x}=$ ____________

h $\frac{7}{9x}-\frac{4}{3x}=$ ____________

i $\frac{3m}{2n}-\frac{m}{6n}=$ ____________

Question 3 Simplify the following products.

a $\frac{x}{4}\times\frac{y}{5}=$ ____________

b $\frac{a}{3}\times\frac{a}{4}=$ ____________

c $\frac{a}{b}\times\frac{m}{n}=$ ____________

d $\frac{ab}{10}\times\frac{5}{a}=$ ____________

e $\frac{x^2}{y^2}\times\frac{y^2}{x^2}=$ ____________

f $\frac{8m}{3a}\times\frac{12a}{16m}=$ ____________

g $\frac{pq}{m}\times\frac{3m}{p}=$ ____________

h $\frac{12m}{5}\times\frac{10m}{9}=$ ____________

i $\frac{2x}{y}\times\frac{4y}{3x}=$ ____________

Question 4 Divide the following.

a $\frac{x}{3}\div\frac{x}{9}=$ ____________

b $\frac{2n}{7}\div\frac{3n}{14}=$ ____________

c $\frac{p}{5}\div\frac{7p}{10}=$ ____________

d $\frac{6}{x}\div\frac{3}{x}=$ ____________

e $\frac{5}{2y}\div\frac{10}{y}=$ ____________

f $\frac{x}{y}\div\frac{3x}{y}=$ ____________

g $\frac{8m}{5n}\div\frac{4m}{15n}=$ ____________

h $\frac{ab}{c}\div\frac{b}{ac}=$ ____________

i $\frac{12a}{b}\div\frac{5a}{7b}=$ ____________

Further algebra and indices

UNIT 10: Negative indices

Question 1 Write the following with positive indices.

a $2^{-5} =$ ______ b $7^{-2} =$ ______ c $3^{-4} =$ ______

d $5^{-6} =$ ______ e $8^{-3} =$ ______ f $10^{-8} =$ ______

g $x^{-4} =$ ______ h $a^{-2} =$ ______ i $9m^{-4} =$ ______

j $(-7)^{-3} =$ ______ k $\frac{1}{2^{-4}} =$ ______ l $\left(\frac{5}{6}\right)^{-2} =$ ______

Question 2 Evaluate the following.

a $3^{-2} =$ ______ b $2^{-3} =$ ______ c $4^{-3} =$ ______

d $5^{-3} =$ ______ e $10^{-5} =$ ______ f $(3^{-1})^3 =$ ______

g $\left(\frac{2}{3}\right)^{-2} =$ ______ h $\left(\frac{3}{2}\right)^{-3} =$ ______ i $\left(\frac{1}{4}\right)^{-2} =$ ______

j $\left(\frac{1}{2}\right)^{-4} =$ ______ k $\left(\frac{1}{3}\right)^{-3} =$ ______ l $\left(\frac{5}{6}\right)^{-2} =$ ______

Question 3 Write the following with negative indices.

a $\frac{1}{9} =$ ______ b $\frac{1}{3^5} =$ ______ c $\frac{1}{a} =$ ______

d $\frac{1}{y^2} =$ ______ e $\frac{4}{x^3} =$ ______ f $\frac{8}{x^5} =$ ______

g $\frac{7}{y} =$ ______ h $\frac{6}{a^4} =$ ______ i $\frac{1}{3x^4} =$ ______

j $\frac{a}{5^3} =$ ______ k $\frac{1}{5m^2} =$ ______ l $\frac{9n}{3m^3} =$ ______

Question 4 Simplify the following, giving your answers as fractions.

a $5^{-2} =$ ______ b $6^{-3} =$ ______ c $2^{-6} =$ ______

d $3^{-2} \times 2^{-1} =$ ______ e $7 \times 2^{-3} =$ ______ f $8^{-1} =$ ______

g $5^{-3} \times 5^0 =$ ______ h $8 \times 10^{-2} =$ ______ i $5 \times 10^{-3} =$ ______

j $2^{-3} \times 3^{-1} =$ ______ k $5^{-7} \div 5^{-9} =$ ______ l $(3^{-1})^3 =$ ______

Question 5 Find the value of x in the following.

a $10^{-3} = 10^x$ ______ b $10^{-3} = \frac{1}{10^x}$ ______ c $\frac{1}{9} = 9^x$ ______

d $5^3 \times 5^{-7} = 5^x$ ______ e $10^7 \div 10^{-4} = 10^x$ ______ f $3^8 \times \frac{1}{3^4} = 3^x$ ______

g $\left(\frac{2}{5}\right)^7 \times \left(\frac{2}{5}\right)^{-3} = \left(\frac{2}{5}\right)^x$ ______ h $5^{-6} \div 5^3 = 5^x$ ______ i $\frac{1}{8^{-3}} = 8^x$ ______

j $6^5 \times 6^{-3} = 6^x$ ______ k $7^8 \div 7^5 = 7^x$ ______ l $5^{-3} \div 5^2 = 5^x$ ______

Question 6 Simplify the following, leaving answers in index form.

a $3^8 \times 3^{-4} \times 3^{-2} =$ ______ b $2^8 \div 2^2 \div 2^3 =$ ______ c $5^7 \div 5^8 \div 5^2 =$ ______

d $(8^2)^{-7} =$ ______ e $7^4 \times 7^8 \div 7^5 =$ ______ f $(6^3)^{-2} \times (6^2)^{-5} =$ ______

g $(7^2)^{-3} \div 7^8 =$ ______ h $(4^7)^{-3} =$ ______ i $8^5 \times 8^3 \div 8^{10} =$ ______

j $(4^5)^{-2} \times 4 =$ ______ k $\frac{(x^3)^2 \div (x^{-1})^3}{x^{-3}} =$ ______ l $a^5 \times a^{-4} =$ ______

Further algebra and indices

UNIT 11: Further indices

QUESTION 1 Use the index laws to simplify:

a $x^4 \times x^5 \times x^8 =$ ____________ b $a^9 \times a^3 \div a^7$ ____________ c $m^6 \div m^2 \div m$ ____________

d $12n^{12} \div 6n^0$ ____________ e $(4p^3q^5)^2$ ____________ f $24k^8 \div 3k^3 \times 2k^2$ ____________

g $14h^{-8} \div 7h^4$ ____________ h $6y^{-2} \times 5y^{-5}$ ____________ i $15q^5 \div 3q^{-6}$ ____________

j $20x^3y^4 \div 5x^2y^{-1}$ ____________ k $3ab^2 \times 4a^2b^3 \div 6ab^{-6}$ ____________ l $8p^3q^5 \div 2p^4q \div 2pq^2$ ____________

QUESTION 2 Simplify, leaving answers in fractional form where necessary.

a $\dfrac{x^3x^5}{x^6}$ ____________ b $\dfrac{a^5b^3}{a^3b^7}$ ____________ c $\dfrac{2p^5q^2}{4p^4q^3}$ ____________

d $\dfrac{n^9 \div n^3}{n^7 \times n^5}$ ____________ e $\dfrac{12z^4}{3z^{-3} \times 2z^2}$ ____________ f $\dfrac{(2q^2)^2}{4q^{-5}}$ ____________

g $\left(\dfrac{x^5}{x^{-2}}\right)^3$ ____________ h $\left(\dfrac{x^4y^3}{x^3y^6}\right)^2$ ____________ i $\left(\dfrac{a^7 \div a^2}{a^3 \times a^2}\right)^4$ ____________

QUESTION 3 Simplify, giving answers without negative indices.

a $\left(\dfrac{x^2}{y^3}\right)^{-4}$ ____________ b $\left(\dfrac{6k^3}{2k^4}\right)^{-1}$ ____________ c $\left(\dfrac{2a^5}{4a}\right)^{-1}$ ____________

d $\left(\dfrac{2x^2}{7}\right)^{-2}$ ____________ e $\left(\dfrac{k^2}{m^3}\right)^{-3}$ ____________ f $\left(\dfrac{8n^3}{4n^5}\right)^{-1}$ ____________

g $\left(\dfrac{a^6}{a^3x^2}\right)^{-5}$ ____________ h $\left(\dfrac{4}{2p^{-3}}\right)^{-1}$ ____________ i $\left(\dfrac{4x^3 \times 3x^2}{6x^4}\right)^{-2}$ ____________

QUESTION 4

a Write down the value of:

i $(\sqrt{4})^2 =$ ________ ii $(\sqrt{9})^2 =$ ________ iii $(\sqrt{25})^2 =$ ________ iv $(\sqrt{100})^2 =$ ________

b Use the index laws to simplify and evaluate:

i $(4^{\frac{1}{2}})^2 =$ ________ ii $(9^{\frac{1}{2}})^2 =$ ________ iii $(25^{\frac{1}{2}})^2 =$ ________ iv $(100^{\frac{1}{2}})^2 =$ ________

c Complete: Finding the square root of a number is the same as raising the number to the power of

____.

d Find the value of: i $81^{\frac{1}{2}} =$ ________ ii $16^{\frac{1}{2}} =$ ________ iii $121^{\frac{1}{2}} =$ ________

iv $36^{\frac{1}{2}} =$ ________ v $144^{\frac{1}{2}} =$ ________ vi $64^{\frac{1}{2}} =$ ________ vii $49^{\frac{1}{2}} =$ ________

Further algebra and indices

TOPIC TEST — PART A

Instructions
- This part consists of 10 multiple-choice questions.
- Fill in only ONE CIRCLE for each question.
- Each question is worth 1 mark.

Time allowed: 15 minutes | **Total marks: 10**

Marks

1 $(2x + 3)(x - 5) =$

(A) $2x^2 + 7x - 15$ (B) $2x^2 - 7x + 15$ (C) $2x^2 + 7x + 15$ (D) $2x^2 - 7x - 15$ | 1

2 The complete factorisation of $3ab - 6a$ is:

(A) $3(ab - 6a)$ (B) $3a(a - 2)$ (C) $3a(b - 2)$ (D) $3a(b - 6)$ | 1

3 $(2x - 5y)^2 =$

(A) $4x^2 - 10xy - 25y^2$ (B) $4x^2 - 10xy + 25y^2$
(C) $4x^2 - 20xy - 25y^2$ (D) $4x^2 - 20x^2 + 25y^2$ | 1

4 $\frac{15 - 5p}{5} =$

(A) $2p$ (B) $3 - p$ (C) $3 - 5p$ (D) $15 - p$ | 1

5 $(7 - 2m)(7 + 2m) =$

(A) $49 - 2m^2$ (B) $49 - 4m^2$ (C) $49 - 28m + 4m^2$ (D) $4m^2 - 49$ | 1

6 When fully factorised, $x^4 - 1 =$

(A) $(x^2 - 1)(x^2 + 1)$ (B) $(x^2 - 1)^2$
(C) $(x^2 + 1)(x + 1)(x - 1)$ (D) $(x - 1)^2(x + 1)^2$ | 1

7 $x^2 - 5x - 6 =$

(A) $(x - 3)(x + 2)$ (B) $(x - 2)(x + 3)$ (C) $(x - 6)(x + 1)$ (D) $(x - 5)(x - 1)$ | 1

8 $\frac{12a}{5b} \div \frac{3a}{10b} =$

(A) 8 (B) 2 (C) $\frac{1}{8}$ (D) $\frac{1}{2}$ | 1

9 $\frac{x^5 \times x^{-3}}{x^8 \div x^{-2}} =$

(A) x^8 (B) $\frac{1}{x^8}$ (C) x (D) $\frac{1}{x}$ | 1

10 $mn - ml - kl + kn$ expressed as a product of factors is:

(A) $(m + k)(n - l)$ (B) $(m - k)(n + l)$ (C) $(m + k)(l - n)$ (D) $(m - k)(n - l)$ | 1

Total marks achieved for PART A ___ /10

Further algebra and indices

TOPIC TEST PART B

Instructions
- This part consists of 4 questions.
- Write only the answer in the answer column.
- For any working use the question column.

Time allowed: 20 minutes **Total marks: 15**

Questions	Answers	Marks
1 Expand and simplify:		
a $(3x-5)(2x+3)$ **b** $(5a-1)(4a-7)$		1 1
c $(3a-2)(3a+2)$ **d** $(x+3)^2$		1 1
2 Factorise fully:		
a m^2-36 **b** $x(a-b)+y(a-b)$		1 1
c $m^2-2m-80$ **d** $a^2-9a+18$		1 1
e $2n^2+16n+30$ **f** $4-4x^2$		1 1
3 Find:		
a $\frac{7x}{3y}-\frac{2x}{5y}$ **b** $\frac{6a^2}{5b}\div\frac{9a^4}{2b^3}$		1 1
4 Simplify:		
a $\frac{5t^3u^2\times 2t^3u}{6tu\times 4u^4}$ **b** $\frac{7x^{-2}\times 2x^4}{3x^9\div x^{-4}}$ **c** $\left(\frac{4x^3}{3x^5}\right)^{-1}$		1 1 1

Total marks achieved for PART B $\frac{\quad}{15}$

CHAPTER 5
Right-angled triangles and trigonometry

UNIT 1: Review of Pythagoras' theorem

QUESTION 1 Find the length of the hypotenuse.

a

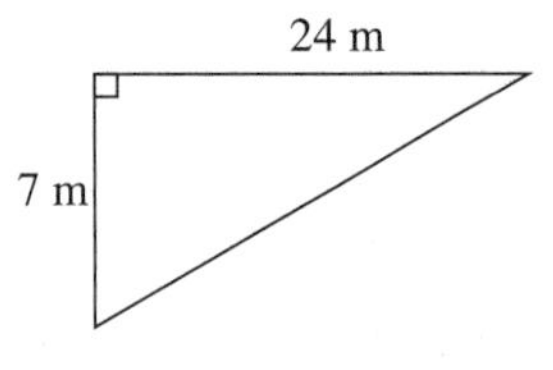

b

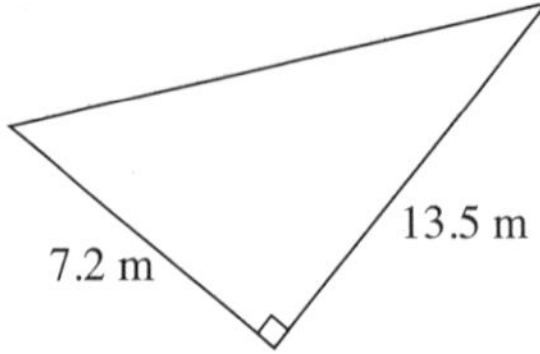

c

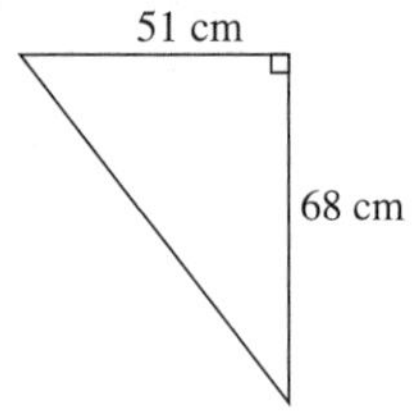

QUESTION 2 Find the length of the unknown side.

a

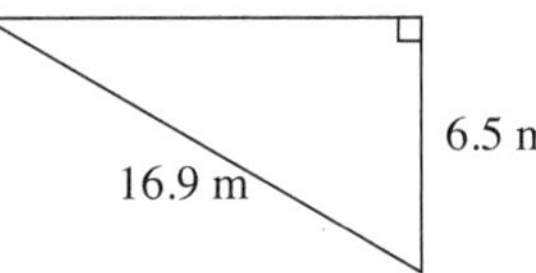

b

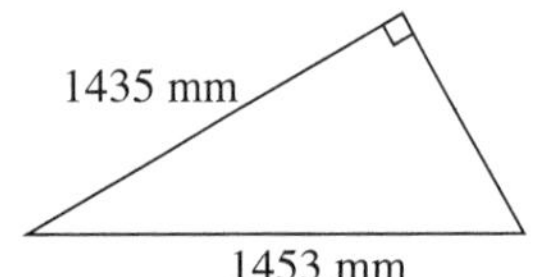

c

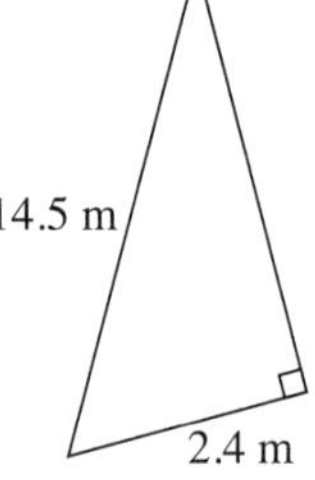

QUESTION 3 Find the value of x. Give your answer correct to one decimal place if necessary.

a

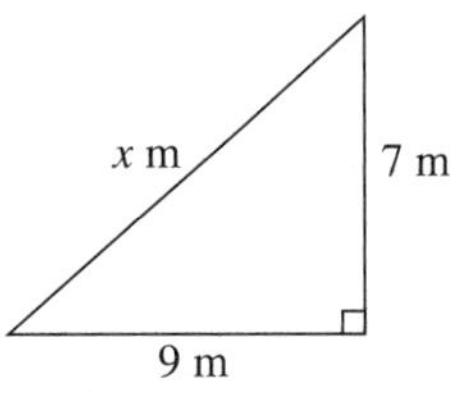

b

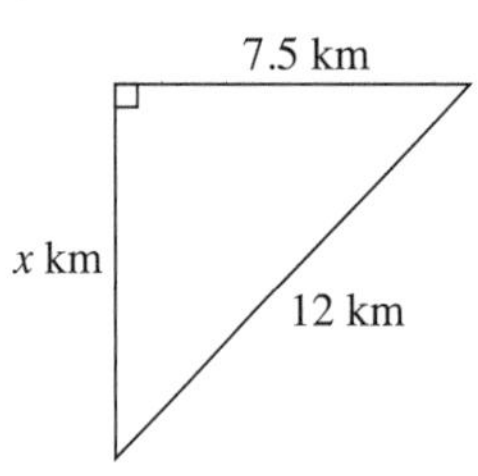

c

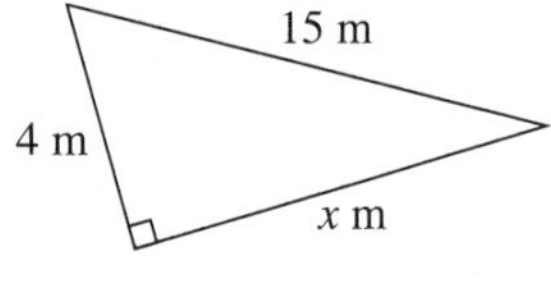

d

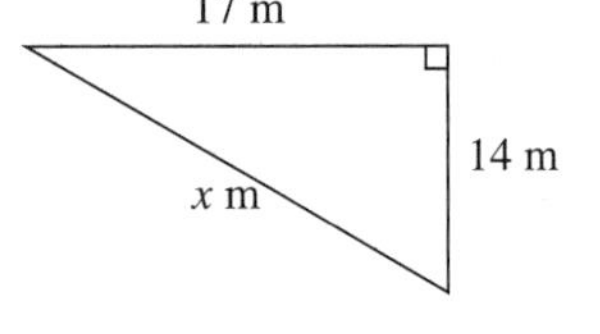

e

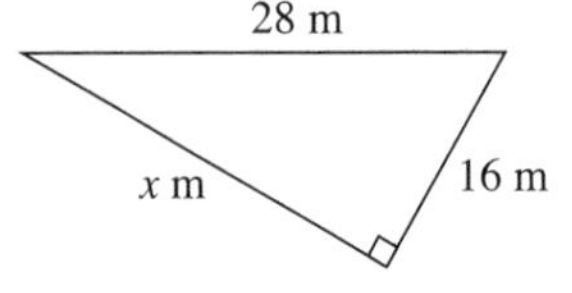

f

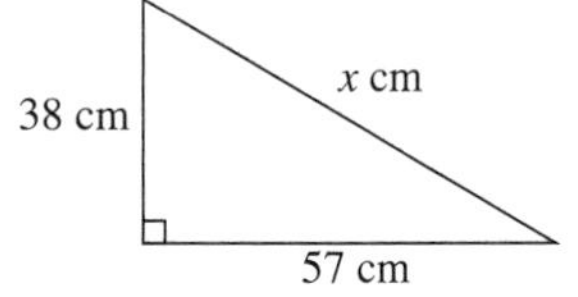

Right-angled triangles and trigonometry

UNIT 2: The trigonometric ratios

QUESTION 1 In each of the following triangles, state whether a, b and c are the opposite side, adjacent side or hypotenuse with reference to the angle marked.

a

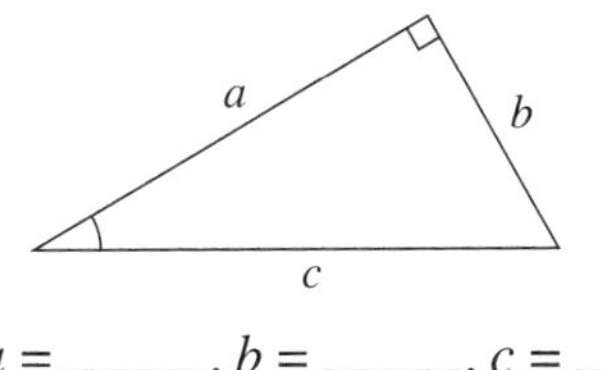

$a =$ ______, $b =$ ______, $c =$ ______

b

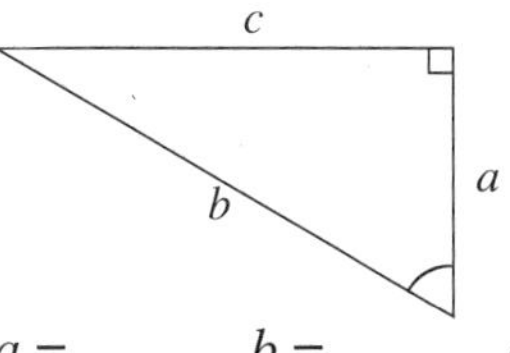

$a =$ ______, $b =$ ______, $c =$ ______

c

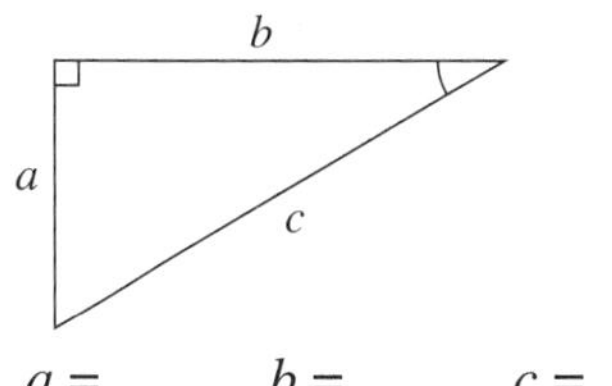

$a =$ ______, $b =$ ______, $c =$ ______

QUESTION 2 Name the hypotenuse in each triangle given below.

a

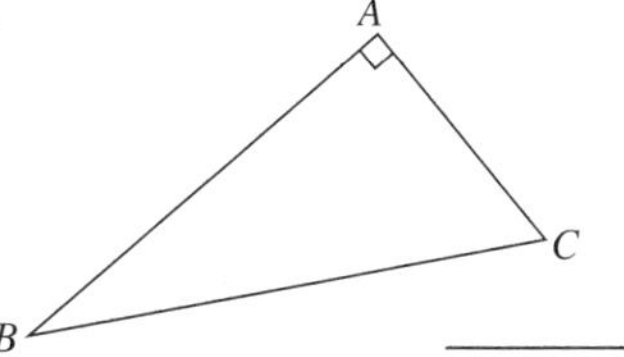

______

b

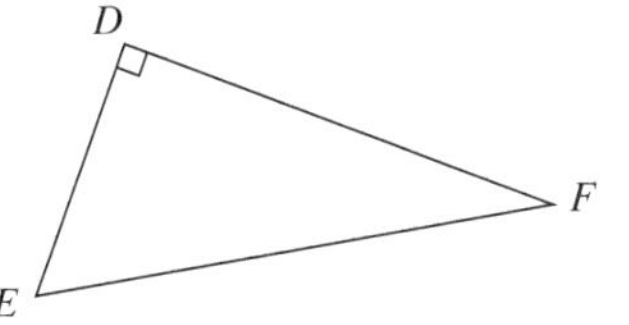

______

c

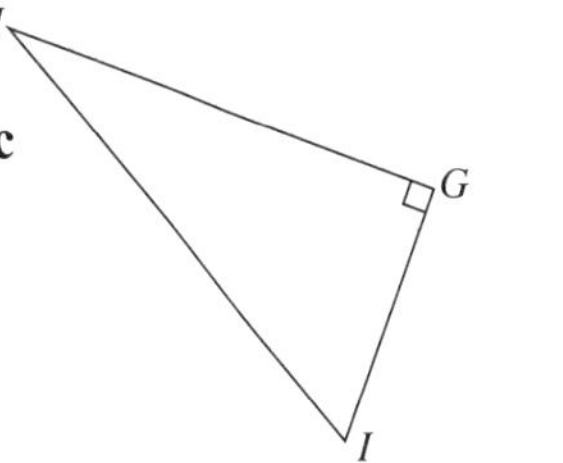

______

QUESTION 3 Write the trigonometric ratios (sine, cosine and tangent) for the following triangles.

a

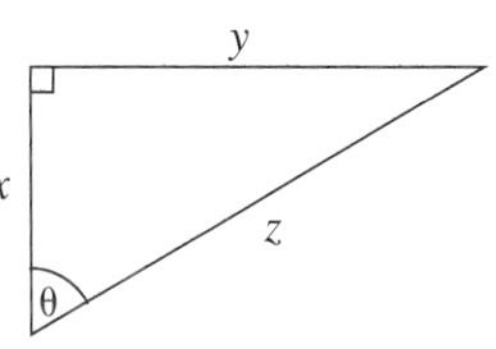

______

______

______

b

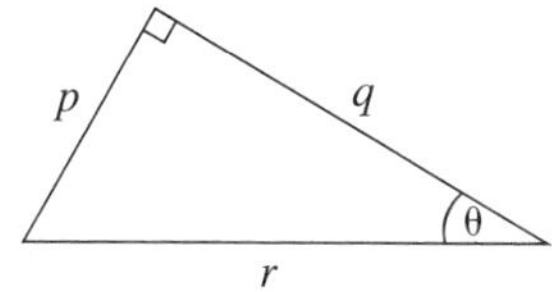

______

______

______

c

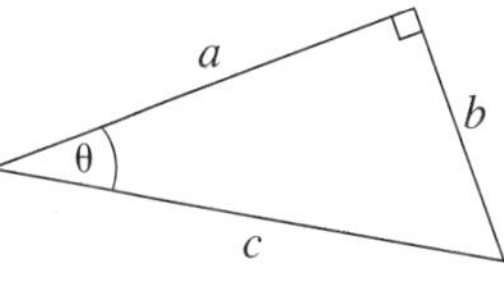

______

______

______

QUESTION 4 Write the fractions equal to $\sin \theta$, $\cos \theta$ and $\tan \theta$ in each of the following triangles.

a

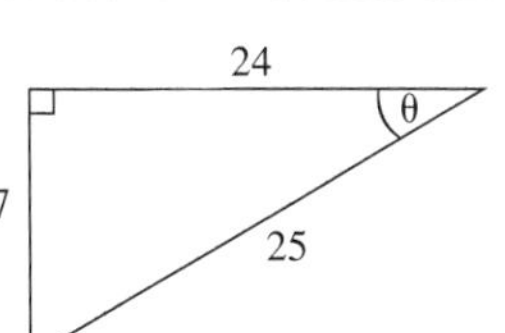

______

______

______

b

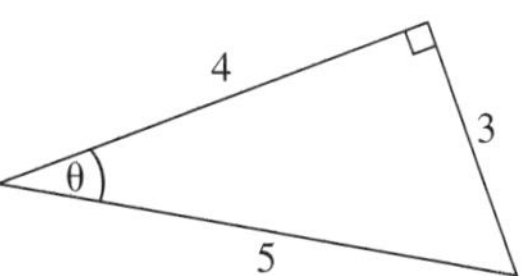

______

______

______

c

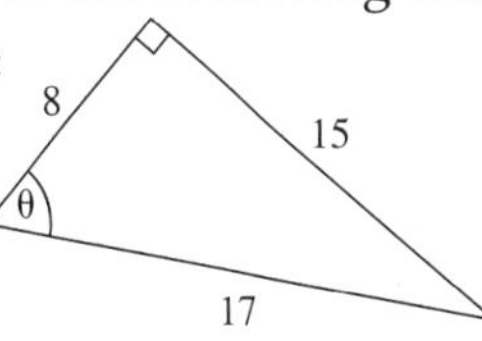

______

______

______

QUESTION 5 Which ratio (sin, cos or tan) could be used to find the angle θ?

a

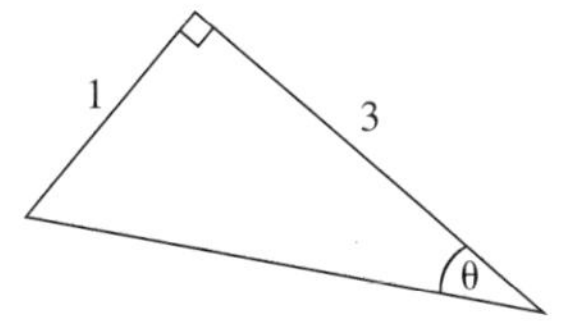

______

b

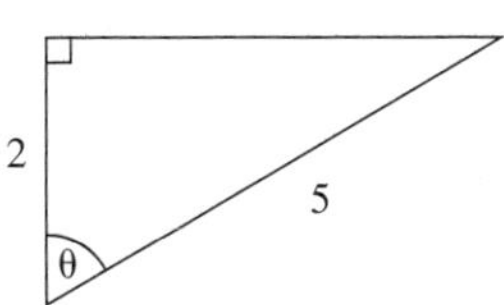

______

c

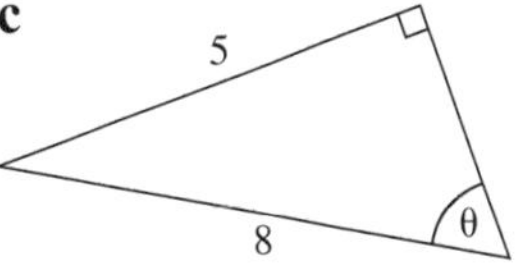

______

Right-angled triangles and trigonometry

UNIT 3: Using a calculator with trigonometric ratios

Question 1 Express to the nearest degree.

a 27°15' ______ **b** 46°19'32" ______ **c** 29°34'8" ______

d 78.325° ______ **e** 77.638° ______ **f** 82.5° ______

g 21°34' ______ **h** 55°18'59" ______ **i** 64°43'32" ______

Question 2 Round off to the nearest minute.

a 83°24'36" ______ **b** 89°34'27" ______ **c** 63°28'18" ______

d 27°15'32" ______ **e** 41°45'26" ______ **f** 30°45'32" ______

g 24.76° ______ **h** 57.349° ______ **i** 54.469° ______

Question 3 Find the value of the following, correct to three decimal places.

a sin 58° = ______ **b** tan 40° ______ **c** cos 38° ______

d cos 82° = ______ **e** sin 60° ______ **f** tan 54° ______

Question 4 Find the value, correct to three significant figures.

a 1.5 sin 36° = ______ **b** tan 68°28' = ______ **c** cos 39°41' ______

d 7 cos 25° = ______ **e** sin 73°25' ______ **f** tan 51°36' ______

g 81.6 cos 60° = ______ **h** 52.63 sin 78° = ______ **i** 8.34 tan 61°25' = ______

Question 5 Use a calculator to find the value, correct to three decimal places.

a $\frac{\sin 56^\circ}{8.3} =$ ______ **b** $\frac{\cos 83^\circ}{2.5} =$ ______ **c** $\frac{25.8}{\sin 23^\circ 8'} =$ ______

d $\frac{\cos 59^\circ 35'}{3.4} =$ ______ **e** $\frac{\sin 81^\circ}{5.4} =$ ______ **f** $\frac{14.932}{\cos 18^\circ 32'} =$ ______

g $\frac{\tan 72^\circ 18'}{5} =$ ______ **h** $\frac{\tan 69^\circ}{3.2} =$ ______ **i** $\frac{120.96}{\tan 65^\circ 28'} =$ ______

Question 6 A is an acute angle. Find its size to the nearest degree.

a $\sin A = 0.5671$ ______ **b** $\cos A = 0.5632$ ______ **c** $\tan A = 3.3815$ ______

d $\cos A = 0.8321$ ______ **e** $\tan A = 2.6815$ ______ **f** $\cos A = 0.6953$ ______

g $\tan A = 1.3654$ ______ **h** $\sin A = 0.3496$ ______ **i** $\sin A = 0.8325$ ______

Question 7 B is an acute angle. Find its size in degrees and minutes.

a $\tan B = 1.6837$ ______ **b** $\sin B = 0.3153$ ______ **c** $\cos B = 0.5673$ ______

d $\sin B = 0.3459$ ______ **e** $\cos B = 0.4567$ ______ **f** $\tan B = 0.8364$ ______

g $\cos B = 0.8621$ ______ **h** $\tan B = 2.8327$ ______ **i** $\sin B = 0.5389$ ______

Right-angled triangles and trigonometry

UNIT 4: Finding a side

QUESTION **1** Find the length of the unknown side correct to one decimal place.

a

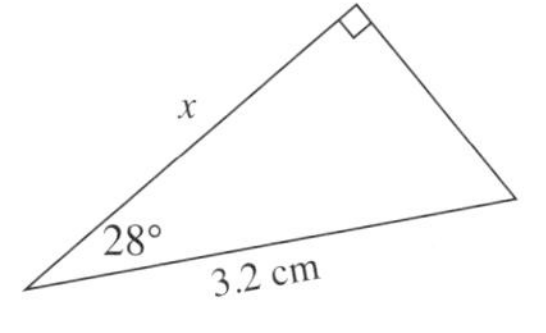

b

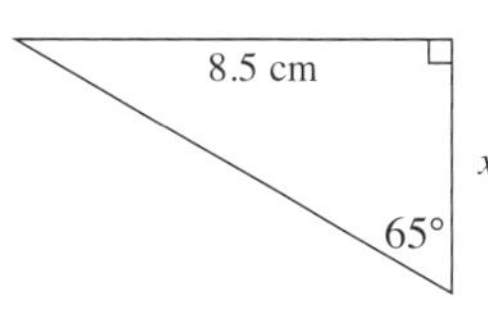

c

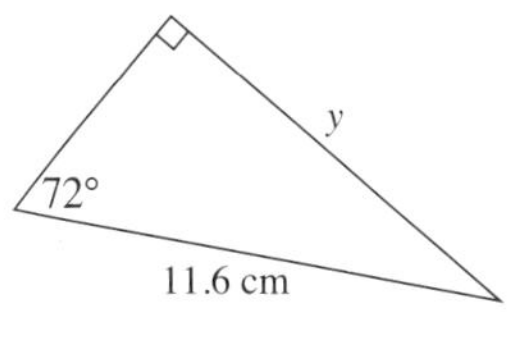

d

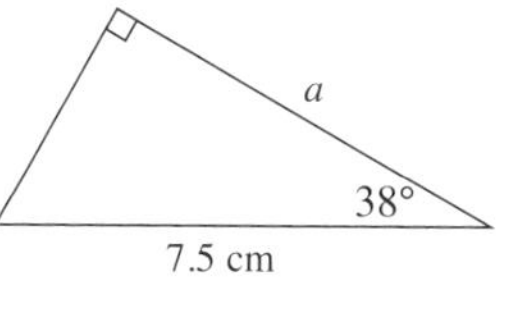

e

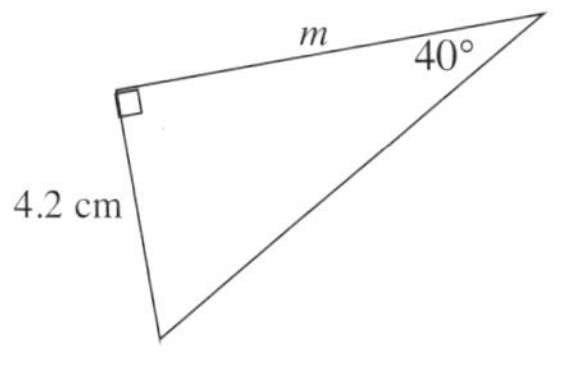

f

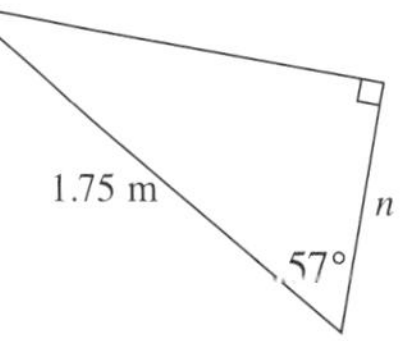

QUESTION **2** Find the value of the pronumeral in each triangle correct to two decimal places.

a

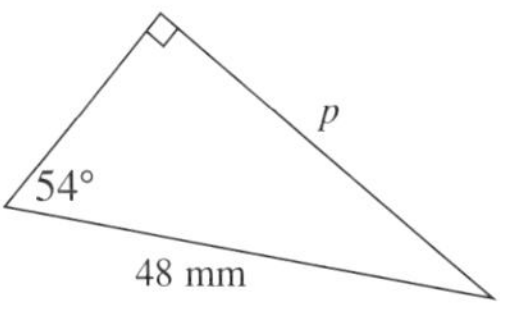

b

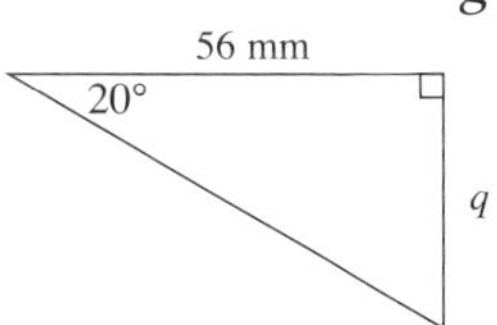

c

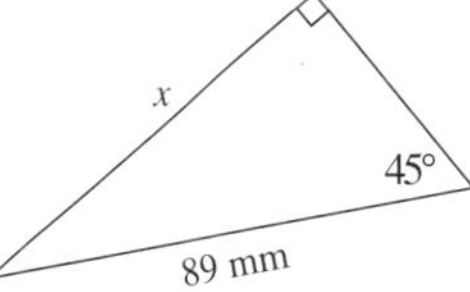

d

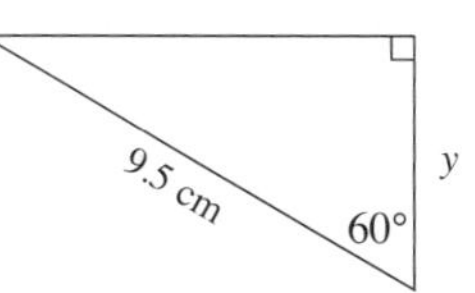

e

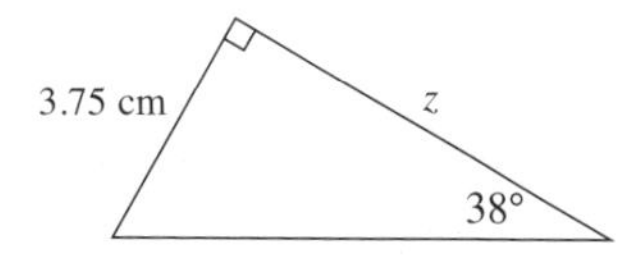

f

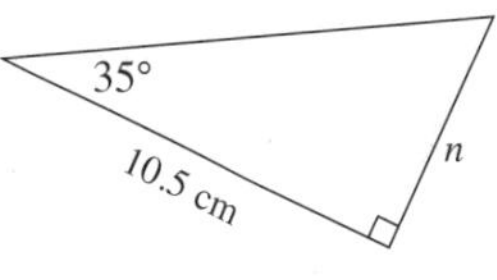

Right-angled triangles and trigonometry

UNIT 5: Finding the hypotenuse

QUESTION 1 Find the length of the hypotenuse correct to two decimal places.

a

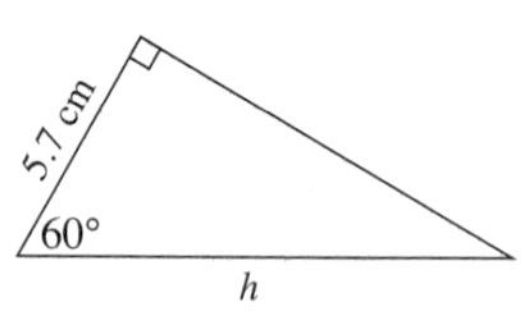

b

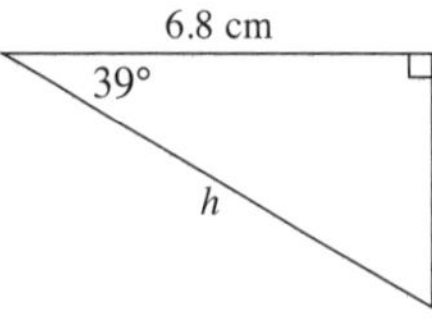

c

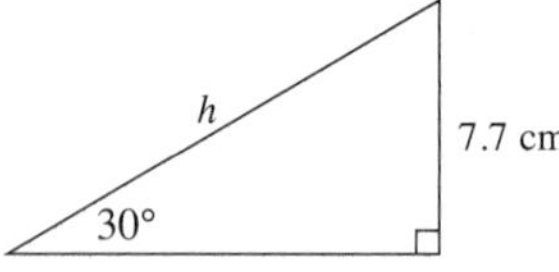

d

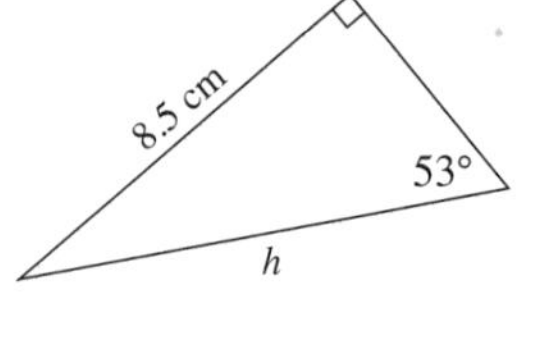

e

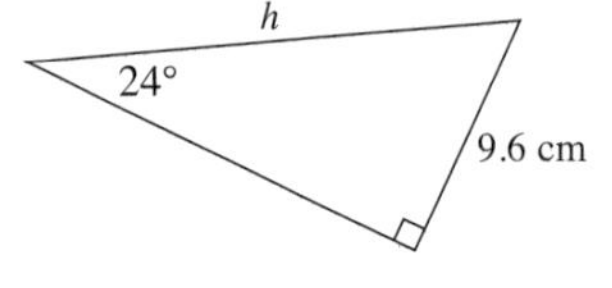

f

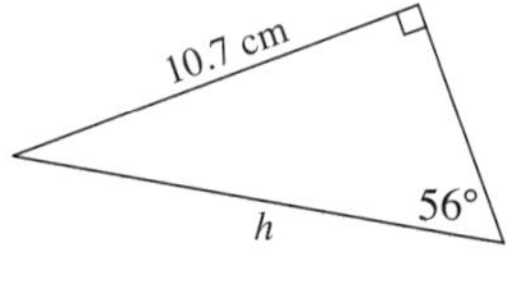

QUESTION 2 Find the length of the hypotenuse correct to one decimal place.

a

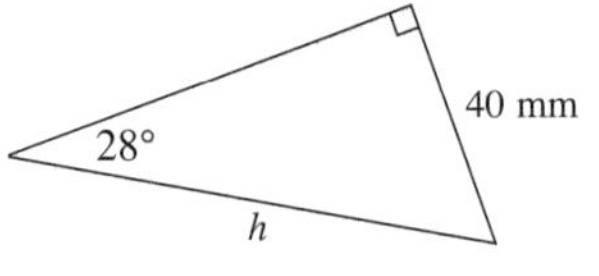

b

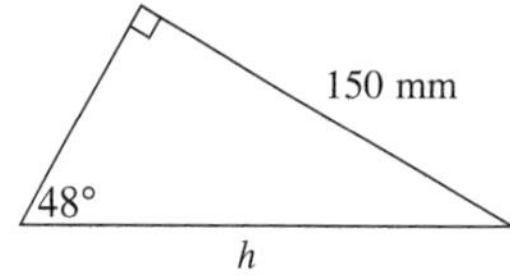

c

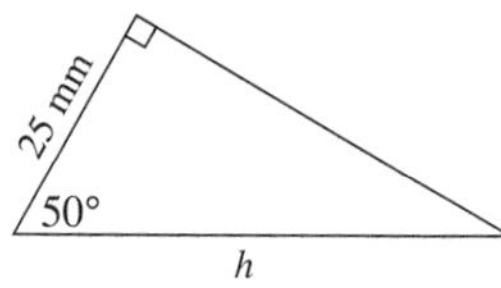

d

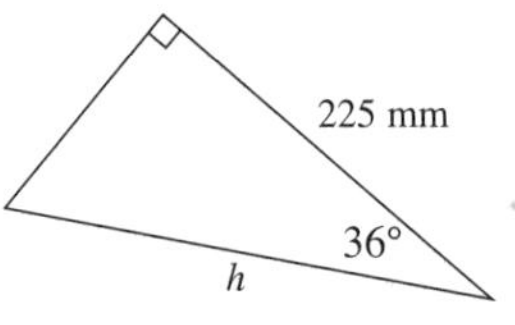

e

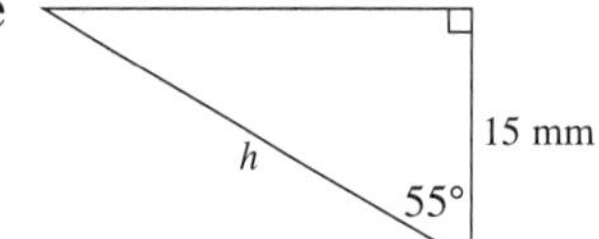

f

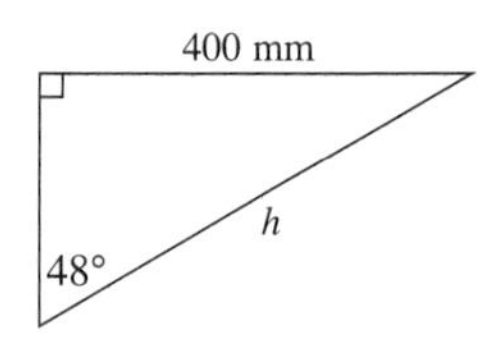

QUESTION 3 Find the length of the hypotenuse correct to two decimal places.

a

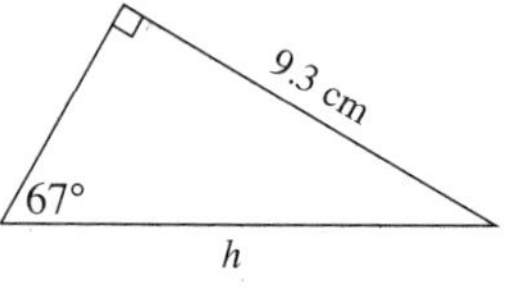

b

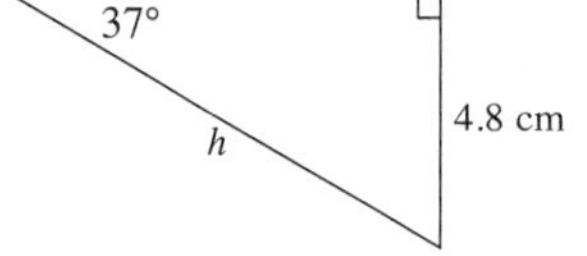

c

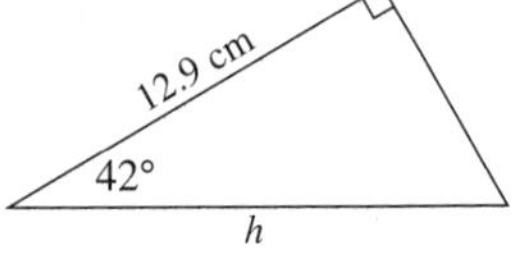

Right-angled triangles and trigonometry

UNIT 6: Finding an unknown angle

QUESTION 1 Find the size of angle B. Give the answer to the nearest degree.

a

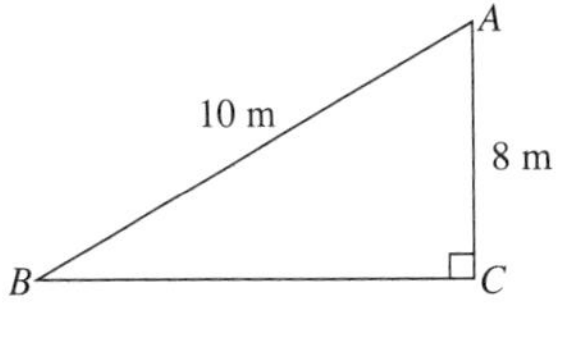

b

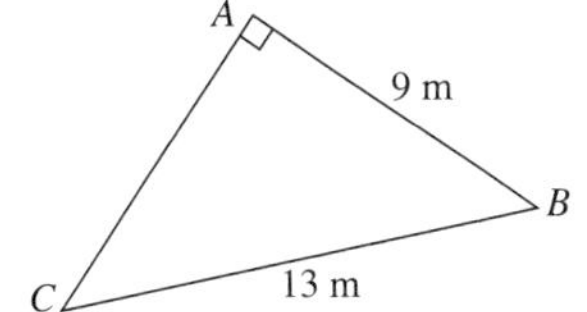

c

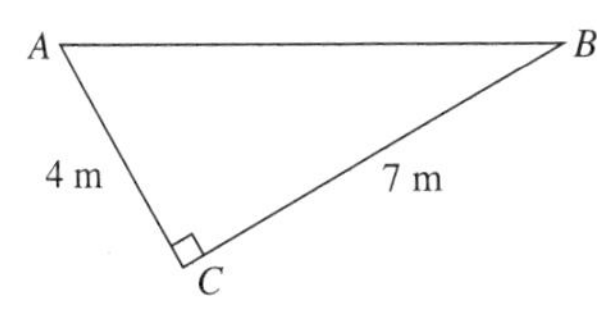

QUESTION 2 Find, in degrees and minutes, the size of the marked angle.

a

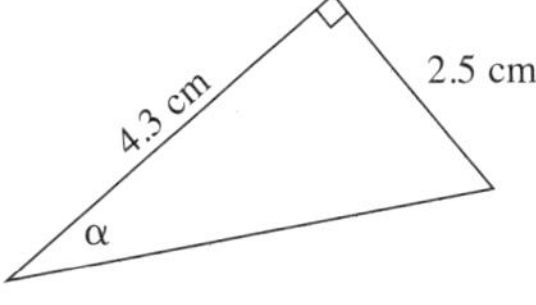

b

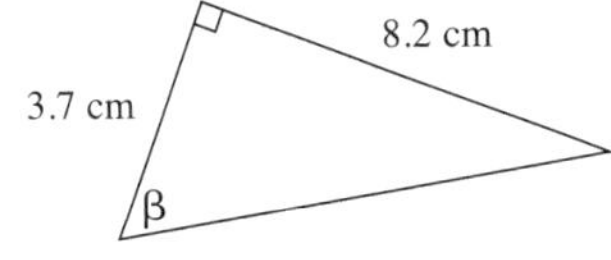

c

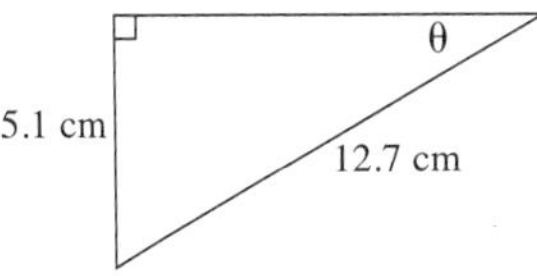

d

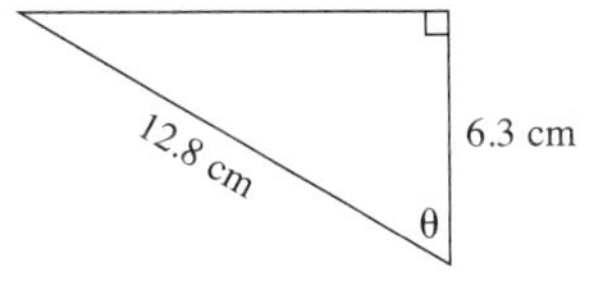

e

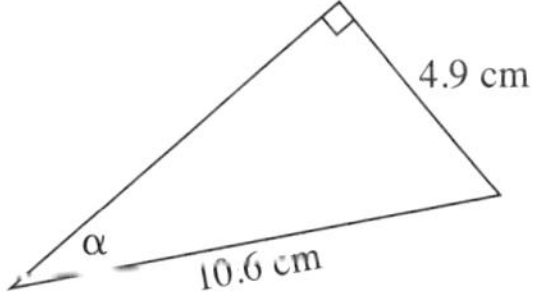

f

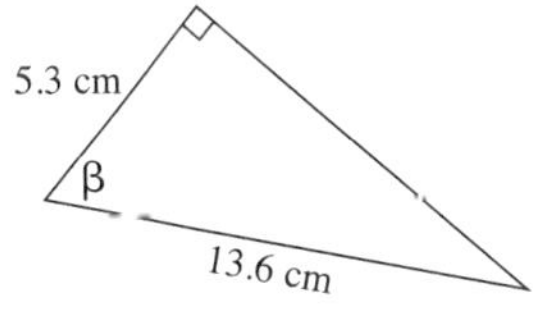

QUESTION 3 Find, in degrees and minutes, the size of the marked angle.

a

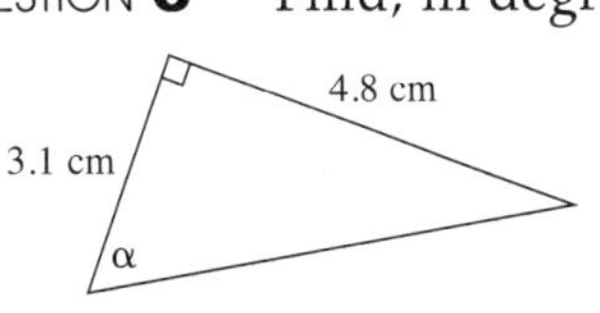

b

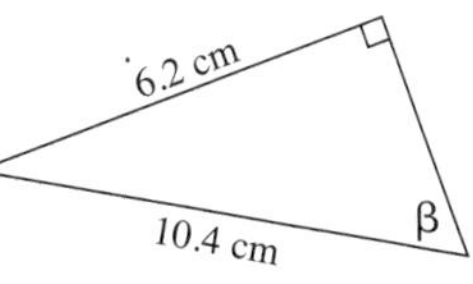

c

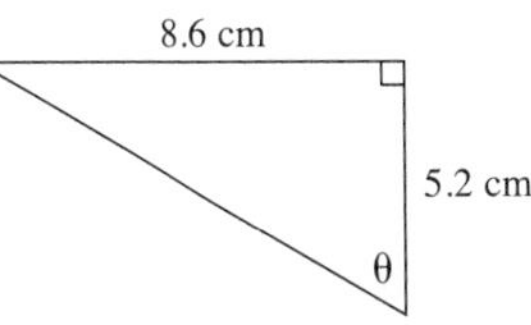

QUESTION 4 Find, in degrees and minutes, the size of the marked angle.

a

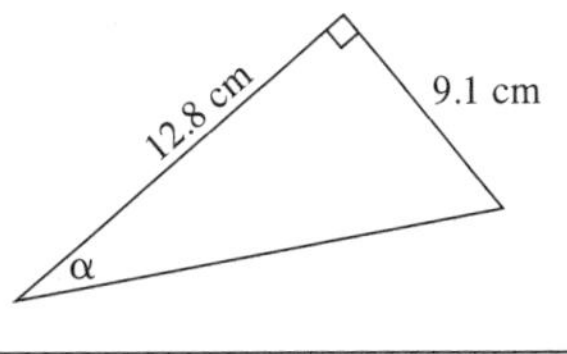

b

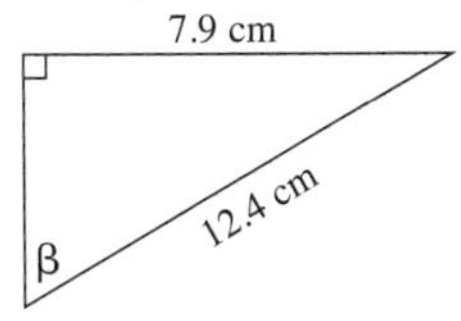

c

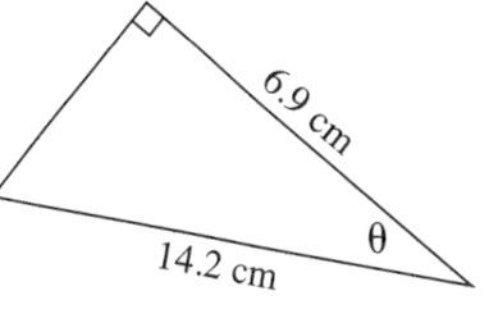

Right-angled triangles and trigonometry

UNIT 7: Mixed problems

Note: Space has been left for you to draw diagrams for some of these questions.

QUESTION 1 In ΔABC, $\angle C = 90°$, $\angle B = 38°40'$ and $AB = 14.6$ cm. Find BC correct to one decimal place.

QUESTION 2 A ladder leans against a vertical wall with its foot 1.5 m from the wall making an angle of 45°36' with the ground. How long is the ladder? Give your answer to the nearest centimetre.

QUESTION 3 A tree 18 m tall casts a shadow 19.5 m long. What angle do the rays of the Sun make with the ground?

QUESTION 4 The height of a ramp is 4.2 m. Given that the ramp is inclined at 30° to the ground, find the length of the ramp to the nearest centimetre.

QUESTION 5 A tree casts a shadow 20.7 m long. If the Sun's rays meet the ground at 29°36', what is the height of the tree to the nearest metre?

QUESTION 6 The diagonal of a square is 24.6 cm long. Find the length of one side to the nearest millimetre.

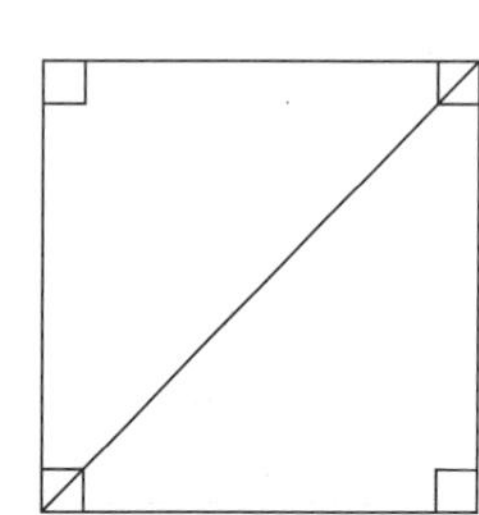

Right-angled triangles and trigonometry

UNIT 8: Angles of elevation and depression (1)

QUESTION 1 The angle of elevation of the top of a tower AB is 62° from a point C on the ground 300 m from the base of the tower. Calculate the height of the tower to the nearest metre.

QUESTION 2 From the top of a building 90 m tall, the angle of depression of a car parked on the ground is 48°. Find the distance of the car from the base of the building. Write your answer correct to two decimal places.

QUESTION 3 A railway track rises uniformly 8.5 m for every 300 m along the track. Find the angle of elevation of this track to the nearest degree.

QUESTION 4 From a point on the ground 20 m from the base of a tree, the angle of elevation of the top of the tree is 53°. Find the height of the tree to the nearest metre.

QUESTION 5 A building that is 45 m tall casts a horizontal shadow 32.3 m long. Find the angle of elevation of the sun to the nearest degree.

QUESTION 6 Anna is 25 m away from a building 38 m high. What is the angle of elevation of the top of the building from her eyes which are 1.7 m above the ground? Answer to the nearest degree.

Right-angled triangles and trigonometry

UNIT 9: Angles of elevation and depression (2)

QUESTION 1 From a point on the ground 27 m from the base of a tree, the angle of elevation of the top of the tree is 56°34'. Find the height of the tree to the nearest metre.

56°34'

27 m

QUESTION 2 A railway track rises uniformly 5.4 m for every 250 m along the track. Find the angle of elevation of this track to the nearest minute.

QUESTION 3 Find the angle of depression from the top of a vertical cliff 80 m high to a boat 388 m from the foot of the cliff. Give your answer correct to the nearest minute.

QUESTION 4 Ryan is sitting in a park and looks towards the top of a 120-m tall tower at an angle of elevation of 31°28'. How far is he sitting from the base of the tower, to the nearest metre?

QUESTION 5 A statue is 25 m tall and casts a horizontal shadow 26.3 m long. Find the angle of elevation of the Sun to the nearest degree.

QUESTION 6 From a point on top of a building that is 98 m tall, the angle of depression of a car is 39°27'. How far is the car from the foot of the building? Give your answer correct to the nearest metre.

Right-angled triangles and trigonometry

UNIT 10: Compass directions

QUESTION **1** What is the size of the angle between each pair of directions?

a N and E ______________ **b** N and S ______________ **c** S and SW ______________

d S and ESE ______________ **e** N and NE ______________ **f** N and ENE ______________

QUESTION **2** Which compass bearing is found between:

a E and NE? ______________ **b** SE and S? ______________ **c** NW and W? ______________

d S and SW? ______________ **e** N and NW? ______________ **f** SE and E? ______________

QUESTION **3** A lighthouse is 10 nautical miles northeast of a ship. How far is the ship west of the lighthouse (correct to two decimal places).

QUESTION **4** Town Q is southwest of town P. Town R is 80 km due south of P and due east of Q.

a Briefly explain why R is the same distance from both P and Q.

b Find the distance from P to Q to the nearest kilometre.

P
80 km
Q
R

QUESTION **5** Town B is SSW of Town A and Town A is WNW of Town C.

a What is the size of $\angle BAC$? ______________

b If $\angle ABC = 40°$ find, to the nearest kilometre, the distance from:

i A to B

ii B to C

A
19 km
C
40°
B

Right-angled triangles and trigonometry

UNIT 11: True bearings

Question 1 For each diagram write down the true bearing of Q from P.

a

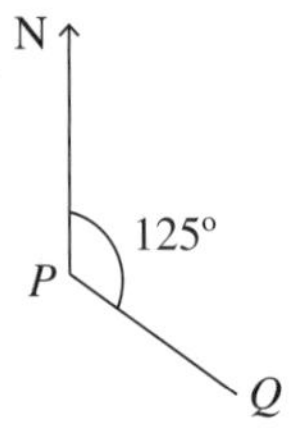

b

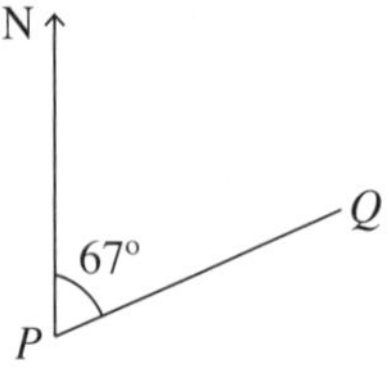

c

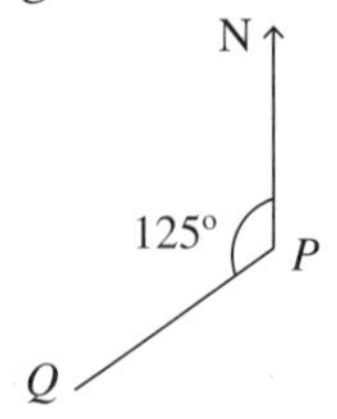

d

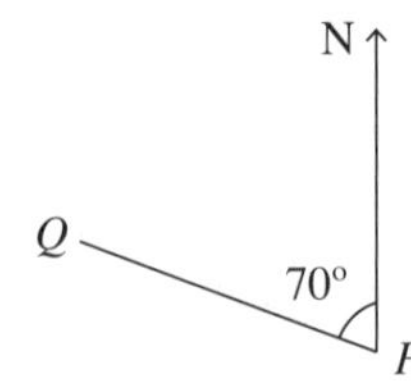

e

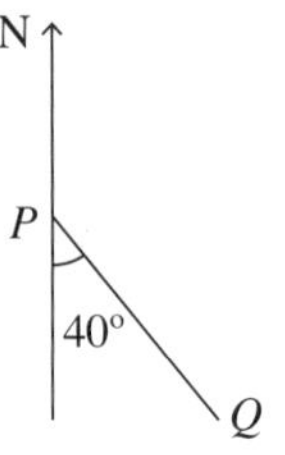

f

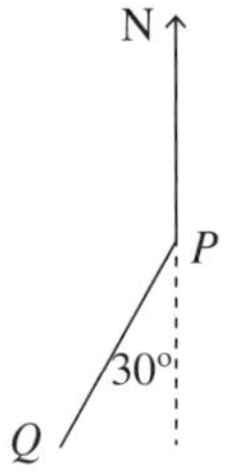

g

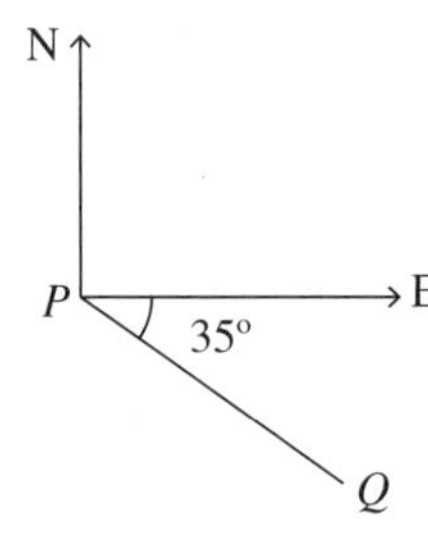

h

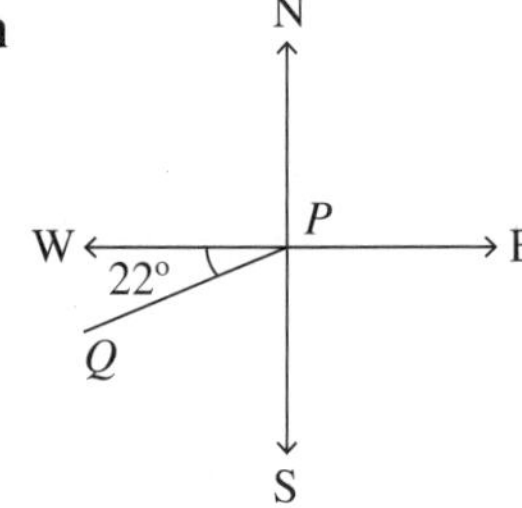

Question 2 Show the position of point Q on the diagram if the bearing of:

a Q from P is 160°

P•

b Q from P is 240°

P•

c Q from P is 080°

P•

Question 3 A ship sailed 12 nautical miles north and then 20 nautical miles west. Find its true bearing (to the nearest degree) from the starting point.

Question 4 A helicopter flies 215 km from P to Q on a bearing of 130°. From Q it flies on a bearing of 220° to R which is due south of P.

a Show this information on a diagram

b What is the size of $\angle PQR$?

c What is the size of $\angle QPR$?

d How far is it, to the nearest kilometre, from Q to R?

Right-angled triangles and trigonometry

UNIT 12: Compass bearings

QUESTION 1 Write down the compass bearing of point Q from P.

a

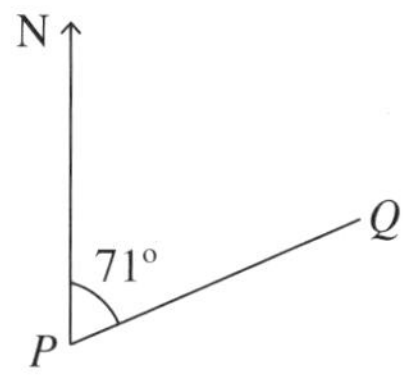

b

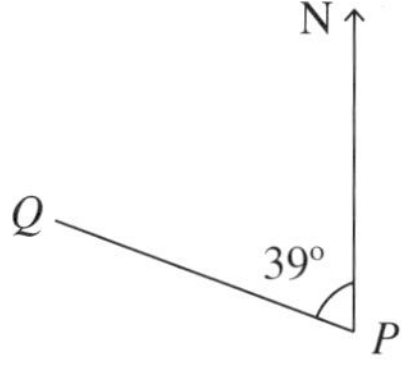

c

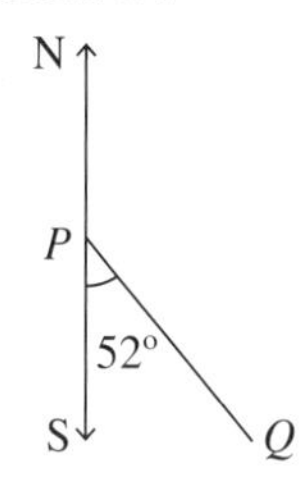

d

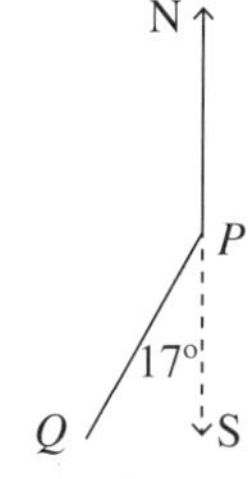

e

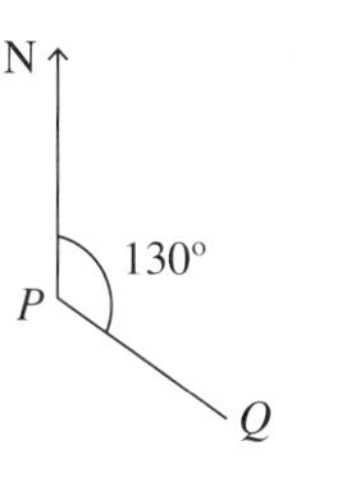

f

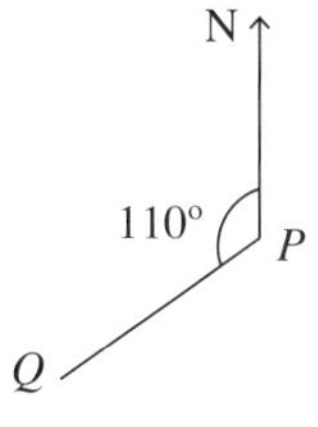

g

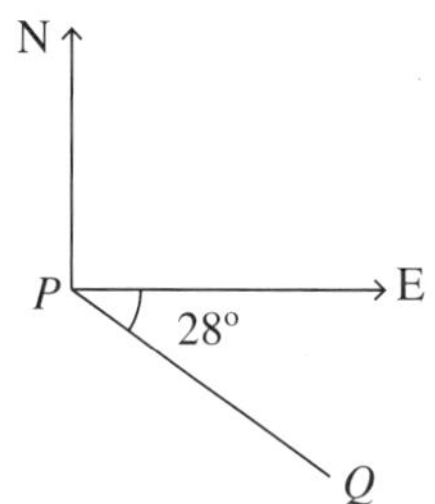

h

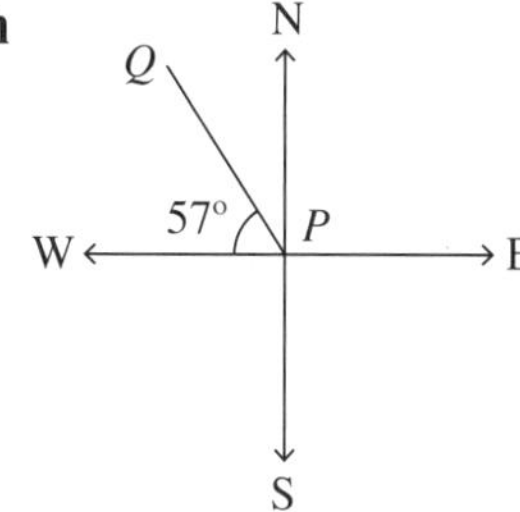

QUESTION 2 A plane left P and flew S60°E to Q. From Q it flew N30°E to R. R is exactly 200 km due East of P. S is due south of P and due West of Q.

a Show this information on a diagram.

b What is the size of $\angle PQR$?

c Find the distance from Q to R.

d Find the distance from S to Q.

QUESTION 3 Change these compass bearings to true bearings.

a N62°E **b** S50°W **c** N40°W **d** S20°E

e N43°E **f** N67°W **g** S34°E **h** S13°W

QUESTION 4 Change these true bearings to compass bearings.

a 055° **b** 120° **c** 340° **d** 190°

e 153° **f** 245° **g** 321° **h** 096°

Right-angled triangles and trigonometry

TOPIC TEST — PART A

Instructions
- This part consists of 10 multiple-choice questions.
- Fill in only ONE CIRCLE for each question.
- Each question is worth 1 mark.

Time allowed: 15 minutes — **Total marks: 10**

Marks

1 The hypotenuse of a right-angled triangle is 25 cm. If one side is 7 cm, the third side is:

Ⓐ 25.96 cm Ⓑ 24 cm Ⓒ 26 cm Ⓓ 26.52 cm

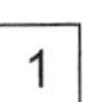

2 Evaluate 15 cos 70° correct to two decimal places.

Ⓐ 0.34 Ⓑ 0.02 Ⓒ 5.13 Ⓓ 43.86

3 If $\sin\theta = \frac{3}{5}$, calculate the size of θ to the nearest degree.

Ⓐ 53° Ⓑ 37° Ⓒ 31° Ⓓ 59°

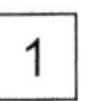

4 In relation to the diagram, which statement is correct?

Ⓐ $\cos\theta = \frac{6}{10}$ Ⓑ $\tan\theta = \frac{8}{6}$

Ⓒ $\sin\theta = \frac{6}{10}$ Ⓓ $\sin\theta = \frac{8}{10}$

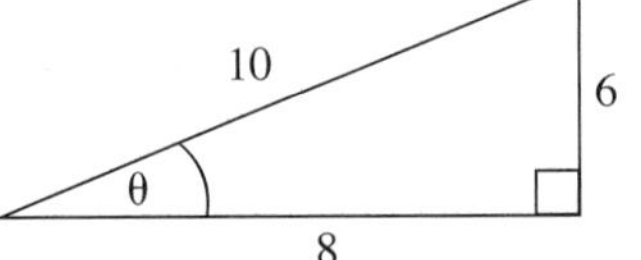

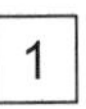

5 If $\cos\theta = 0.5$, find the size of angle θ.

Ⓐ 30° Ⓑ 45° Ⓒ 55° Ⓓ 60°

6 The value of sin 49°28' is closest to:

Ⓐ 0.650 Ⓑ 0.760 Ⓒ 1.169 Ⓓ 0.482

7 A compass bearing of S25°W is a true bearing of:

Ⓐ 155° Ⓑ 205° Ⓒ 215° Ⓓ 245°

8 The value of x in the diagram is given by:

Ⓐ $36 \times \cos 18°$ Ⓑ $36 \times \sin 18°$

Ⓒ $\frac{36}{\cos 18°}$ Ⓓ $\frac{36}{\sin 18°}$

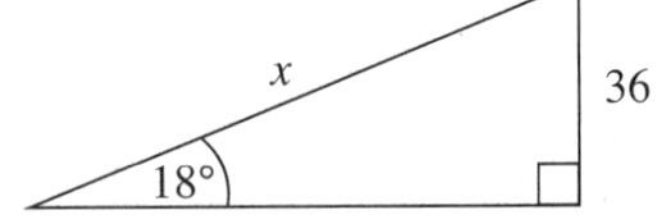

1

9 In ΔABC, the angle B is 90°, AB is 6 m and AC is 10 m. Find the size of angle A correct to the nearest degree.

Ⓐ 27° Ⓑ 30° Ⓒ 37° Ⓓ 53°

10 From the diagram the correct expression for h is:

Ⓐ $h = 30 \tan 25°$ Ⓑ $h = 25 \tan 30°$

Ⓒ $h = \frac{\tan 25°}{30}$ Ⓓ $h = \frac{30}{\tan 25°}$

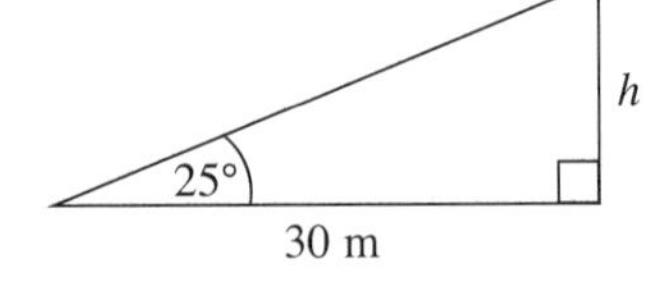

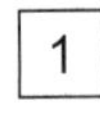

Total marks achieved for PART A

Right-angled triangles and trigonometry

TOPIC TEST PART B

Instructions
- This part consists of 5 questions.
- Write only the answer in the answer column.
- For any working use the question column.

Time allowed: 20 minutes **Total marks: 15**

Questions	Answers	Marks
1 The angle of depression of a car from the top of a building is 64°. The building is 40 m tall. How far is the car from the base of the building? Give the answer correct to one decimal place. (diagram: 40 m)		1
2 Point B is due east of A and northeast of C. (diagram: A, B, C)		
a What is the true bearing of B from C?		1
b What is the compass bearing of C from B?		1
c How far is it from B to C if it is 70 m from A to B?		1
3 The bearing of R from P is 240° and the bearing of R from Q is 270°. It is 6 km from R to Q. P is due north of Q.		
a Show the information on a diagram.		1
b What is the size of: **i** $\angle PQR$?		1
ii $\angle QPR$?		1
c Find the distance from P to R.		1
4 The angle of elevation of A from C is 60° and the angle of elevation of A from D is 40°. C is 250 m from B. (diagram: A, B, C, D, 40°, 60°, 250 m)		
a Using ΔACB, find the length of AB.		1
b Using the answer to part **a** and ΔADB, find the length of DB.		1
c What is the length of DC?		1
5 Town Y is 40 km due south of town X and due west of town Z. The bearing of Z from X is 110°.		
a Show the information on a diagram.		1
b What is the size of **i** $\angle XYZ$?		1
ii $\angle ZXY$?		1
c What is the distance from X to Z (to the nearest kilometre)?		1

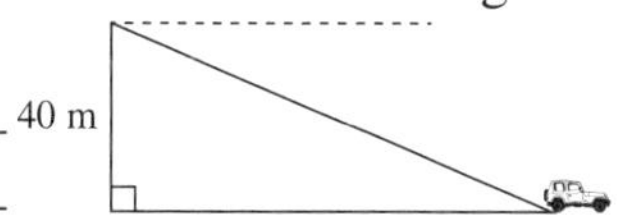

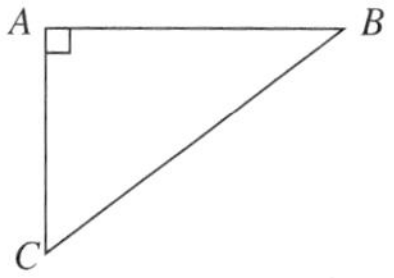

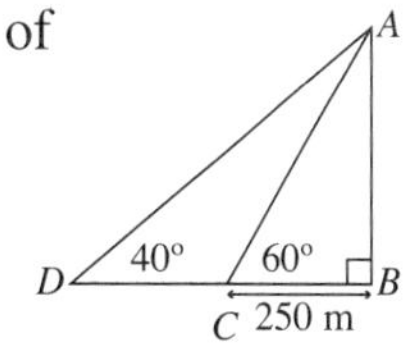

Total marks achieved for PART B /15

CHAPTER 6
Linear relationships

UNIT 1: Review of midpoint, gradient and distance

QUESTION 1 The diagram shows the points $A(-2, 1)$, $B(6, 7)$ and $C(2, -2)$.

a Find the gradient of the line joining:

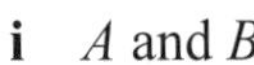

i A and B **ii** A and C

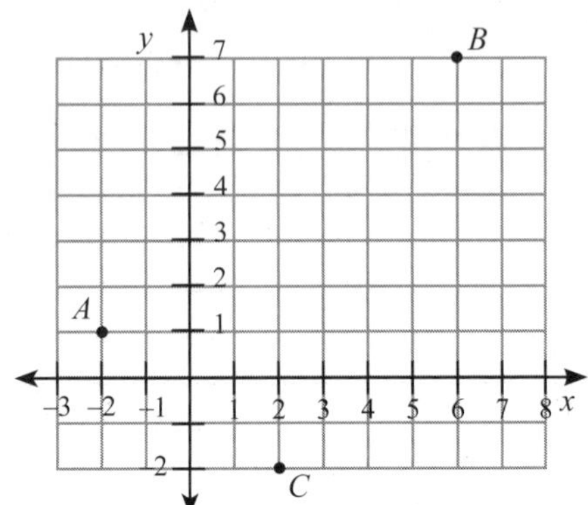

b Find the midpoint of:

i AB **ii** BC

c Find the distance between:

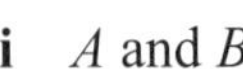

i A and B **ii** A and C

QUESTION 2 A is the point $(7, 9)$ and B the point $(-9, -3)$. Find:

a the midpoint of AB **b** the gradient of AB **c** the distance from A to B

QUESTION 3 Find the midpoint of:

a $(1, -3)$ and $(5, -5)$ **b** $(-6, 4)$ and $(8, -2)$ **c** $(5, -3)$ and $(-2, 0)$

QUESTION 4 Find the gradient of the line joining:

a $(2, 1)$ and $(4, 9)$ **b** $(-1, 6)$ and $(5, 0)$ **c** $(-4, -3)$ and $(-7, -1)$

QUESTION 5 Find the distance (leaving the answer as a square root if necessary) between:

a $(-1, 8)$ and $(4, -4)$ **b** $(2, 3)$ and $(5, 5)$ **c** $(7, -2)$ and $(2, -6)$

Linear relationships

UNIT 2: Gradient and y-intercept

QUESTION 1 The graph shows the lines $y = 2x + 1$, $y = x - 4$ and $y = -x + 2$. For each line write down **i** the gradient and **ii** the y-intercept.

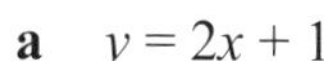

a $y = 2x + 1$

i ______________ **ii** ______________

b $y = x - 4$

i ______________ **ii** ______________

c $y = -x + 2$

i ______________ **ii** ______________

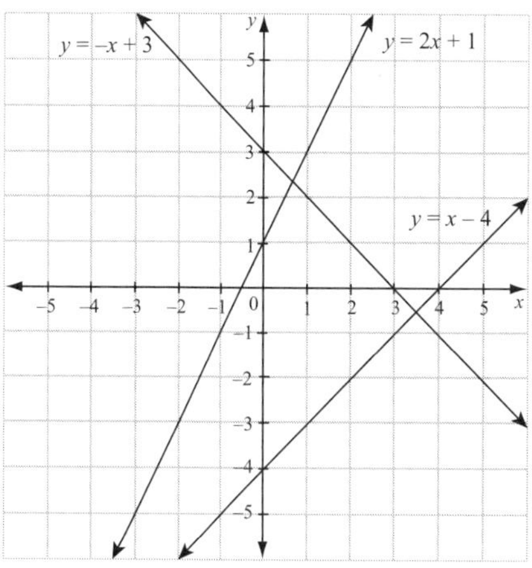

QUESTION 2 Complete:

In the equation of a line $y = mx + c$, m gives the ______________

and c is the ______________.

QUESTION 3 Write the equation of the line with;

a gradient 2 and y-intercept 3 ______________

b gradient $\frac{1}{2}$ and y-intercept 4 ______________

c gradient −2 and y-intercept 5 ______________

d gradient 3 and y-intercept −2 ______________

e gradient −3 and y-intercept −1 ______________

f gradient $-\frac{3}{4}$ and y-intercept 0 ______________

g gradient $\frac{2}{3}$ and y-intercept $\frac{1}{2}$ ______________

h gradient 0 and y-intercept 1 ______________

QUESTION 4 Write the gradient (m) and y-intercept (c) for each line.

a $y = 4x + 1$

m __________ c __________

b $y = 2x - 5$

m __________ c __________

c $y = -3x + 2$

m __________ c __________

d $y = x - 4$

m __________ c __________

e $y = \frac{3}{2}x + \frac{1}{2}$

m __________ c __________

f $y = -8$

m __________ c __________

g $y = -x - 7$

m __________ c __________

h $y = 6 - x$

m __________ c __________

i $y = 2 - 5x$

m __________ c __________

QUESTION 5

a Plot the points P (−4, 3) and Q (6, −2) and show the line that passes through those two points.

b Find the gradient of PQ.

____

____

c What is the y-intercept? ______________

d Write down the equation of the line. ______________

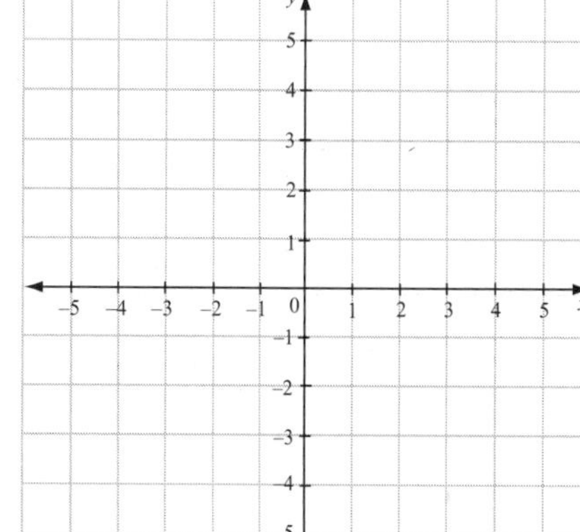

Linear relationships

UNIT 3: Graphing lines

QUESTION **1** For each equation, mark the y-intercept on the graph, use the gradient to mark a second point and then graph the line.

a $y = x + 2$

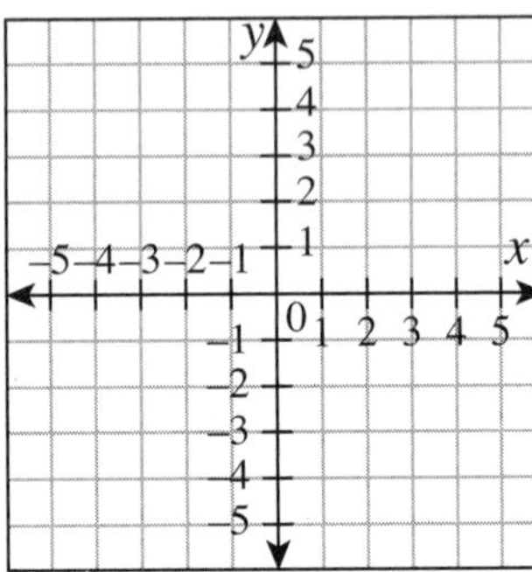

b $y = 2x - 3$

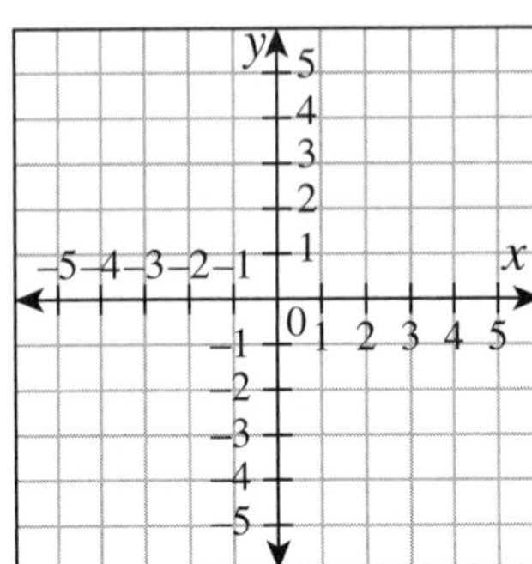

c $y = \frac{1}{2}x + 1$

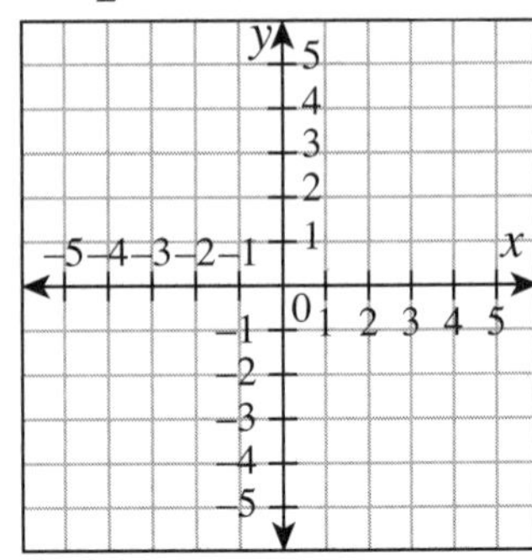

d $y = -2x + 4$

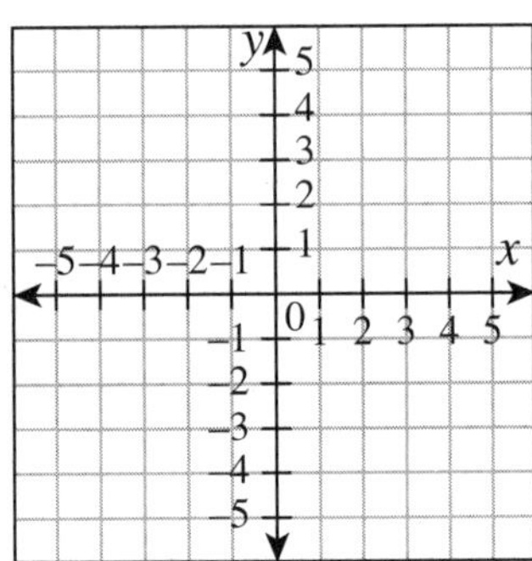

e $y = -x - 1$

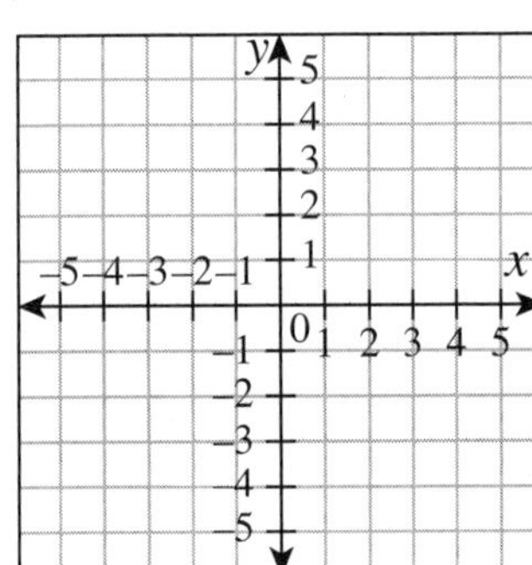

f $y = \frac{4}{3}x$

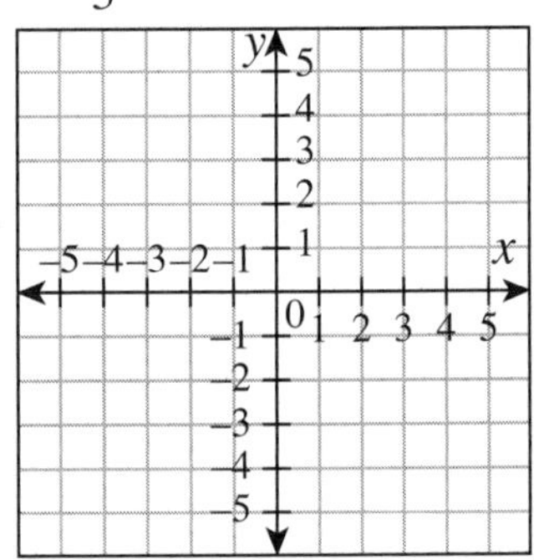

QUESTION **2** Graph each of these lines on the grid supplied.

a $y = 3x - 1$

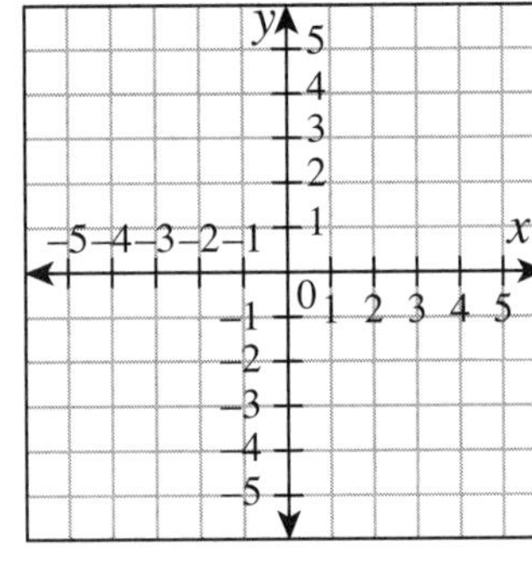

b $y = x - 5$

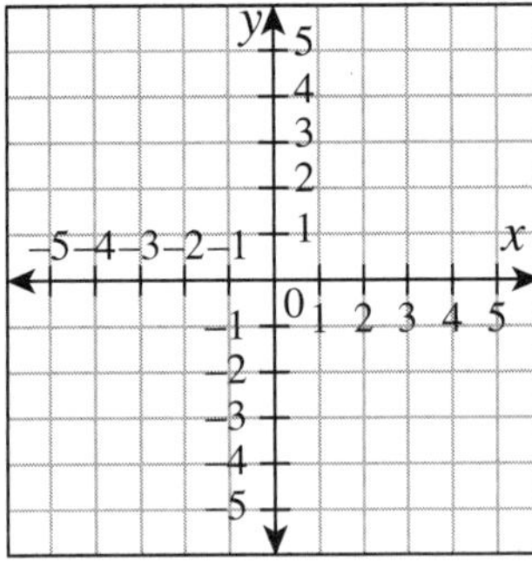

c $y = -x + 3$

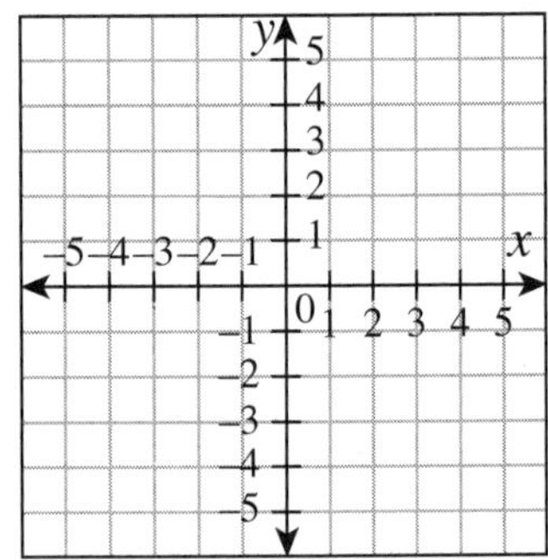

d $y = -2x$

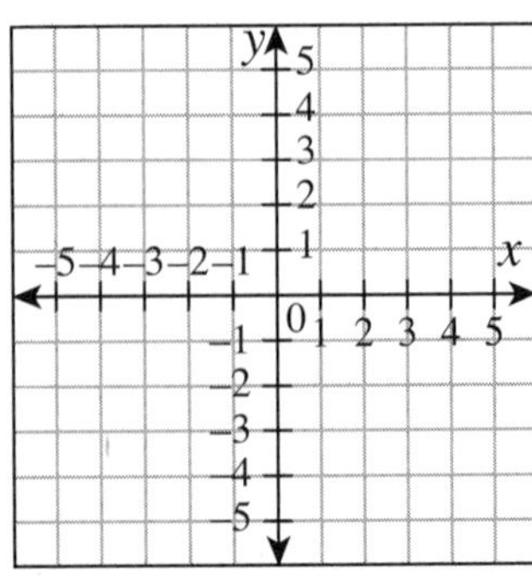

e $y = 2x + \frac{1}{2}$

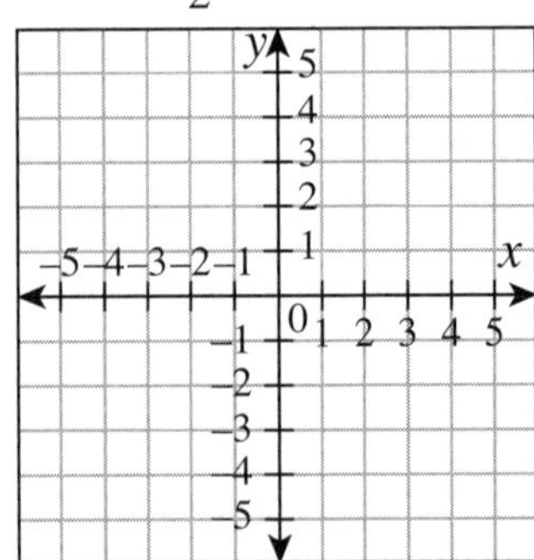

f $y = \frac{2}{3}x + 2$

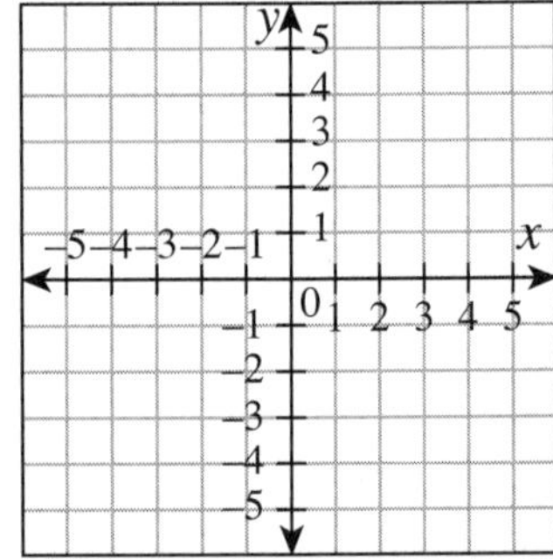

Linear relationships

UNIT 4: Meaning for gradient and *y*-intercept

QUESTION 1 Andrew receives a fixed amount of pocket money each week. In addition, if Andrew chooses to help his mother, she gives him an extra amount per hour for the time worked. The graph shows the amount of money Andrew might receive in pocket money each week.

a What is the intercept on the vertical axis?

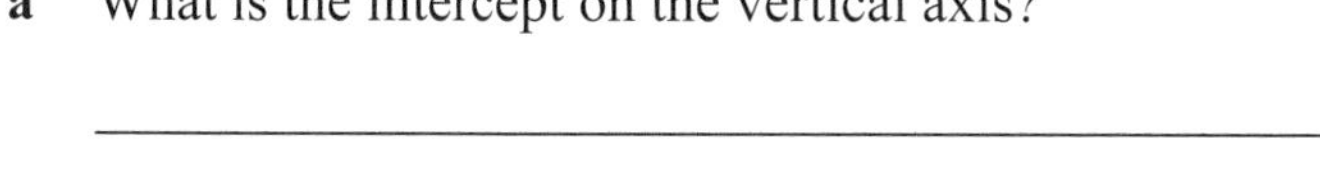

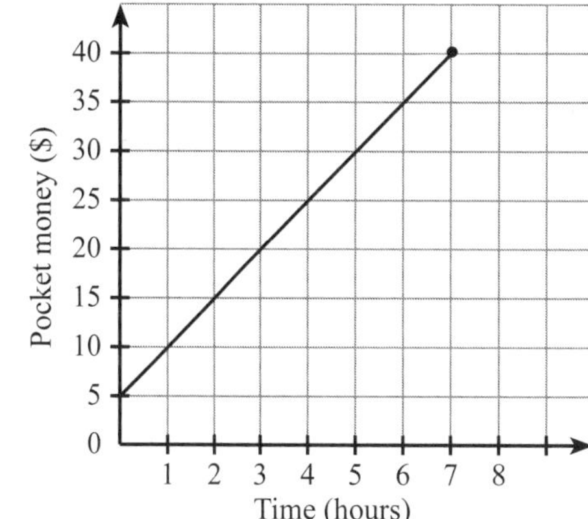

b What does the intercept on the vertical axis represent?

c What is the gradient of this line?

d What does the gradient represent?

QUESTION 2 Melissa intends to ride a bicycle from Baxton to Clair to raise money for the local hospital. The graph shows her expected distance from Clair in kilometres over time (in hours).

a What is the intercept on the vertical axis?

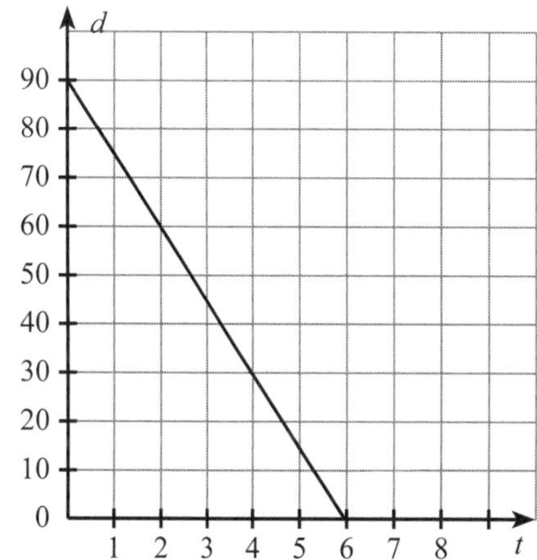

b What information does this intercept tell us?

c What is the gradient of the line?

d What information does the gradient tell us?

e What is the equation of the line?

Linear relationships

UNIT 5: Lines with the same gradient

QUESTION 1 On the same number plane, draw the graphs of the following.

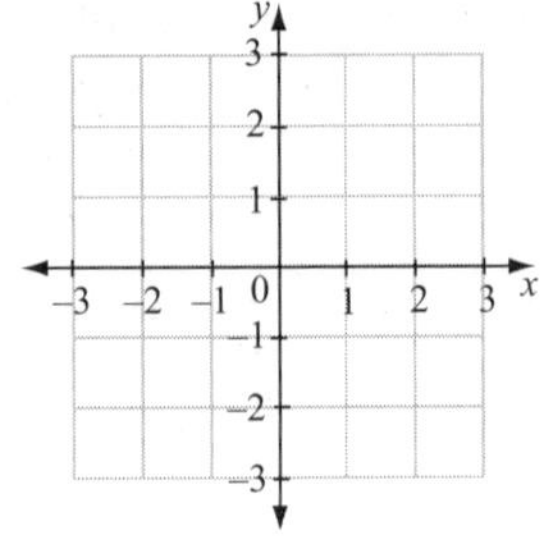

a $y = x$

b $y = x + 1$

c $y = x - 1$

d $y = x - 3$

QUESTION 2 On the same number plane, draw the graphs of the following.

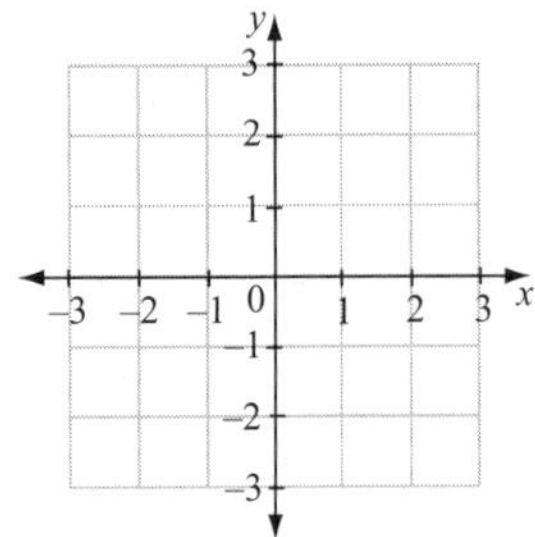

a $y = 2x$

b $y = 2x + 1$

c $y = 2x - 2$

QUESTION 3 On the same number plane, draw the graphs of the following.

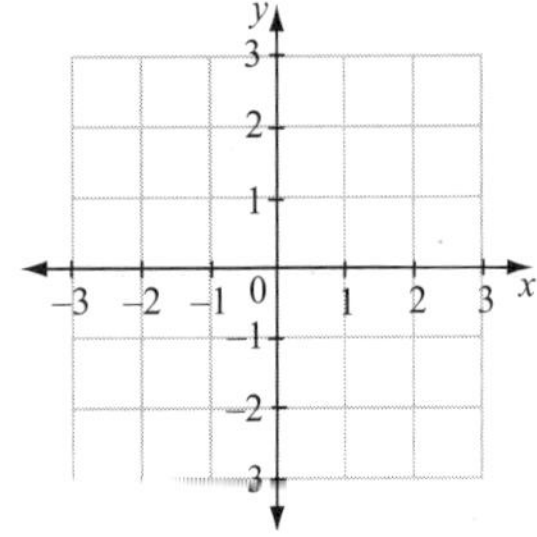

a $y = -x$

b $y = -x - 2$

c $y = 1 - x$

QUESTION 4 On the same number plane, draw the graphs of the following.

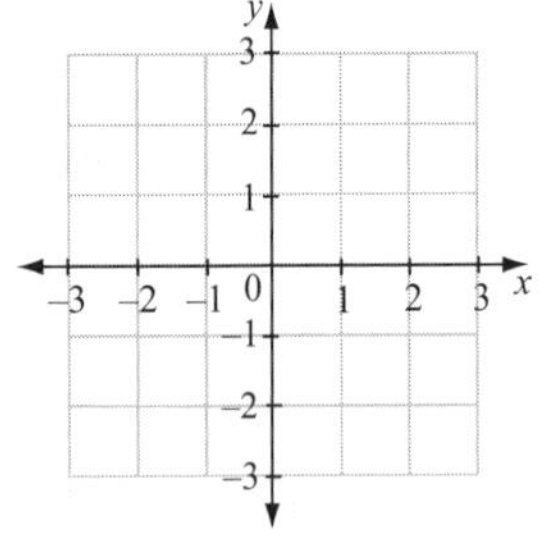

a $y = \frac{1}{3}x$

b $y = \frac{1}{3}x + 2$

c $y = \frac{1}{3}x - 1$

QUESTION 5 On the same number plane, sketch the graphs of the following.

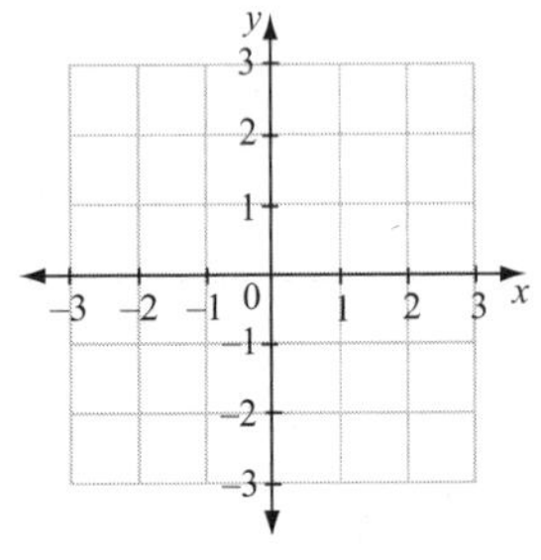

a $y = -\frac{2}{3}x$

b $y = -\frac{2}{3}x + 1$

c $y = -\frac{2}{3}x - 1$

QUESTION 6 Complete: Lines that have the same gradient are ____________________.

Linear relationships

UNIT 6: Lines with gradients that are negative reciprocals

QUESTION 1 Find these products.

a $2 \times -\frac{1}{2} =$ ________ **b** $-3 \times \frac{1}{3} =$ ________ **c** $\frac{1}{5} \times -5 =$ ________ **d** $-\frac{1}{7} \times 7 =$ ________

e $\frac{2}{3} \times -\frac{3}{2} =$ ________ **f** $-\frac{3}{4} \times \frac{4}{3} =$ ________ **g** $\frac{8}{5} \times -\frac{5}{8} =$ ________ **h** $-\frac{5}{3} \times \frac{3}{5} =$ ________

QUESTION 2 Complete:

The product of any number and its negative reciprocal is always ____________________.

QUESTION 3 On the same number plane, draw the graphs of the following.

a $y = x$

b $y = -x + 1$

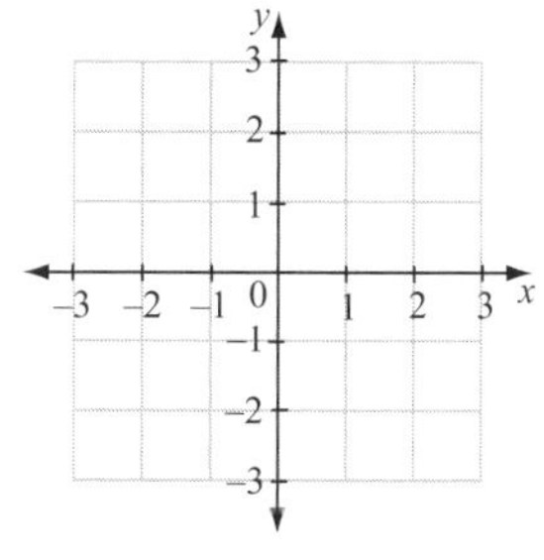

QUESTION 4 On the same number plane, draw the graphs of the following.

a $y = -2x$

b $y = \frac{1}{2}x - 1$

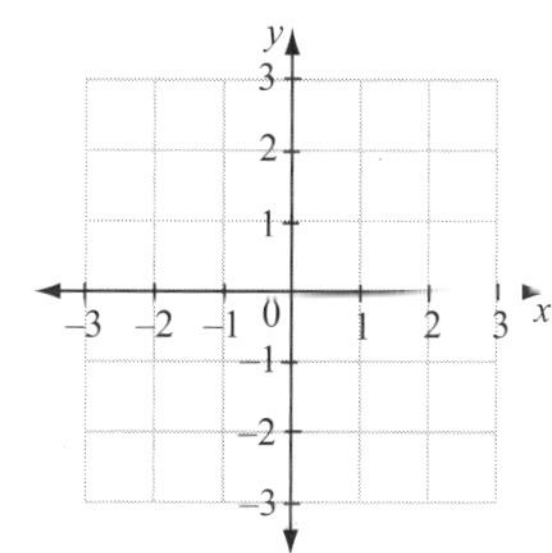

QUESTION 5 On the same number plane, draw the graphs of the following.

a $y = \frac{2}{3}x + 1$

b $y = -\frac{3}{2}x - 1$

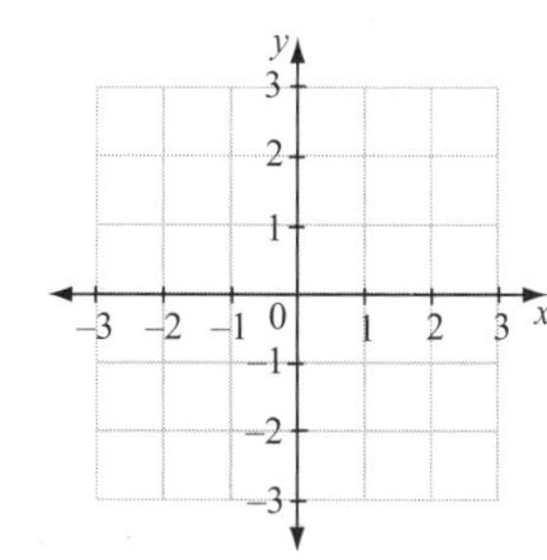

QUESTION 6 On the same number plane, draw the graphs of the following.

a $y = -\frac{3}{4}x$

b $y = \frac{4}{3}x$

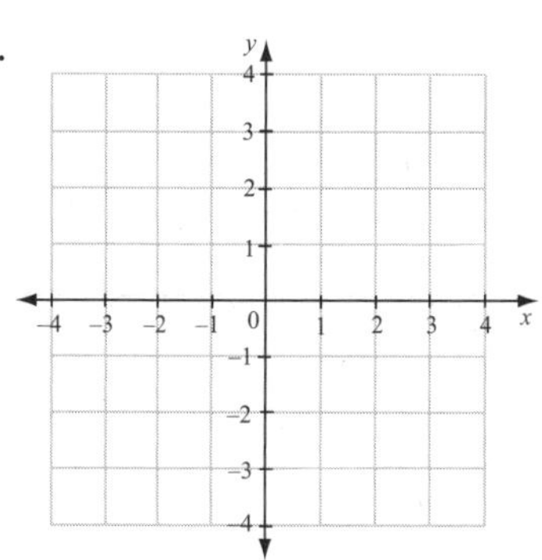

QUESTION 7 Complete.

Lines whose gradients are negative reciprocals are always ____________________.

Linear relationships

UNIT 7: Parallel and perpendicular lines (1)

QUESTION 1 Complete.

a Lines that are parallel have gradients that are ______________________.

b Lines that are perpendicular have gradients that are ______________________.

QUESTION 2 Determine whether the two given lines are parallel or perpendicular or neither.

a $y = 2x + 3$
$y = 2x - 1$

____

b $y = \frac{1}{2}x + 5$
$y = 2x - 3$

____

c $y = \frac{3}{2}x + 1$
$y = -\frac{2}{3}x - 1$

____

d $y = \frac{x}{4} + \frac{2}{3}$
$y = -4x + 2$

____

e $y = 6 - x$
$y = -x + 3$

____

f $y = 1 - 2x$
$y = \frac{x}{2} + 5$

____

g $y = -\frac{3}{4}x - 2$
$y = -\frac{4}{3}x + 2$

____

h $y = \frac{5x}{6} - 7$
$y = -\frac{6}{5}x + 1$

____

QUESTION 3 Write down the equation of the line that passes through the point (0, 2) and which is parallel to the given line.

a $y = 3x + 1$

____

b $y = -2x - 3$

____

c $y = \frac{1}{2}x + \frac{1}{4}$

____

d $y = -\frac{5x}{3} - 4$

____

QUESTION 4 Write down the equation of the line that passes through the origin and which is perpendicular to the given line.

a $y = 4x + 3$

____

b $y = -\frac{x}{3} + 7$

____

c $y = 9 - 2x$

____

d $y = \frac{3}{2}x + 4$

____

QUESTION 5 The diagram shows the graph of line l.

a What is the gradient of l? ______________

b What is the gradient of any line parallel to l? ______________

c What is the gradient of any line perpendicular to l? ______________

d Line m has equation $y = -\frac{3}{2}x - 12$. It intersects line l at P. If l meets the y-axis at Q and m meets the y-axis at R, what is the size of $\angle QPR$? ______________

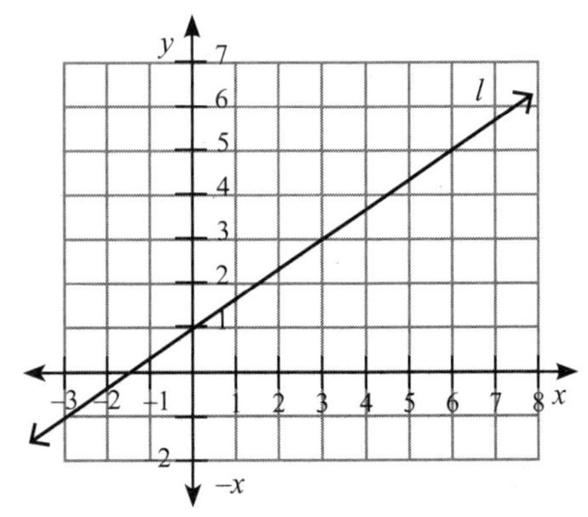

Linear relationships

UNIT 8: General form of linear equations

QUESTION 1 Write each of the following linear equations in general form.

a $2x - 5y = 9$ b $3x + 4y = 8$ c $5x - 7 = 2y$

d $8y - 3 = 4x$ e $2x = 9 - y$ f $y = 8x + 7$

g $3y - 2x = 6$ h $9y = 8x + 12$ i $2y = \frac{x}{3} + 1$

QUESTION 2 Each of the following equations is in general form. Change it to gradient-intercept form, then write down its gradient and y-intercept.

a $2x + 3y - 8 = 0$ b $x + 5y - 7 = 0$ c $3x - 2y - 3 = 0$

d $x - y + 7 = 0$ e $2x + y - 9 = 0$ f $5x - 6y + 11 = 0$

g $3x - 2y - 6 = 0$ h $4x + 5y + 3 = 0$ i $2x - y + 6 = 0$

QUESTION 3 Write the equation of each line in gradient-intercept form and then change it to general form.

a $m = 4$, $c = 3$ b $m = 2$, $c = -5$ c $m = 3$, $c = 7$

d $m = \frac{1}{2}$, $c = 4$ e $m = \frac{2}{3}$, $c = 6$ f $m = -\frac{5}{6}$, $c = 3$

Linear relationships

UNIT 9: Parallel and perpendicular lines (2)

QUESTION 1 State whether the following pairs of lines are parallel or not.

a $x + 3y + 9 = 0$ and $x + 3y - 7 = 0$ ________

b $2x + y = 6$ and $3x - 7y = 9$ ________

c $3x - 7y + 8 = 0$ and $3x - 7y = 2$ ________

d $x + 2y = 6$ and $x + 2y - 5 = 0$ ________

e $x + y - 2 = 0$ and $x + y - 7 = 0$ ________

f $y = 4x + 3$ and $y = 4x - 5$ ________

g $y = 2x + 1$ and $y = 2x + 8$ ________

h $y = 3x - 1$ and $y = -5x + 7$ ________

QUESTION 2 State whether the following pairs of lines are perpendicular or not.

a $x - 3y = 7$ and $3x - y - 2 = 0$ ________

b $5x - 3y + 7 = 0$ and $3x + 5y - 6 = 0$ ________

c $2x + 7y = 8$ and $3x - 4y + 7 = 0$ ________

d $8x - 3y = 2$ and $3x + 8y = 9$ ________

e $5x - 6y = 15$ and $6x - 5y + 3 = 0$ ________

f $2x - 3y + 7 = 0$ and $3x + 2y + 5 = 0$ ________

g $2x - 9y = 7$ and $3x + 6y = 8$ ________

h $x - 2y = 6$ and $2x + y = 7$ ________

QUESTION 3 State whether the following pairs of lines are parallel, perpendicular or neither.

a $x - 2y + 5 = 0$ and $2x - 4y - 8 = 0$ ________

b $3x - y - 3 = 0$ and $9x - 3y + 1 = 0$ ________

c $x + 7y = 0$ and $2x - 9y = 0$ ________

d $x + y - 7 = 0$ and $3x - 3y + 3 = 0$ ________

e $3x - 4y + 2 = 0$ and $8x + 6y - 3 = 0$ ________

f $4x - 8y = 8$ and $2x + 9y = 6$ ________

g $x + 3y - 2 = 0$ and $2x + 6y - 5 = 0$ ________

h $x - 5y - 2 = 0$ and $10x + 2y + 3 = 0$ ________

QUESTION 4 Find the general form of the equation of the straight line passing through:

a (2, 5) parallel to $3x - y + 7 = 0$

____

____

____

b (0, 0) parallel to the line $4x - 5y + 6 = 0$

____

____

____

c (–2, 3) perpendicular to $2x + y = 9$

____

____

____

d the point (3, –4) and perpendicular to the line $x - y + 5 = 0$

____

____

____

QUESTION 5 Show that the lines $x - 2y + 7 = 0$ and $2x + y - 16 = 0$ are perpendicular to each other.

____

____

____

Linear relationships

UNIT 10: Graphical solutions of simultaneous equations

Question 1 Solve, by drawing graphs, the following pairs of simultaneous equations.

a $y = x + 5$
$y = -3x + 9$

____

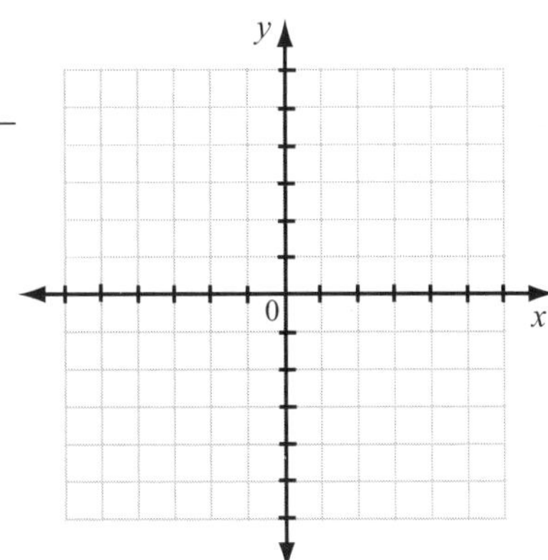

b $y = x + 3$
$y = 2x + 5$

____

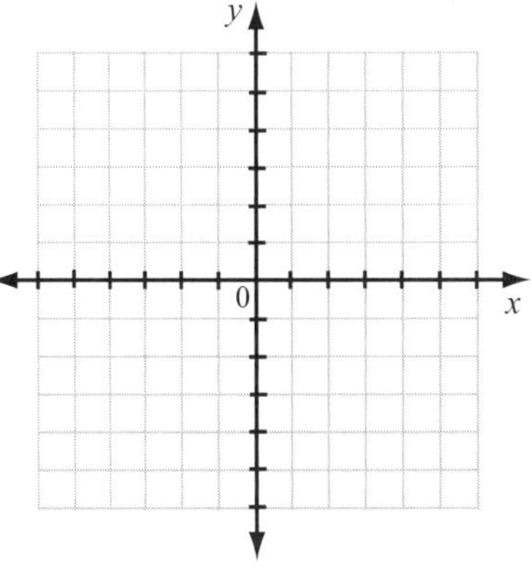

Question 2 Solve graphically the following pairs of simultaneous equations.

a $y = x + 1$
$y - 2x + 3$

____

b $y = x - 4$
$y = 3x - 6$

____

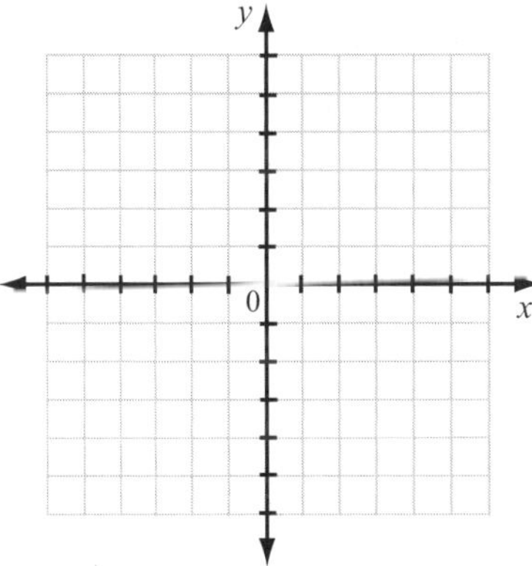

Question 3 Graph each pair of equations on the same number plane to find their solution.

a $y = -x + 2$
$y = 2x - 7$

____

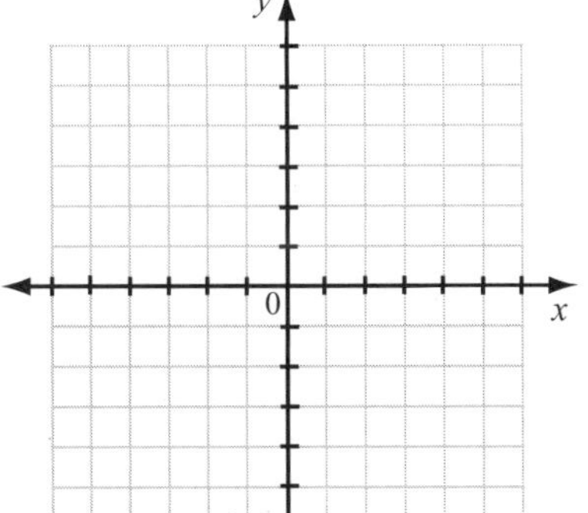

b $x + y = 3$
$2x - y = -9$

____

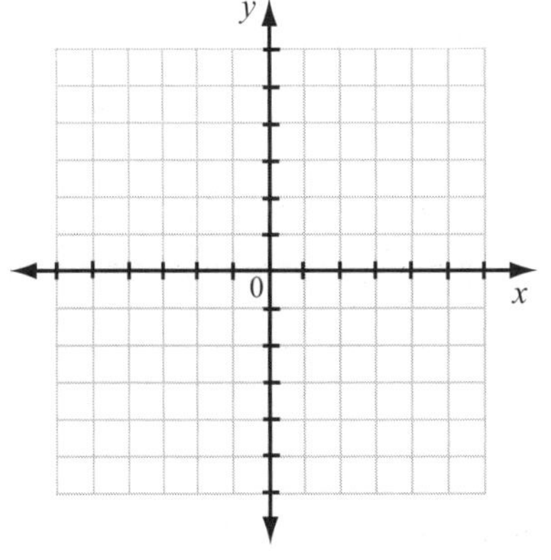

Linear relationships

UNIT 11: Problem solving with coordinates

QUESTION 1 Write the equation, in gradient-intercept form, of the line with the given gradient that passes through point A.

a $A(-3, 2)$, $m = 4$ ______________

____

____

b $A(4, -1)$, $m = 2$ ______________

____

____

c $A(1, 5)$, $m = \frac{1}{2}$ ______________

____

____

d $A(1, 8)$, $m = \frac{2}{3}$ ______________

____

____

e $A(2, 5)$, $m = -\frac{1}{3}$ ______________

____

____

f $A(0, 8)$, $m = -3$ ______________

____

____

QUESTION 2 If the midpoint of $(x, 5)$ and $(9, y)$ is $(1, 6)$, what are the values of x and y?

____

____

QUESTION 3 G is the point $(-2, 1)$ and H is the point $(1, 3)$.

a Find the gradient of GH.

____

____

____

b Find the equation of the line GH.

____

____

QUESTION 4 P is the point $(1, 2)$. Write the coordinates of P' under these transformations.

P (1, 2)

a a translation of 4 units to the right and 2 units up

____

b a translation of 2 units to the left and 5 units down

____

c a reflection in the x-axis ______________________

d a reflection in the y-axis ______________________

e a clockwise rotation of 90° about the origin ______________________

f an anticlockwise rotation of 90° about the origin

____

QUESTION 5 Consider the line $3x - 2y - 6 = 0$.

a Find the x-intercept.

____

b Find the y-intercept.

____

c Hence graph the line $3x - 2y - 6 = 0$.

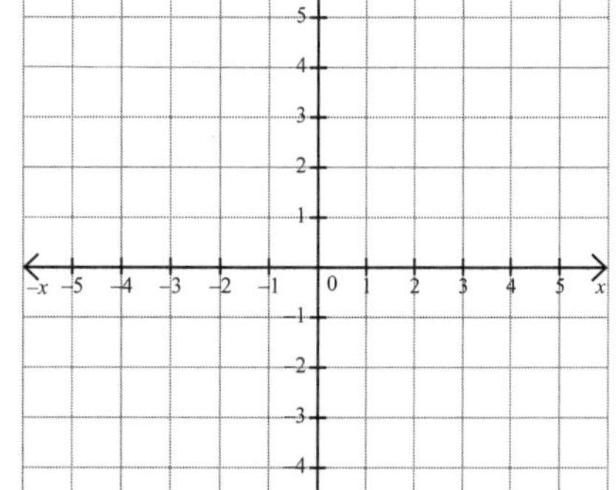

Linear relationships

TOPIC TEST — PART A

Instructions
- This part consists of 10 multiple-choice questions.
- Fill in only ONE CIRCLE for each question.
- Each question is worth 1 mark.

Time allowed: 15 minutes **Total marks: 10**

Marks

1 The straight line $y = 3x - 2$ passes through one of the following points. Which one?

Ⓐ $(0, -2)$ Ⓑ $(0, 2)$ Ⓒ $(-2, 0)$ Ⓓ $(2, 0)$ 1

2 What is the equation of the line parallel to the x-axis passing through P(2, 4)?

Ⓐ $x = 4$ Ⓑ $y = 2$ Ⓒ $x = 2$ Ⓓ $y = 4$ 1

3 Which one of the following is a linear equation?

Ⓐ $y = x^2 + 7$ Ⓑ $y = 5 - \frac{7}{x}$ Ⓒ $y = 6 - 7x$ Ⓓ $y = \sqrt{x} - 3$ 1

4 The gradient of the line $2x - 5y = 10$ is:

Ⓐ 2 Ⓑ 5 Ⓒ $\frac{2}{5}$ Ⓓ $\frac{5}{2}$ 1

5 The point $(9, -1)$ lies on which of these lines?

Ⓐ $3x + y - 6 = 0$ Ⓑ $3x - y + 6 = 0$ Ⓒ $x + 3y - 6 = 0$ Ⓓ $x + 3y + 6 = 0$ 1

6 What is the equation of the line which passes through the point $(-2, 3)$ and has a gradient of -2?

Ⓐ $y = 2x - 1$ Ⓑ $y = -2x - 1$ Ⓒ $y = -2x - 7$ Ⓓ $y = 2x + 7$ 1

7 What is the equation of the line on the graph shown on the right?

Ⓐ $x + y + 1 = 0$ Ⓑ $x + y - 1 = 0$

Ⓒ $x - y + 1 = 0$ Ⓓ $x - y - 1 = 0$ 1

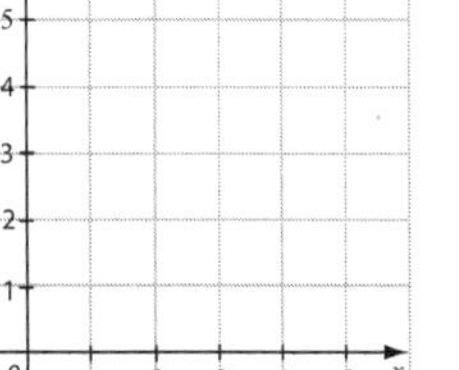

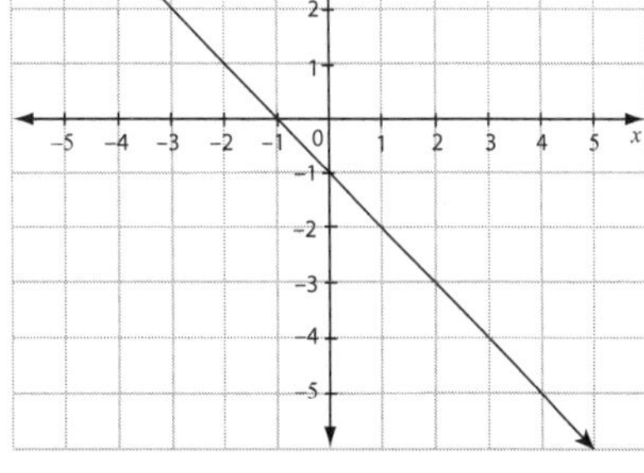

8 Write $4y - 3x = 12$ in gradient-intercept form.

Ⓐ $y = \frac{3}{4}x - 3$ Ⓑ $y = \frac{3}{4}x + 3$

Ⓒ $y = -\frac{3}{4}x - 3$ Ⓓ $y = -\frac{3}{4}x + 3$ 1

9 The point $(3, 6)$ lies on the line:

Ⓐ $x + 2y + 12 = 0$ Ⓑ $x + 2y - 12 = 0$ Ⓒ $2x + y + 12 = 0$ Ⓓ $2x + y - 12 = 0$ 1

10 What is the gradient of any line perpendicular to $y = \frac{x}{4} - 2$?

Ⓐ $\frac{1}{4}$ Ⓑ $-\frac{1}{4}$ Ⓒ 4 Ⓓ -4 1

Total marks achieved for PART A /10

Linear relationships

TOPIC TEST — PART B

Instructions
- This part consists of 3 questions.
- Write only the answer in the answer column.
- For any working use the question column.

Time allowed: 20 minutes **Total marks: 15**

Questions	Answers	Marks
1 The equation of a line is $2x - y - 3 = 0$.		
a Make y the subject of this equation.		1
b What is the gradient of this line?		1
c What is the y-intercept of this line?		1
d Is this line parallel to the line $y = 2x + 1$?		1
2 From the diagram opposite:		
a what is the gradient of AB?		1
b what is the gradient of BC?		1
c is AB perpendicular to BC? Justify your answer.		1
d what is the midpoint, M, of AB?		1
e is the line joining M to $O(0, 0)$ parallel to BC? Justify your answer.		1
3 P is the point $(-2, 5)$ and Q is the point $(4, -3)$.		
a Find the gradient of the line joining P to Q.		1
b What is the equation, in gradient-intercept form, of the line PQ?		1
c What is the equation of PQ in general form?		1
d Does the point $(16, -19)$ lie on the line PQ?		1
e What is the equation, in gradient-intercept form, of the line through the origin parallel to PQ?		1
f What is the distance between P and Q?		1

Total marks achieved for PART B

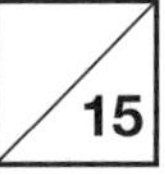

Chapter 7
Non-linear relationships

UNIT 1: Quadratic graphs

Question 1 Complete the table of values and then, on the same number plane, draw the graphs of the following.

a $y = x^2$

b $y = 2x^2$

c $y = \frac{1}{2}x^2$

x	–3	–2	–1	0	1	2	3
$y = x^2$							
$y = 2x^2$							
$y = \frac{1}{2}x^2$							

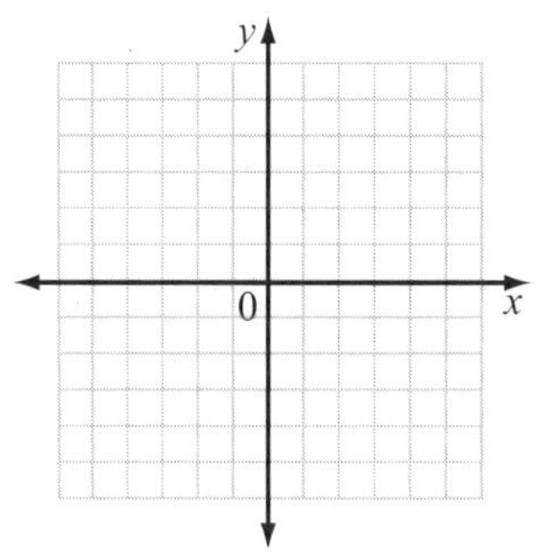

Question 2 Complete the table of values and then, on the same number plane, draw the graphs of the following.

a $y = x^2$

b $y = x^2 + 1$

c $y = x^2 - 1$

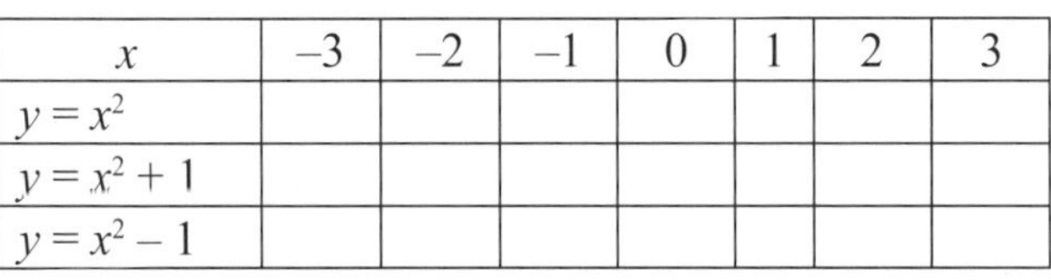

x	–3	–2	–1	0	1	2	3
$y = x^2$							
$y = x^2 + 1$							
$y = x^2 - 1$							

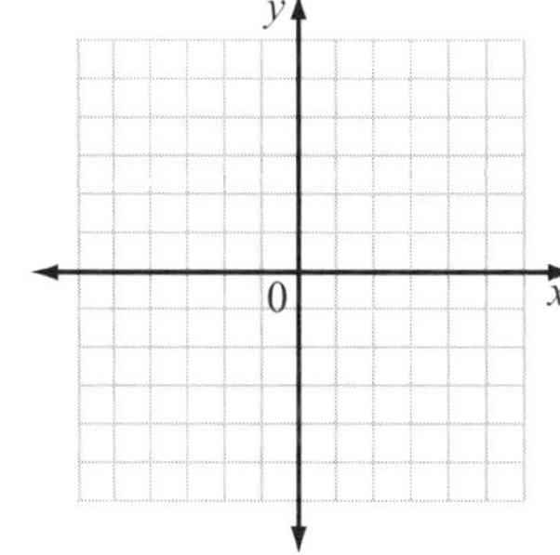

Question 3 Complete the table of values for $y = 1 - x^2$ and sketch its graph.

x	–3	–2	–1	0	1	2	3
$1 - x^2$							

a What is the equation of its axis of symmetry? ______________________

b What are the coordinates of its vertex? ______________________

c What is the maximum value for $y = 1 - x^2$? ______________________

d Find the x-intercepts. ______________________

e Find the y-intercept. ______________________

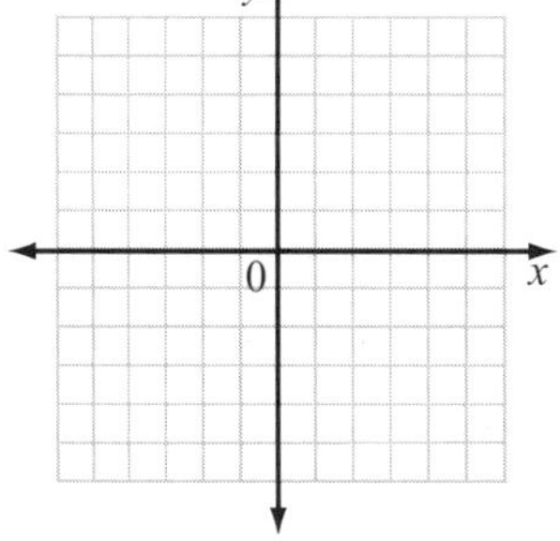

Question 4 Sketch the graphs of the following.

a $y = x^2$

b $y = x^2 + 2$

c $y = x^2 - 2$

d Explain how the graphs of $y = x^2 + 2$ and $y = x^2 - 2$ can be drawn using $y = x^2$.

____

____

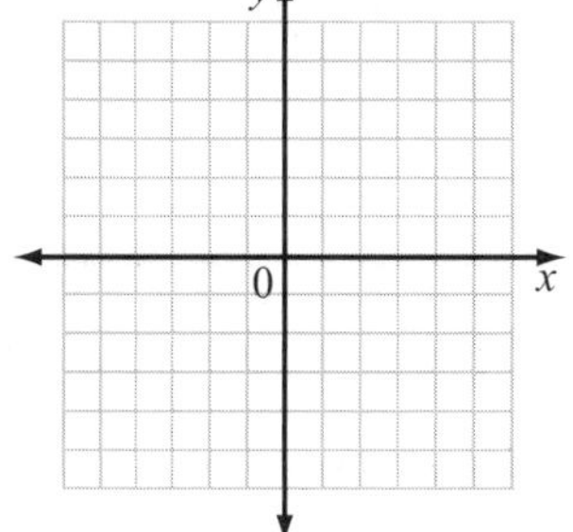

Non-linear relationships

UNIT 2: Features of parabolas

QUESTION 1 The graphs of $y = x^2$, $y = 2x^2$, $y = \frac{1}{4}x^2$, $y = -x^2$, $y = -\frac{1}{2}x^2$ and $y = -4x^2$ are shown.

a Do the curves have the same axis of symmetry?
If so, what is the equation of that axis?

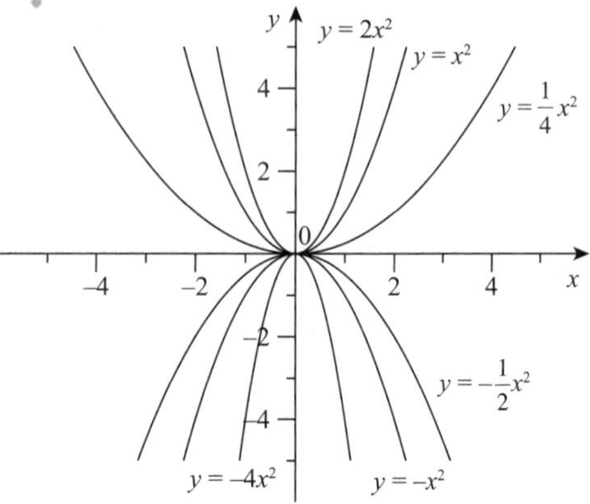

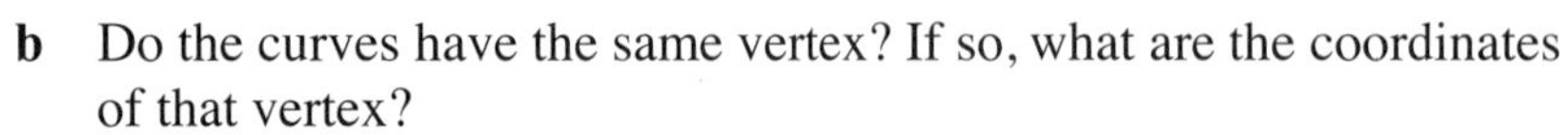

b Do the curves have the same vertex? If so, what are the coordinates of that vertex?

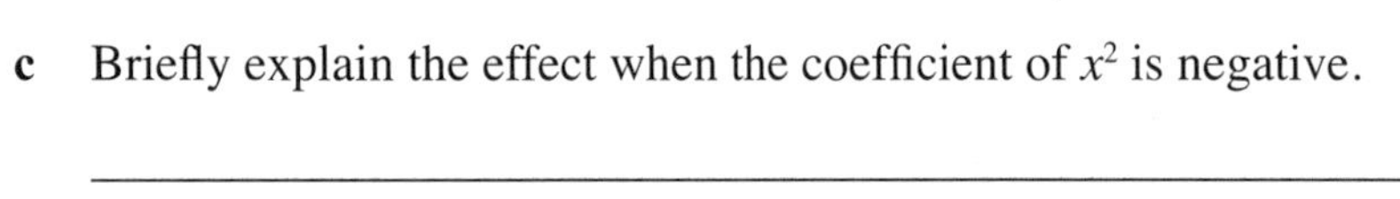

c Briefly explain the effect when the coefficient of x^2 is negative.

d Briefly explain any other effects of the value of a in the equation $y = ax^2$.

QUESTION 2 The graphs of $y = x^2 + 2$, $y = x^2 - 2$, $y = -x^2 - 2$ and $y = 2 - x^2$ are shown.

a Do the curves have the same axis of symmetry?
If so, what is the equation of that axis?

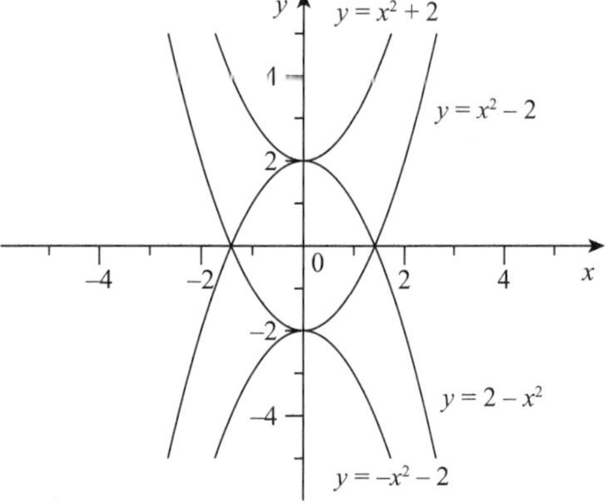

b Do the curves have the same vertex? If so, what are the coordinates of that vertex?

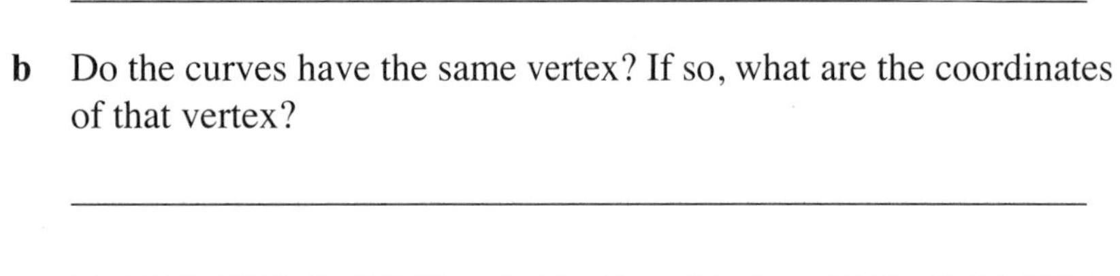

c Briefly explain the effect of the value of c in the equation $y = x^2 + c$.

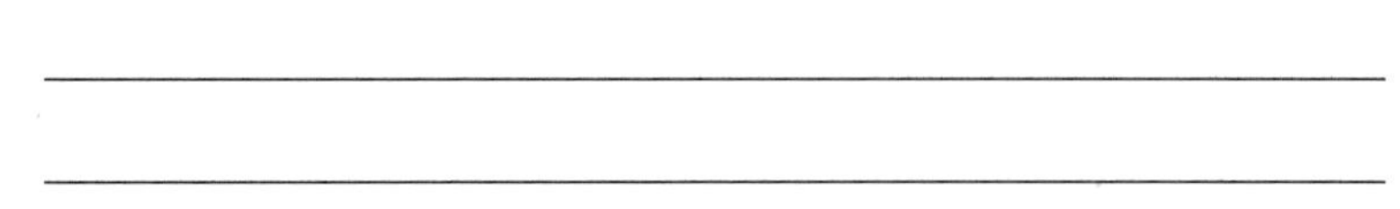

QUESTION 3 The graph of $y = 2x^2 - 2$ is shown.

a What is the y-intercept? ______________________

b What are the x-intercepts? ______________________

c Show the graph of $y = -2x^2 + 2$ on the same diagram.

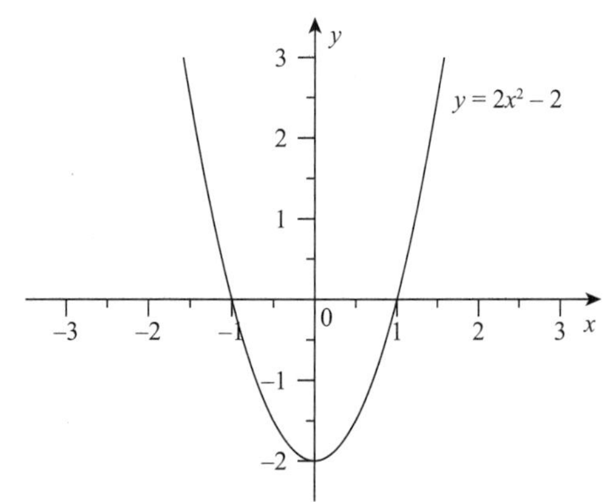

Non-linear relationships

UNIT 3: The parabola $y = ax^2 + c$

QUESTION 1 Consider the parabola $y = 2x^2 + 3$.

a What is the equation of the axis of symmetry? ____________________

b What are the coordinates of the vertex? ____________________

c If the parabola passes through the point $(1, p)$, what is the value of p?

____

d If the parabola passes through the point $(q, 11)$, where $q > 0$, what is the value of q?

____

e Explain why, if we know the coordinates of any point on a parabola of the form $y = ax^2 + c$, other than the vertex, we can immediately write down the coordinates of another point. Give two examples for the given parabola.

____

____

QUESTION 2

a A parabola of the form $y = ax^2 + c$ has vertex at $(0, -2)$ and passes through the point $(3, -29)$.

i Write down the value of c. ____________________

ii Find the value of a. ____________________

____

____

iii Will the parabola intercept the x-axis? If so, find the x-intercepts. If not, explain why not.

____

____

____

b A parabola of the form $y = ax^2 + c$ has vertex at $(0, -8)$ and passes through the point $(2, -6)$.

i Write down the value of c. ____________________

ii Find the value of a. ____________________

____

____

iii Will the parabola intercept the x-axis? If so, find the x-intercepts. If not, explain why not.

____

____

____

QUESTION 3 Determine the equation of each of these parabolas.

a

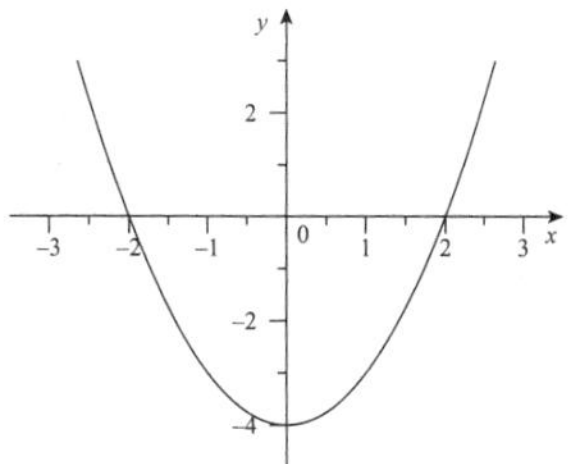

____

____

____

b

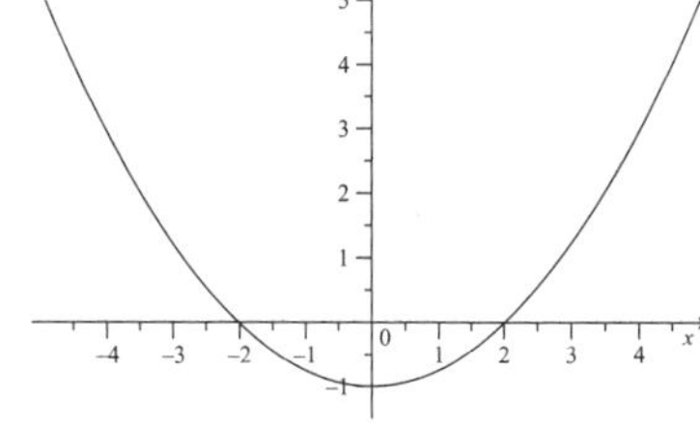

____

____

____

c

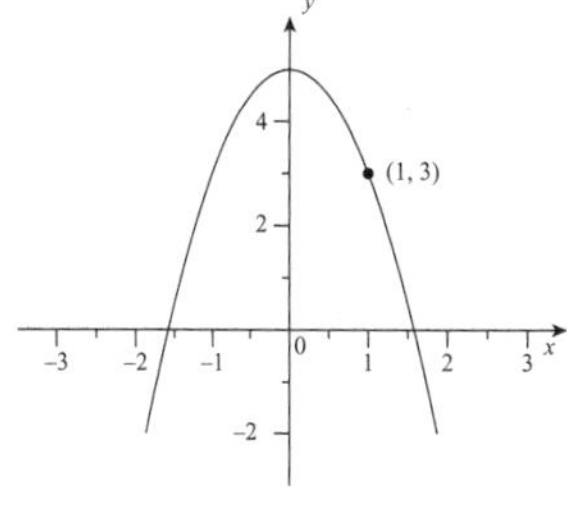

____

____

____

Non-linear relationships

UNIT 4: Exponential graphs

QUESTION 1 Make a table of values and then draw the graphs of the following exponential functions on the same set of axes.

a $y = 2^x$

b $y = 2^{-x}$

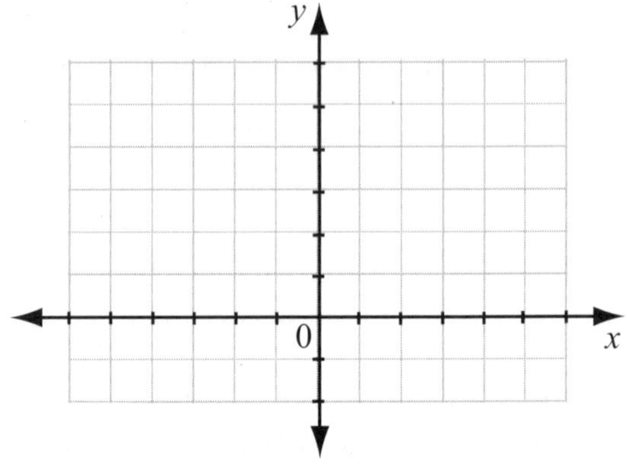

QUESTION 2 Make a table of values and then draw the graphs of the following exponential functions on the same set of axes.

a $y = 2^x$

b $y = 3^x$

c $y = 5^x$

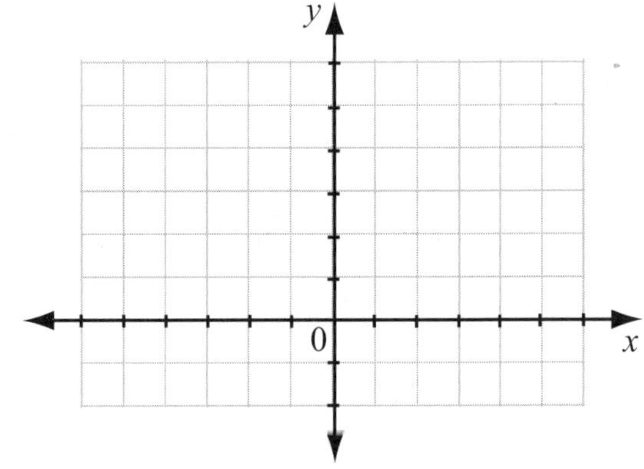

QUESTION 3 Complete the table of values and then draw the graph of $y = \dfrac{3^x + 3^{-x}}{2}$

x	–3	–2	–1	0	1	2	3
$y = 3^x$							
$y = 3^{-x}$							
$y = \dfrac{3^x + 3^{-x}}{2}$							

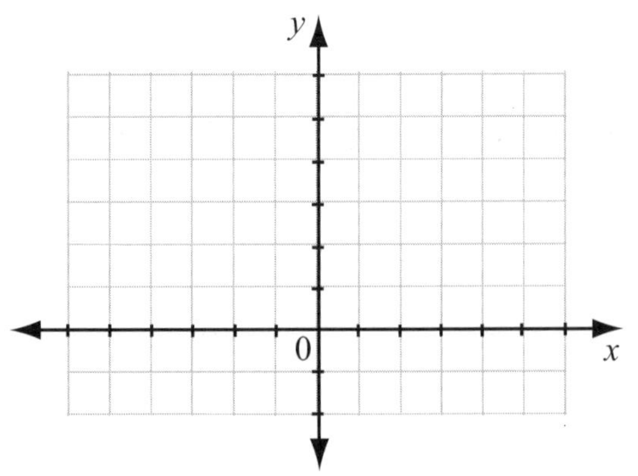

Non-linear relationships

UNIT 5: Exponential curves of the form $y = a^x$

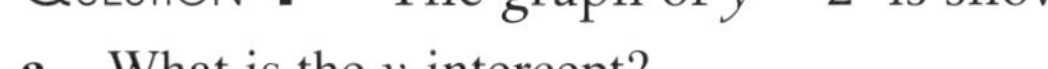

Question 1 The graph of $y = 2^x$ is shown.

a What is the y-intercept? ______________________

b Does the curve intercept the x-axis? If so, what is the x-intercept? If not, why not?

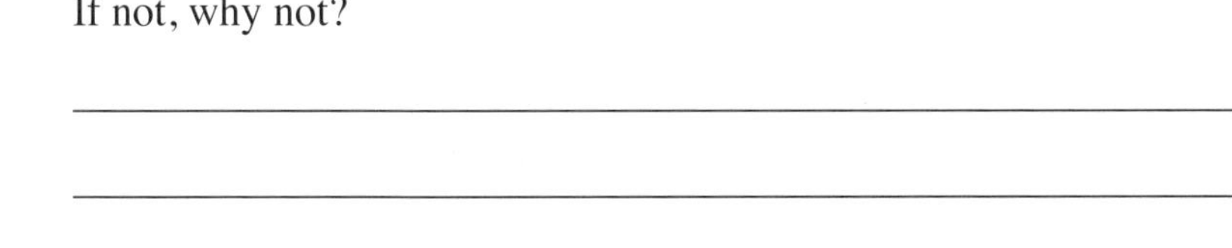

____

____

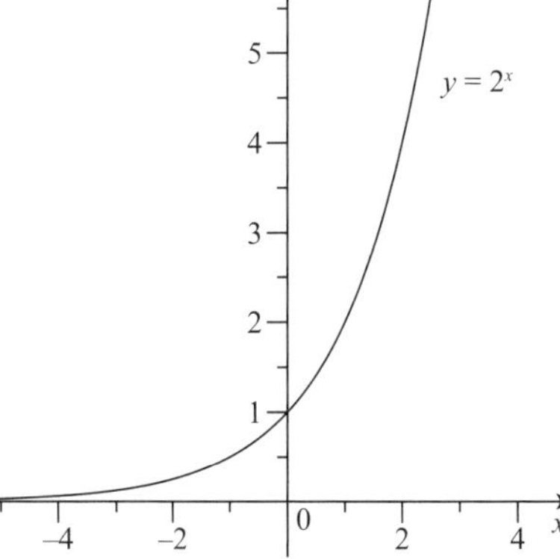

c What is the equation of the line that is an asymptote to this curve?

____

d What is the value of 2^x when:

i $x = 3$ ________ **ii** $x = 4$ ________ **iii** $x = 5$ ________ **iv** $x = 10$ ________ **v** $x = 20$ ________ **vi** $x = 30$ ________

e How do the results to part **d** show the 'exponential' nature of the curve? Briefly explain.

____

Question 2 Consider values of $a > 0$.

a What is the value of a^0? ______________________

b What is the y-intercept of any curve with the equation $y = a^x$? ______________________

c Explain why all curves of the form $y = a^x$ must have the same y-intercept.

____

d Is there any value of x for which a^x is negative? Briefly explain why or why not.

____

e What feature does this mean that all exponential curves of the form $y = a^x$ must have?

____

f What other similarities do all curves of the form $y = a^x$ have?

____

Question 3

a Briefly explain why the curves $y = 2 \times 3^x$ and $y = 6^x$ are not the same curve.

____

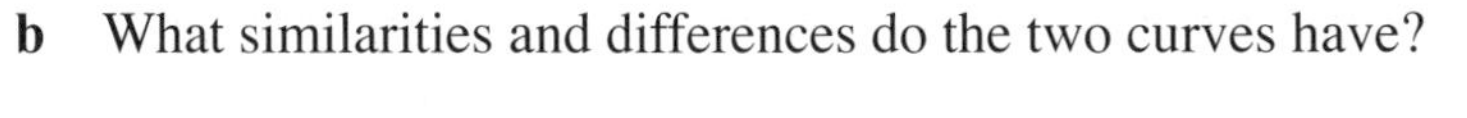

b What similarities and differences do the two curves have?

____

c Sketch both curves on the same set of axes, clearly labelling each.

Non-linear relationships

UNIT 6: The circle

QUESTION 1 Write the coordinates of the centre and the length of the radius for each of the following circles.

a $x^2 + y^2 = 4$ ____________ **b** $x^2 + y^2 = 49$ ____________

c $x^2 + y^2 = \frac{4}{9}$ ____________ **d** $x^2 + y^2 = 81$ ____________

QUESTION 2 Write the equation of each of the following circles, whose centre and radius are given.

a Centre $(0, 0)$, radius = 3 units

b Centre $(0, 0)$, radius = 7 units

c Centre $(0, 0)$, radius = 2 units

d Centre $(0, 0)$, radius = 10 units

QUESTION 3 Write the equation of each of the following circles.

a

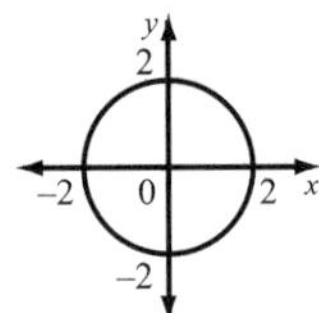

b

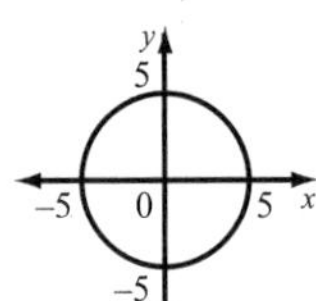

c

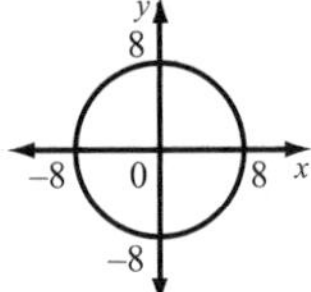

QUESTION 4 Graph each of the following circles, stating the radius and the centre.

a $x^2 + y^2 = 16$

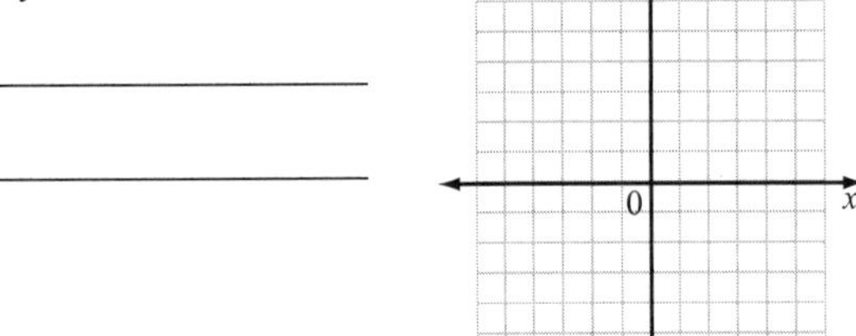

b $x^2 + y^2 = 1$

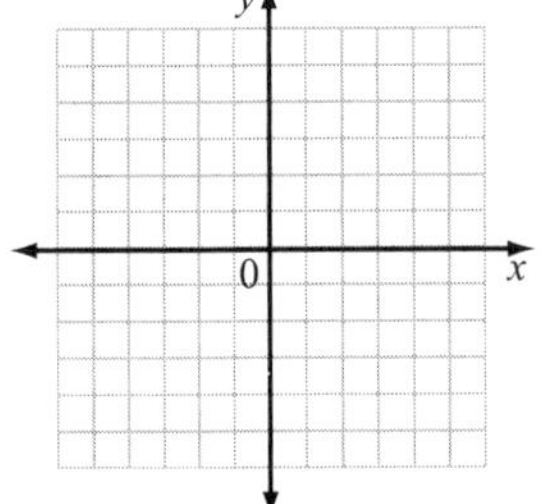

c $x^2 + y^2 = 9$

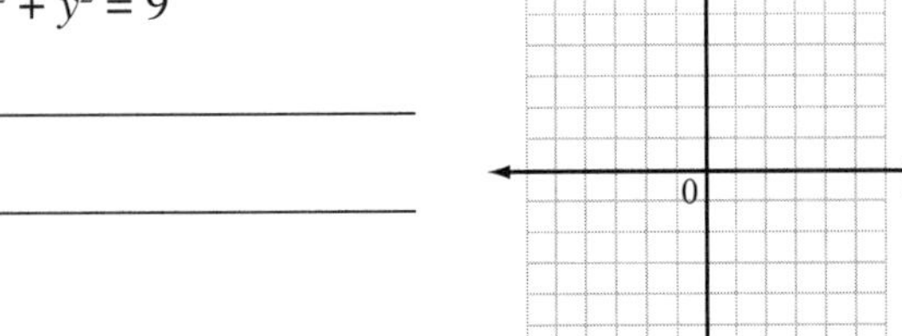

d $x^2 + y^2 = 36$

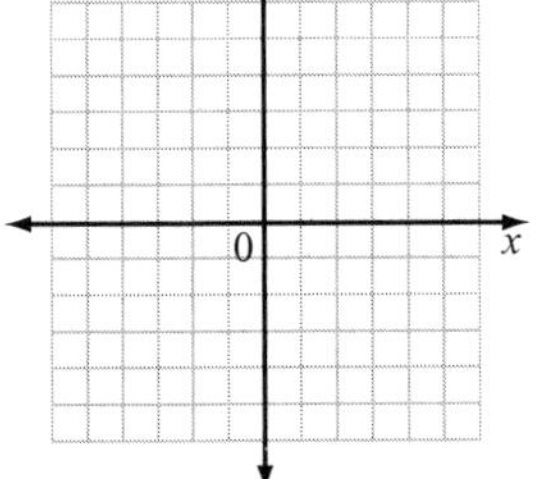

Non-linear relationships

UNIT 7: Miscellaneous graphs

QUESTION 1 For the following equations, write whether the graphs are straight lines, parabolas, circles, exponential functions or none of these.

a $y = x$ ____________ **b** $y = -x^2$ ____________ **c** $y = 0$ ____________

d $y = x^2 - 5x + 6$ ____________ **e** $y = x^2$ ____________ **f** $y = 3 - x$ ____________

g $y = x^2 - 1$ ____________ **h** $y = -10^x$ ____________ **i** $y = x^3$ ____________

j $x^2 + y^2 = 16$ ____________ **k** $y = 2^x$ ____________ **l** $x^2 + y^2 = 64$ ____________

QUESTION 2 Match the equations with the graphs sketched below.

a $y = 2x + 1$ ________ **b** $y = 1 - x^2$ ________ **c** $y = 2^{-x}$ ________ **d** $x^2 + y^2 = 1$ ________

e $y = x^2 + 2$ ________ **f** $y = -2^x$ ________ **g** $y = x$ ________ **h** $y = 2^x$ ________

i $y = x^2$ ________ **j** $y = x^2 - 4x + 3$ ________ **k** $y = -x$ ________ **l** $x^2 + y^2 = 25$ ________

A

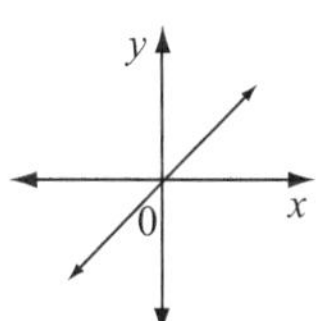

B

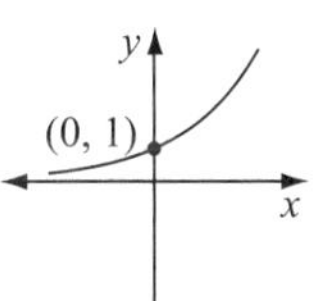

C

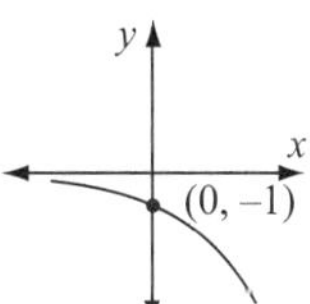

D

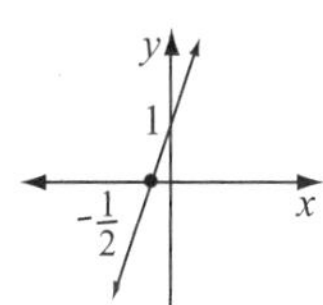

E

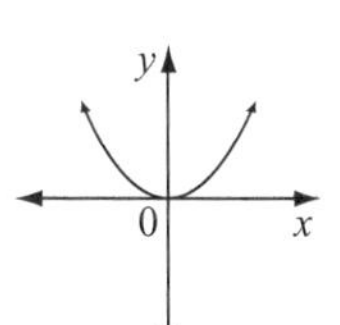

F

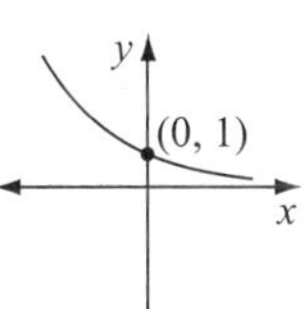

G

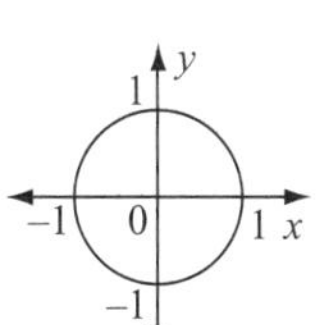

H

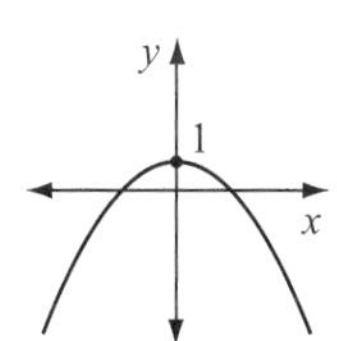

I

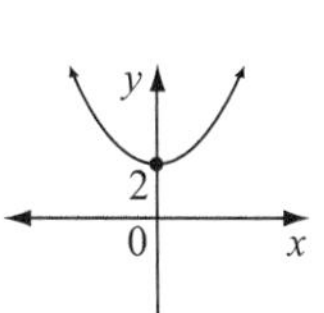

J

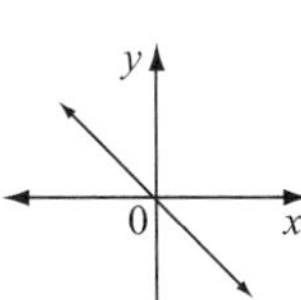

K

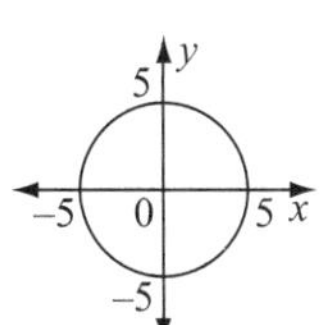

L

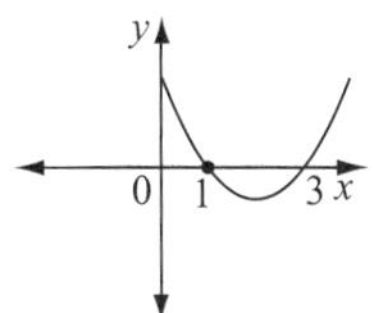

QUESTION 3 Draw a separate sketch for each of the following.

a $y = 2x + 3$ **b** $y = 2x^2$ **c** $x^2 + y^2 = 9$

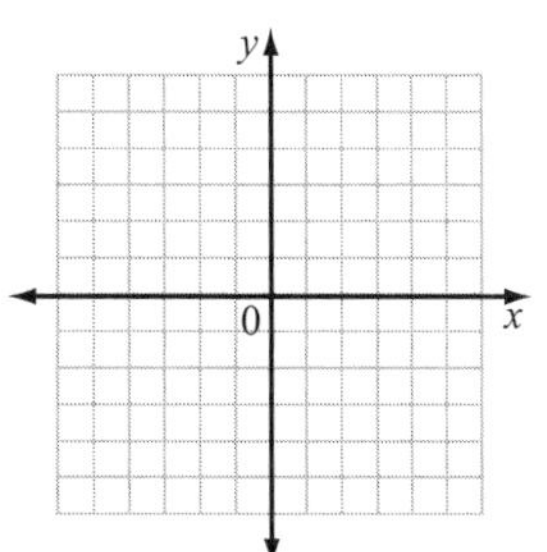

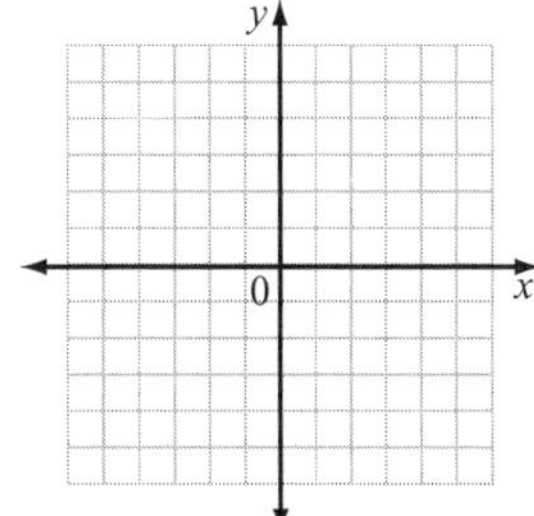

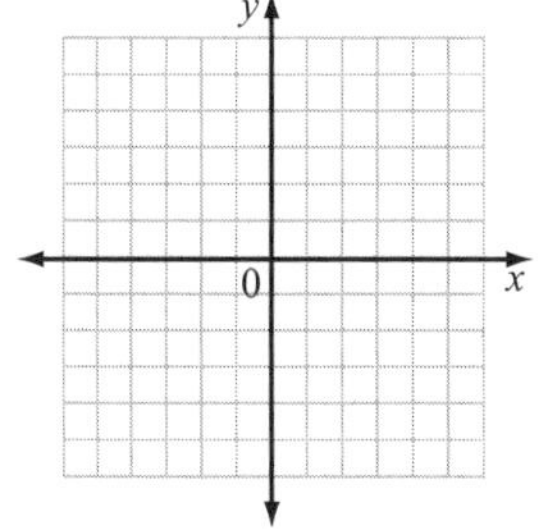

Non-linear relationships

TOPIC TEST — PART A

Instructions
- This part consists of 10 multiple-choice questions.
- Fill in only ONE CIRCLE for each question.
- Each question is worth 1 mark.

Time allowed: 15 minutes — **Total marks: 10**

Marks

1 Which is **not** a linear relationship? 1

Ⓐ $y = 7$ Ⓑ $y = \frac{x}{7}$ Ⓒ $y = 7x$ Ⓓ $y = 7^x$

2 Which graph best represents $y = x^2$? 1

Ⓐ 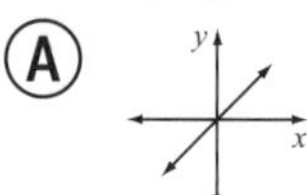
Ⓑ 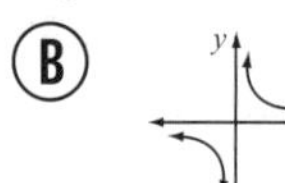
Ⓒ 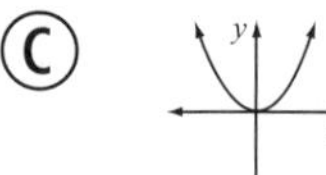
Ⓓ  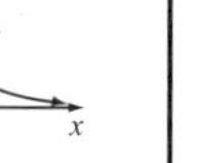

3 What is the equation of the axis of symmetry of $y = ax^2 + c$? 1

Ⓐ $x = 0$ Ⓑ $y = 0$ Ⓒ $x = a$ Ⓓ $y = c$

4 The radius of the circle $x^2 + y^2 = 4$ is equal to: 1

Ⓐ 2 units Ⓑ 4 units Ⓒ 16 units Ⓓ none of these

5 The graph shown could be part of the graph with equation: 1

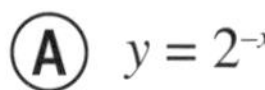

Ⓐ $y = 2^{-x}$ Ⓑ $y = 2^x$ Ⓒ $y = -2^x$ Ⓓ $y = -2^{-x}$

6 What is the vertex of $y = 3 - 2x^2$? 1

Ⓐ (0, 0) Ⓑ (0, –2) Ⓒ (0, 3) Ⓓ (3, –2)

7 Which of these equations is that of a curve that is concave down? 1

Ⓐ $y = 4x - 3$ Ⓑ $y = 6 - 2x^2$ Ⓒ $y = 4^x$ Ⓓ $y = -5 + 3x^2$

8 Which is **not** the equation of an exponential curve? 1

Ⓐ $y = 1^x$ Ⓑ $y = 2^x$ Ⓒ $y = 8^x$ Ⓓ $y = 25^x$

9 A parabola of the form $y = ax^2 + c$ passes through the point (2, 5). Which other point must it also pass through? 1

Ⓐ (0, 0) Ⓑ (1, 2) Ⓒ (–2, 5) Ⓓ (2, –5)

10 Which of these curves will pass through (–1, 0.5)? 1

Ⓐ $y = x^2 - 1$ Ⓑ $y = \frac{1}{2}x^2 + 1$ Ⓒ $y = 2^x$ Ⓓ $y = 5^x$

Total marks achieved for PART A /10

Non-linear relationships

TOPIC TEST PART B

Instructions
- This part consists of 4 questions.
- Write only the answer in the answer column.
- For any working use the question column.

Time allowed: 20 minutes **Total marks: 15**

Questions	Answers	Marks
1 Consider $y = 3 - \frac{1}{3}x^2$.		
a Is the curve concave up or concave down? ______	______	1
b What are the coordinates of the vertex? ______	______	1
c What are the x-intercepts? ______	______	1
d Briefly explain why the curve $y = 3 + \frac{1}{3}x^2$ will not have x-intercepts. ______	______	1
2 Write down the equation of each of these curves.		
a 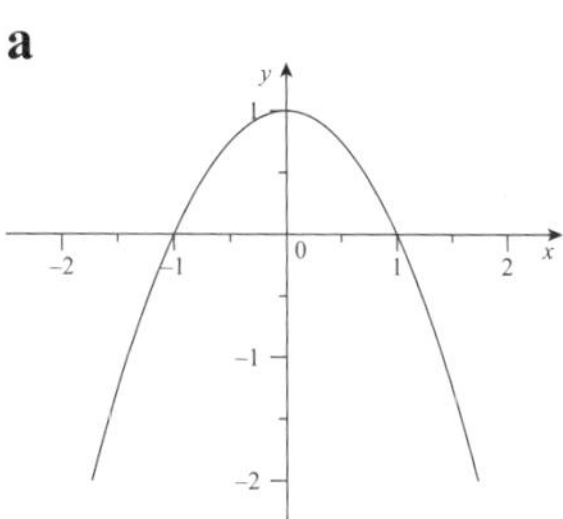	______	1
b 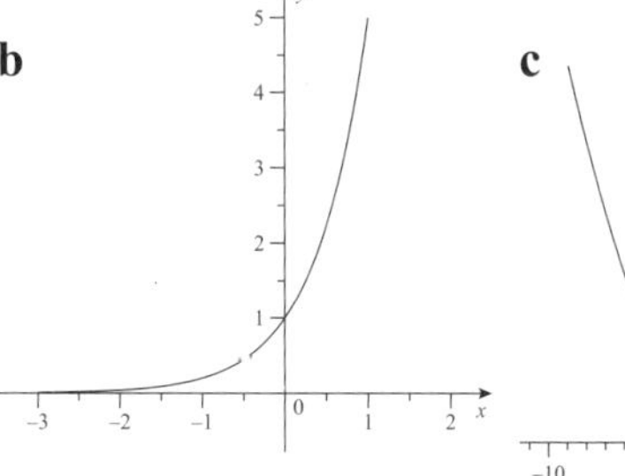	______	1
c 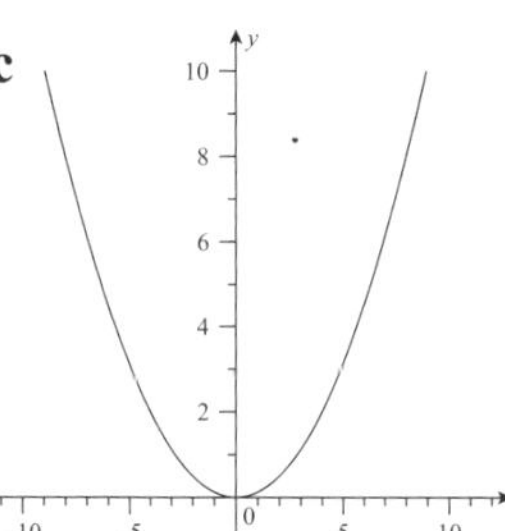	______	1
3 A straight line intersects with the curve $y = 3^x$ when $x = 0$ and $x = 1$.		
a What is the y-intercept of the line? ______	______	1
b Write the coordinates of both points of intersection. ______	______	1
c What is the equation of the line? ______	______	1
d Will the line intersect the curve in any other points? ______	______	1
4 The graph shows the curve $y = ax^2 + c$.		
a What name is given to this type of curve? ______	______	1
b What is the value of c? ______	______	1
c Find the value of a. ______	______	1
d Find y when $x = 6$. ______	______	1

Total marks achieved for PART B ___ / 15

CHAPTER 8
Area, surface area and volume

UNIT 1: Area of plane shapes

QUESTION **1** Complete the following table by writing the formula of the given plane shape.

	Shape	Area
a	Triangle	$A =$
b	Square	$A =$
c	Rectangle	$A =$
d	Parallelogram	$A =$

	Shape	Area
e	Trapezium	$A =$
f	Rhombus	$A =$
g	Kite	$A =$
h	Circle	$A =$

QUESTION **2** Find the area of each shape:

a

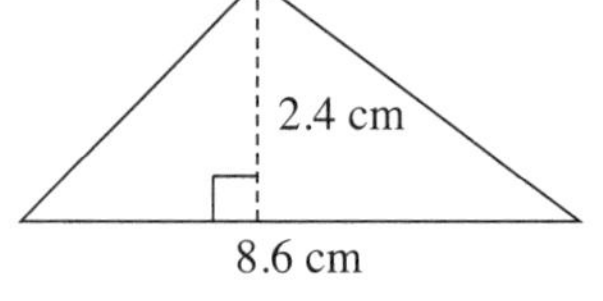

b

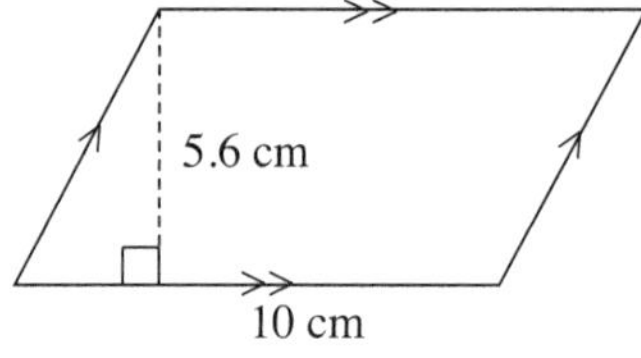

c

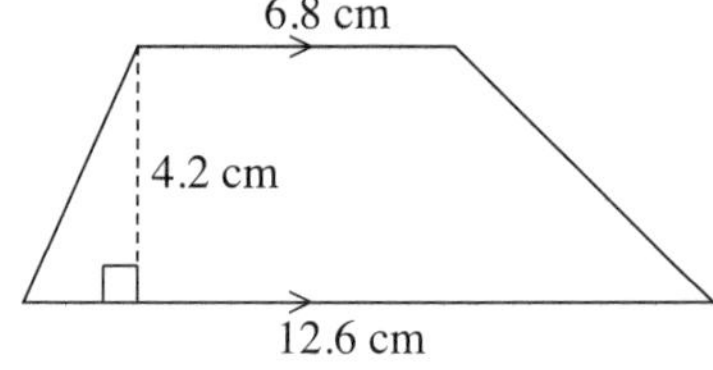

d

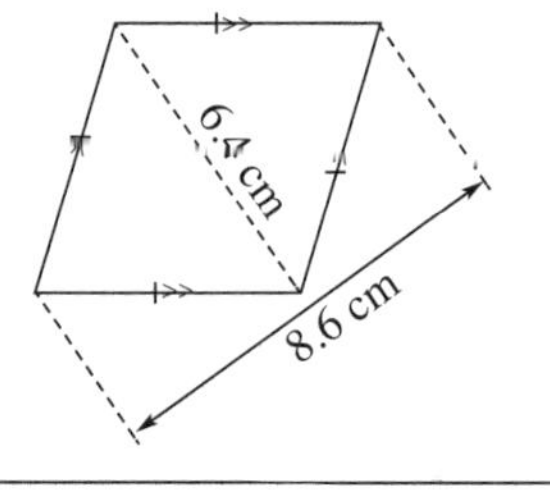

e

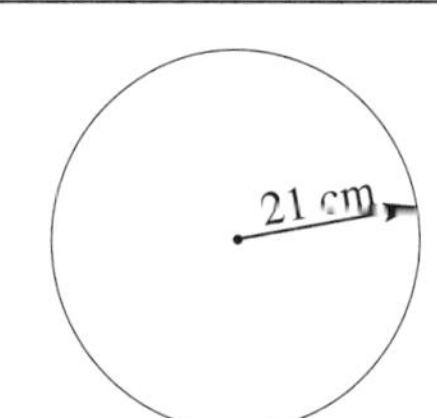

f

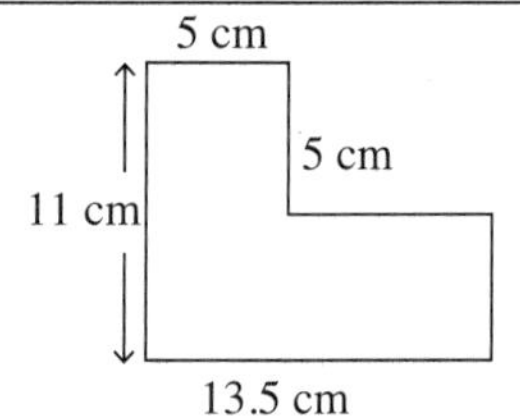

g

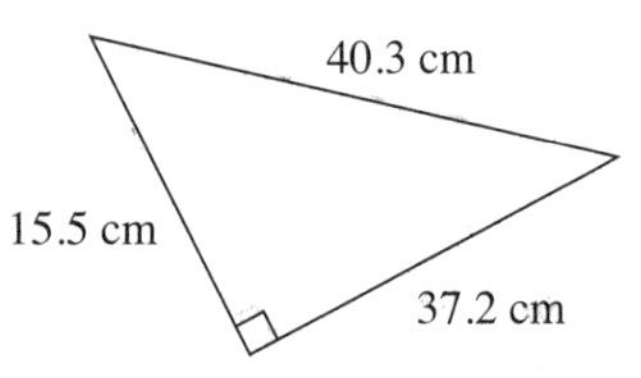

h

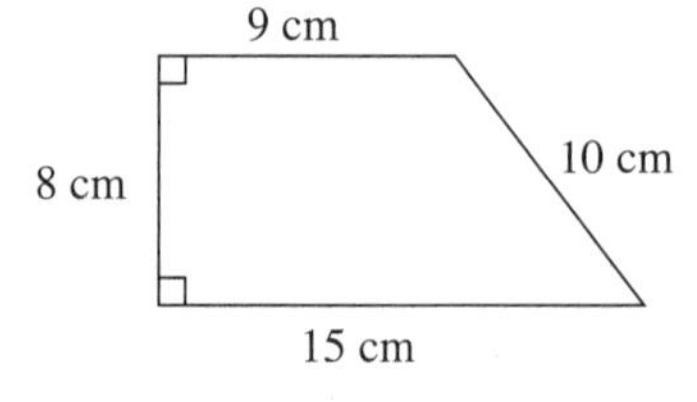

i

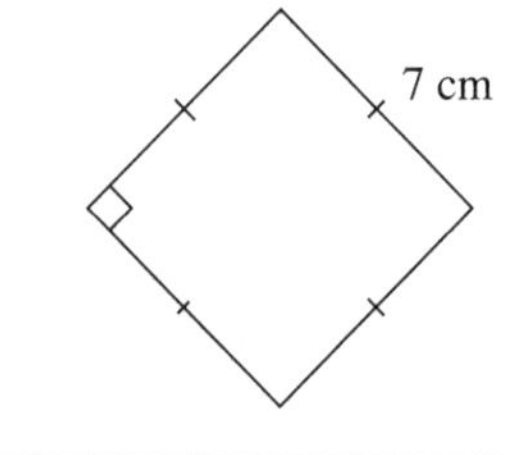

Area, surface area and volume

UNIT 2: Area of composite shapes (1)

QUESTION 1 Find the area of each shape. All measurements are in centimetres, and all angles are right angles.

a

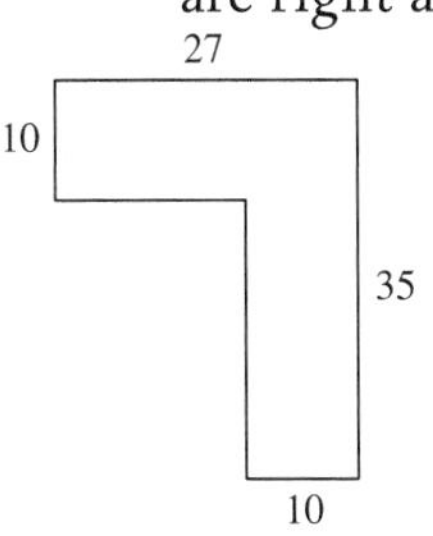

b

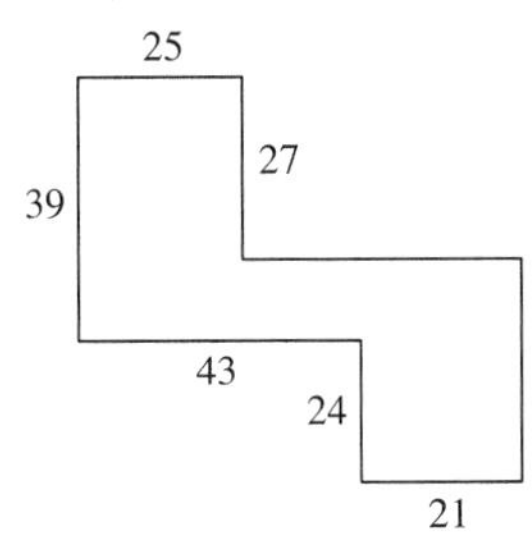

c

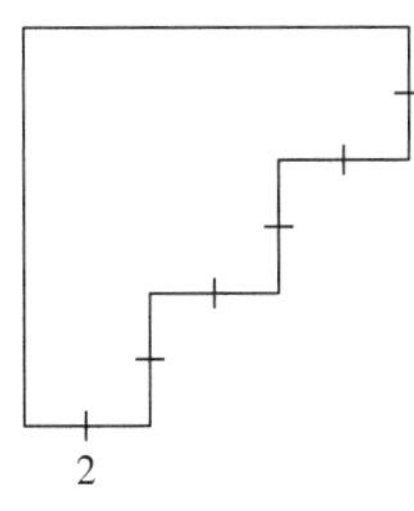

QUESTION 2 Find each shaded area.

a

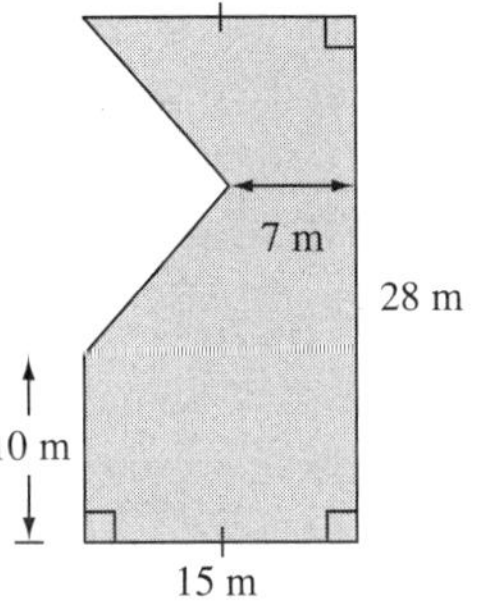

b

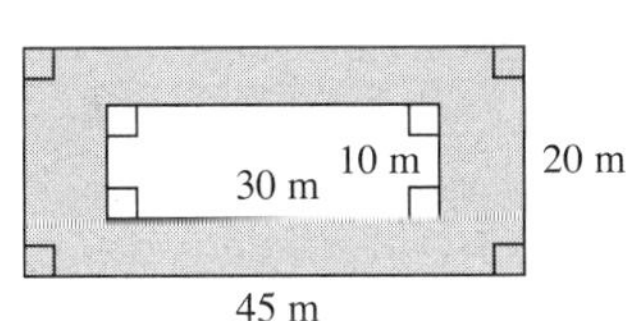

c

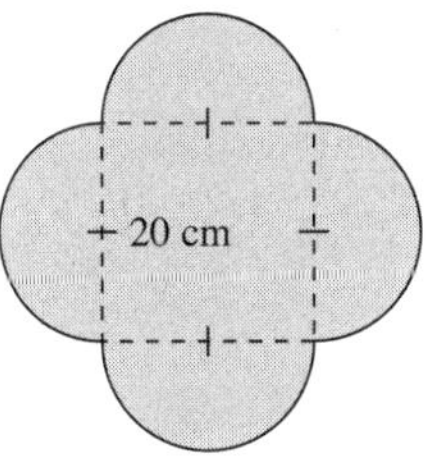

d

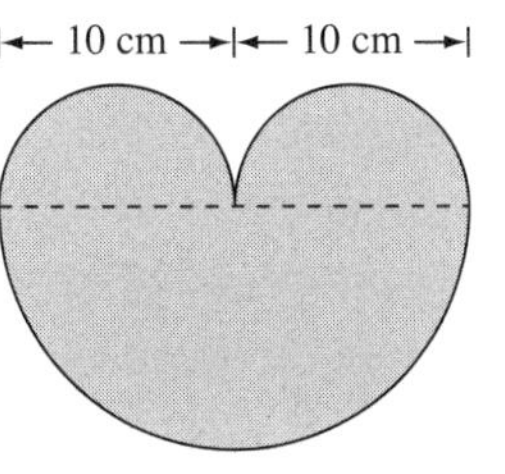

e

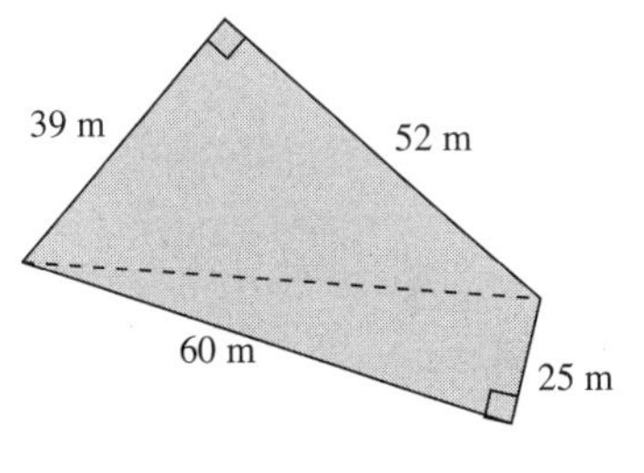

f

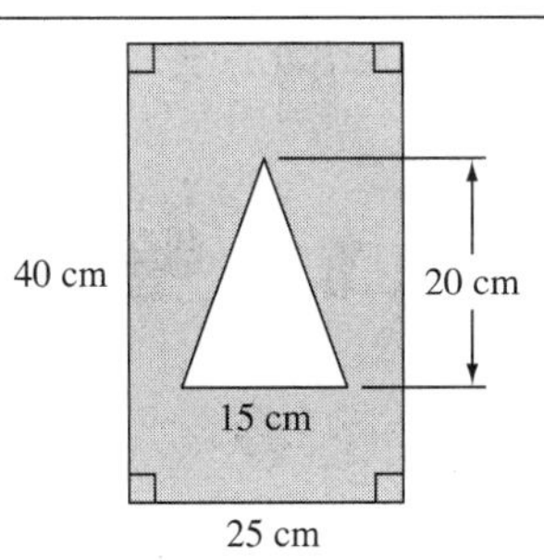

UNIT 3: Area of composite shapes (2)

QUESTION **1** Find each shaded area (correct to one decimal place where necessary).

a

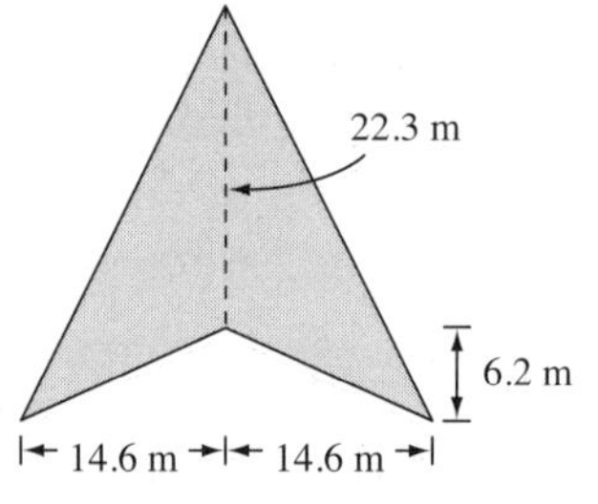

b

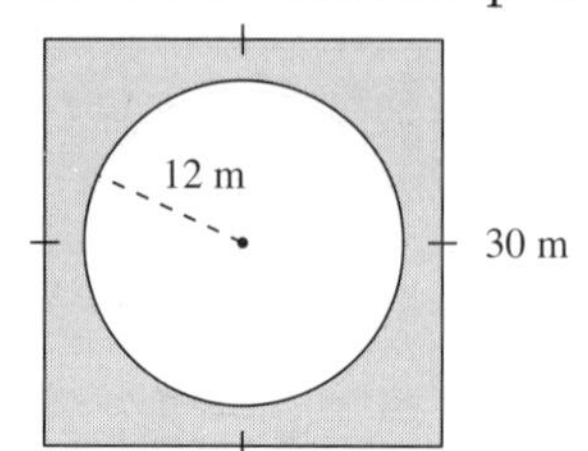

c

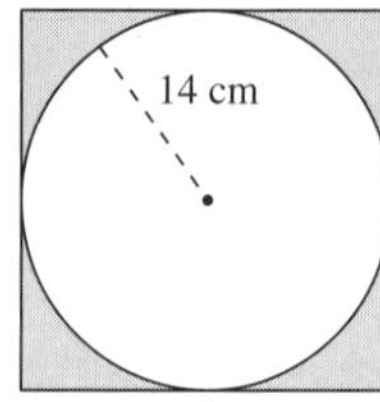

d

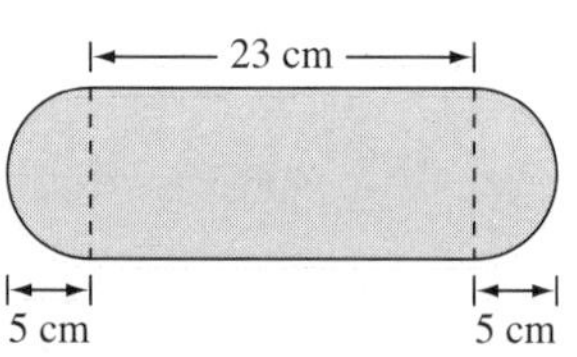

e

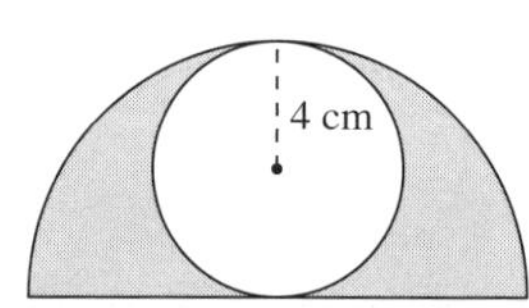

f

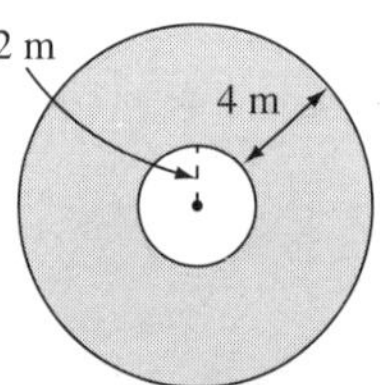

g

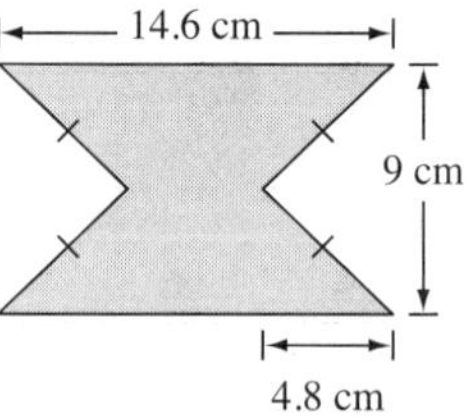

h

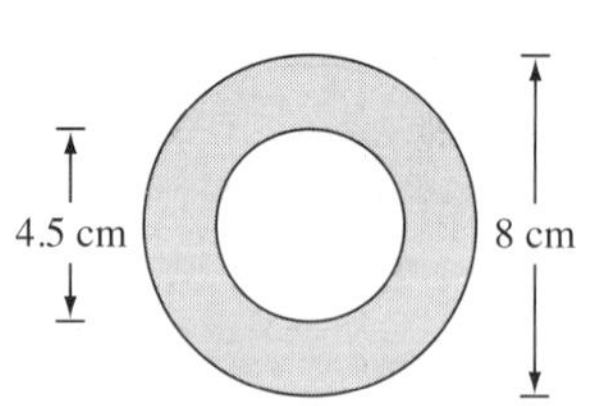

i

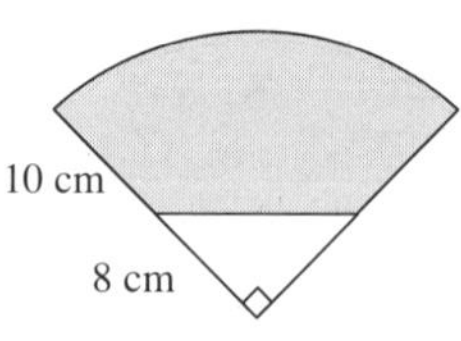

Area, surface area and volume

UNIT 4: Surface areas of right prisms (1)

QUESTION 1 Find the surface area of each shape.

a

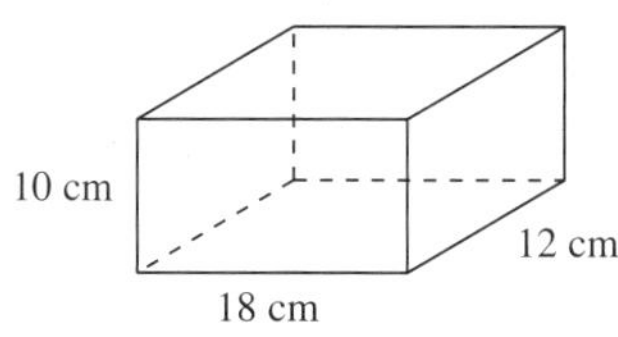

b

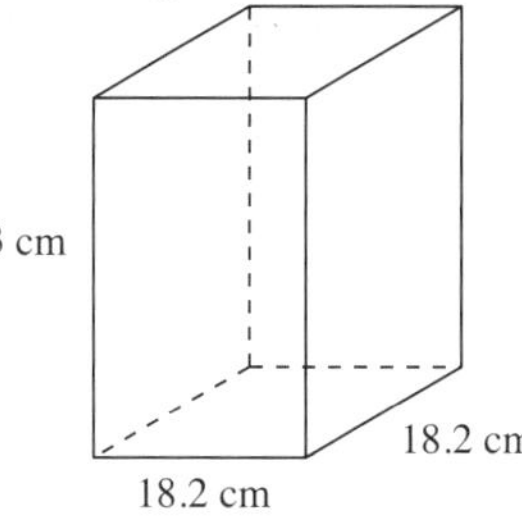

c

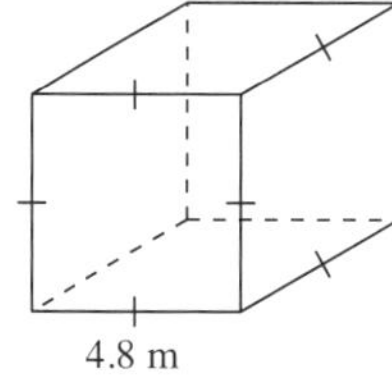

QUESTION 2 Find the surface area of each solid (correct to two decimal places), given its net.

a

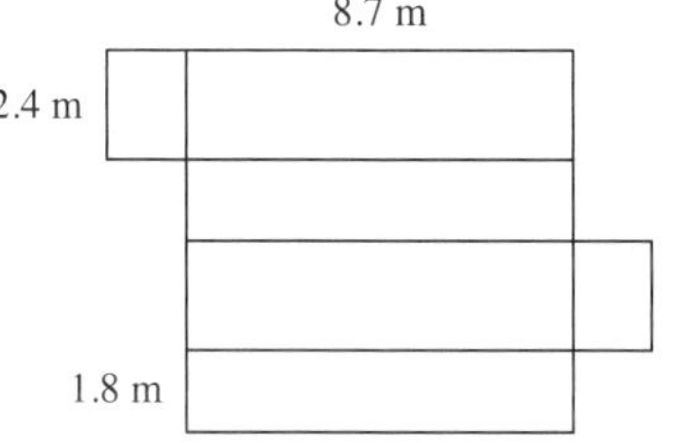

b

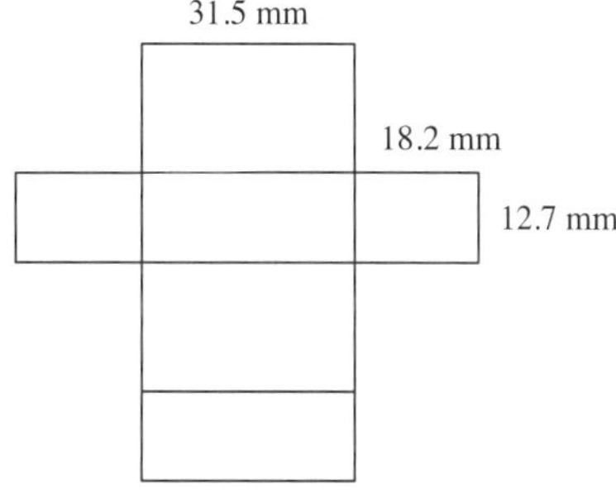

QUESTION 3 Find the surface area of each prism.

a

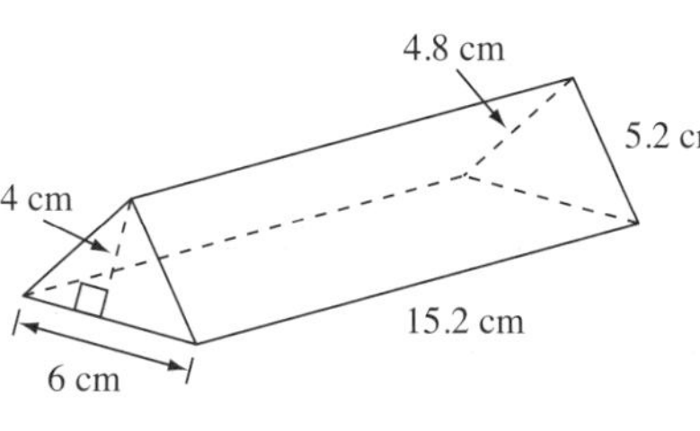

b

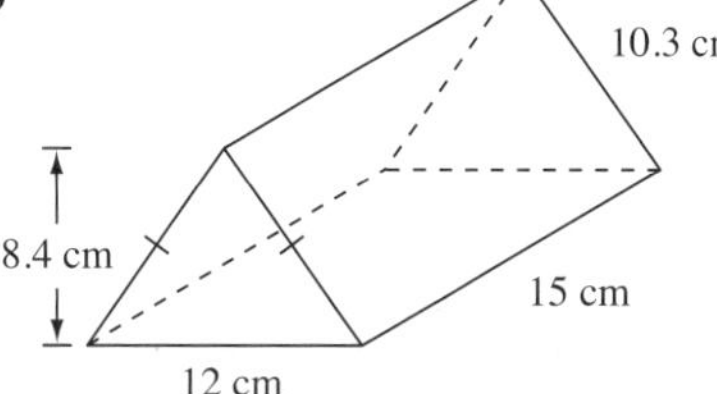

c

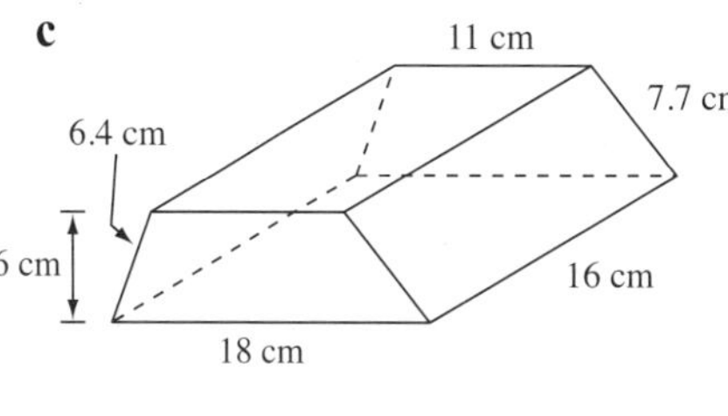

UNIT 5: Surface areas of right prisms (2)

QUESTION **1** Calculate the surface area of each shape (correct to one decimal place where necessary). You will need to use Pythagoras' theorem to calculate an unknown length.

a

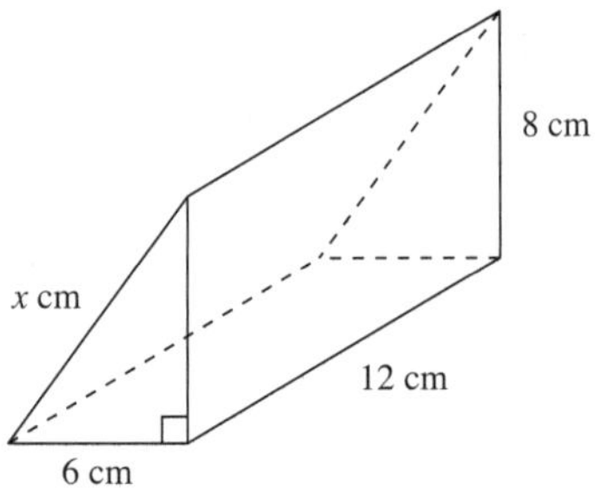

b

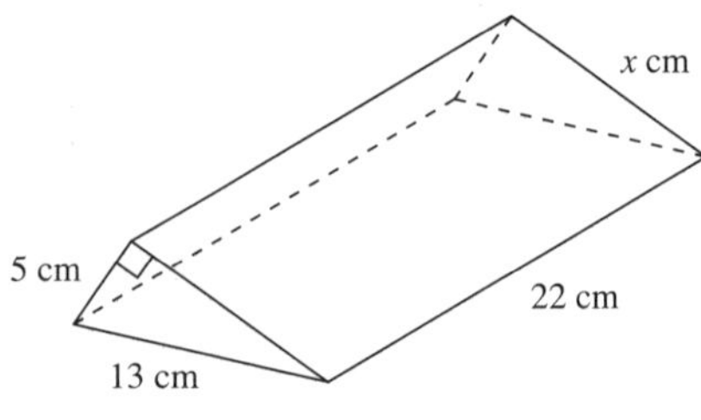

c

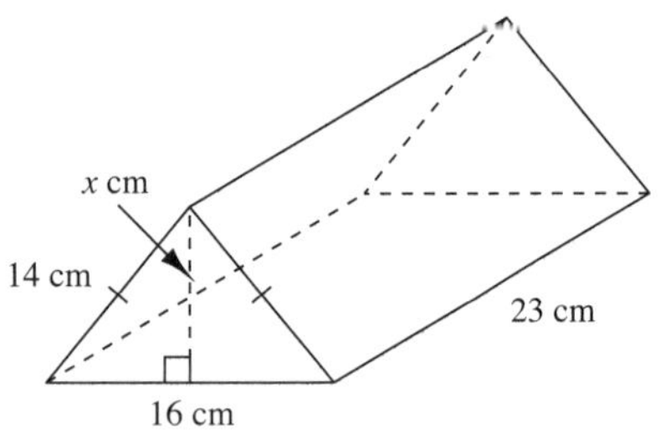

d

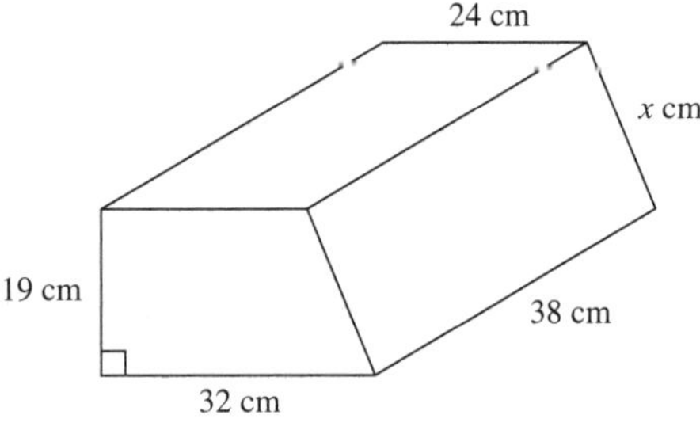

Area, surface area and volume

UNIT 6: Surface area of composite solids

QUESTION 1 Calculate the surface area of each shape (correct to one decimal place where necessary). You might need to use Pythagoras' theorem to calculate an unknown length.

a

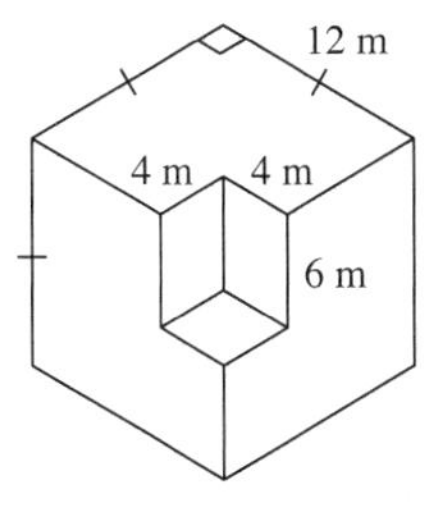

b

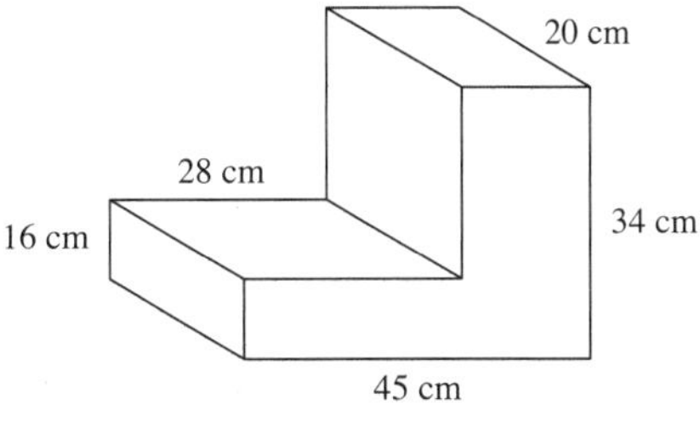

c

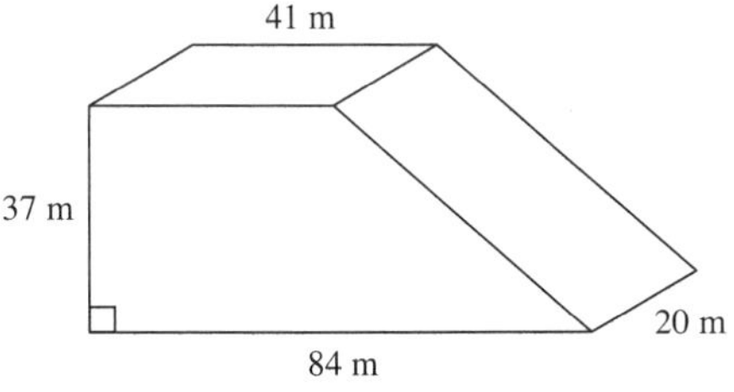

d

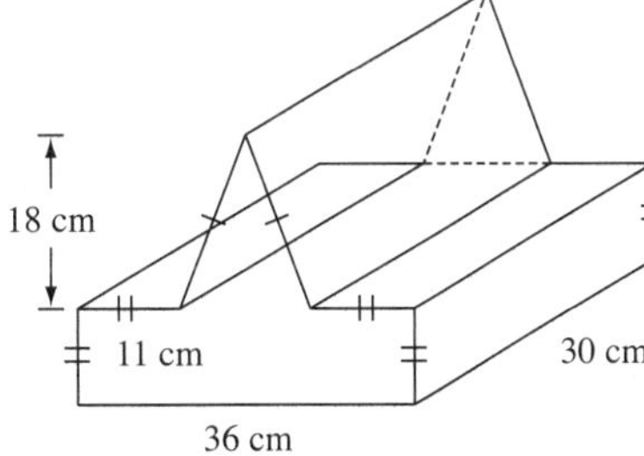

Area, surface area and volume

UNIT 7: Surface area of right cylinders (1)

QUESTION 1 For each cylinder, find **i** the area of a circular base **ii** the area of the curved surface, correct to two decimal places.

a

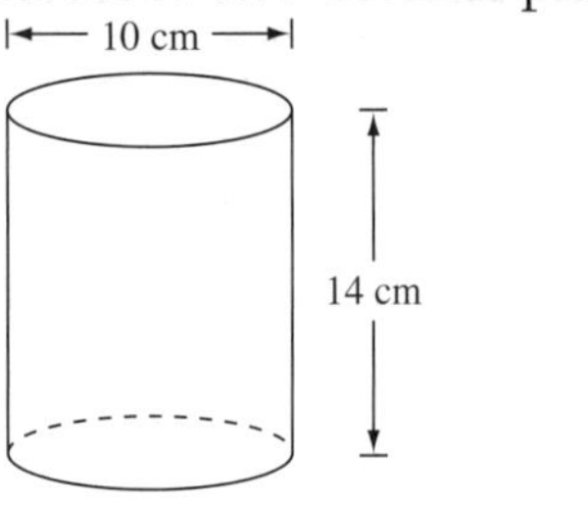

i ______________________________

____

ii ______________________________

____

b

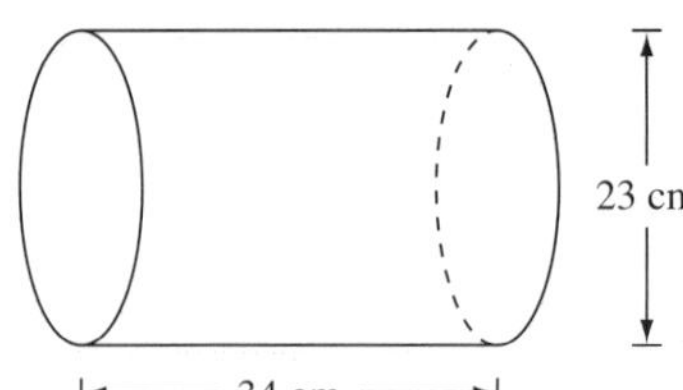

i ______________________________

____

ii ______________________________

____

c

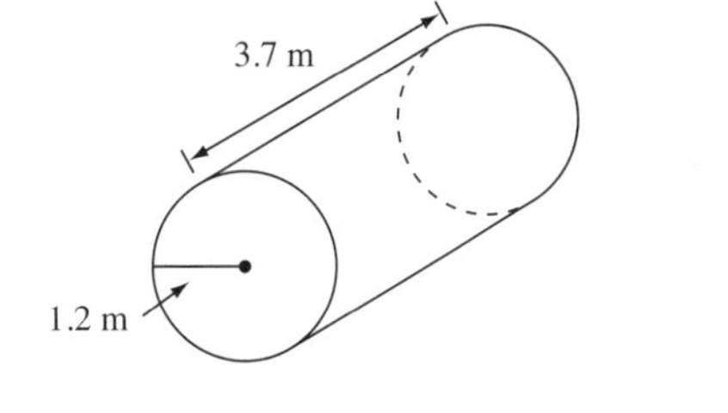

i ______________________________

____

ii ______________________________

____

d

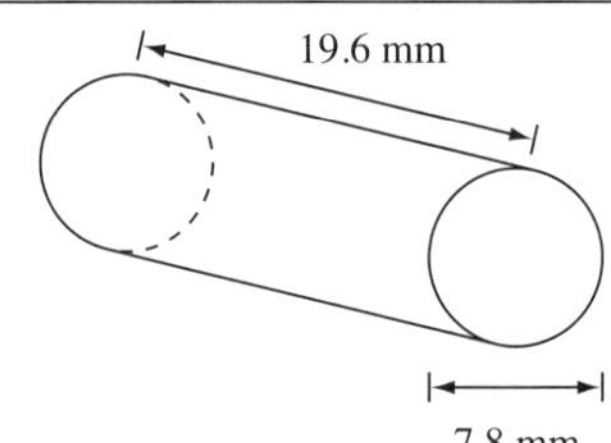

i ______________________________

____

ii ______________________________

____

QUESTION 2 Find the curved surface area of each cylinder, leaving your answers in terms of π.

a

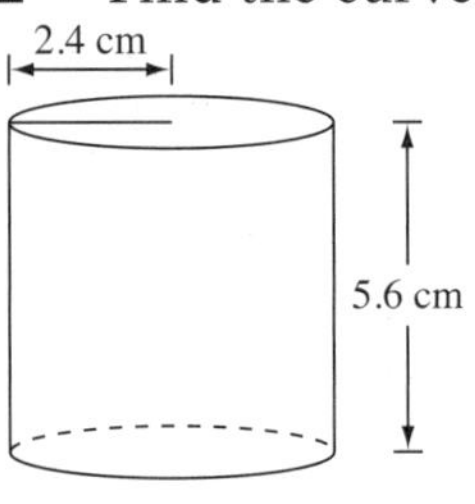

____

____

____

____

b

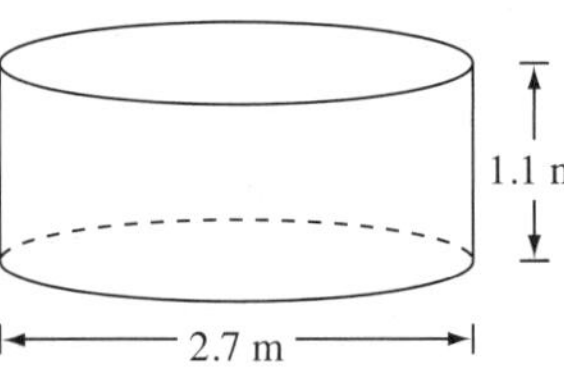

____

____

____

____

c

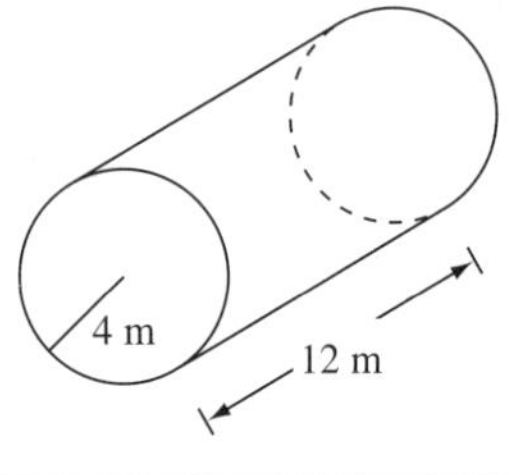

____

____

____

____

d

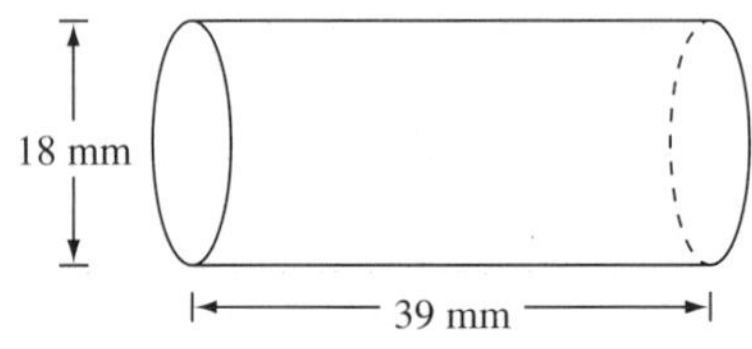

____

____

____

____

Area, surface area and volume

UNIT 8: Surface area of right cylinders (2)

Question 1 For each cylinder, find to three significant figures **i** the area of the two circular ends **ii** the area of the curved surface **iii** the total surface area.

a

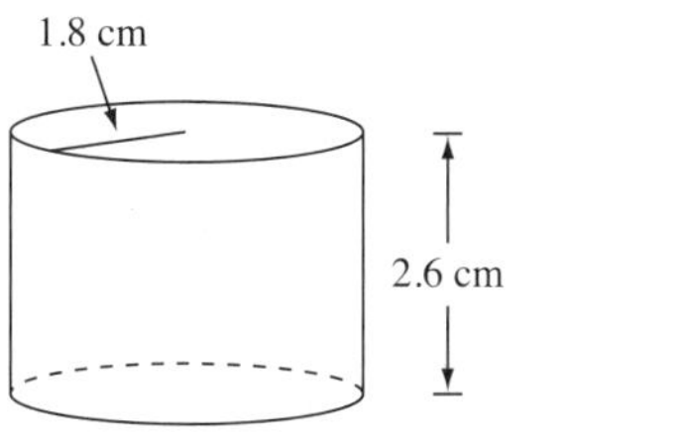

i ______________________

____

____

ii ______________________

____

____

iii ______________________

b

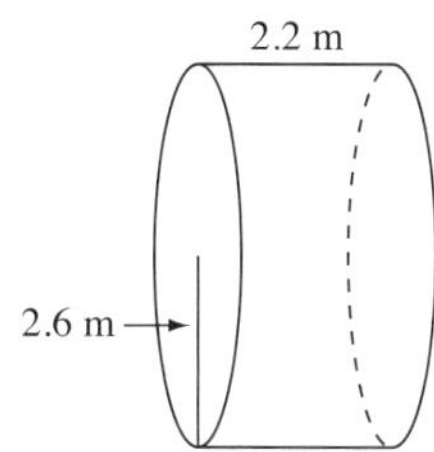

i ______________________

____

____

ii ______________________

____

____

iii ______________________

c

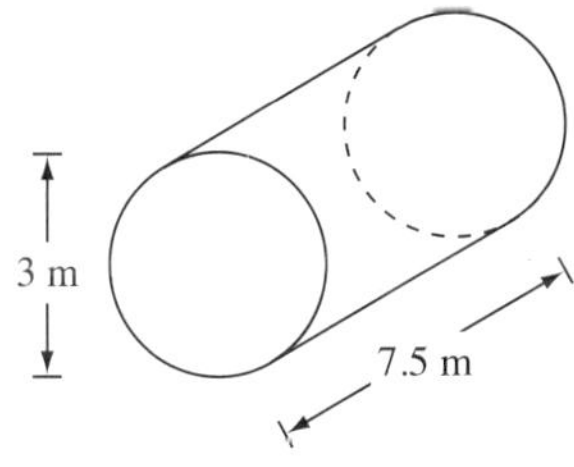

i ______________________

____

____

ii ______________________

____

____

iii ______________________

d

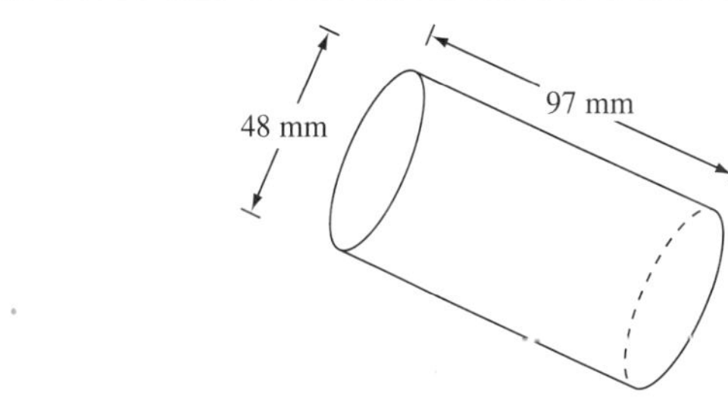

i ______________________

____

____

ii ______________________

____

____

iii ______________________

Question 2 A pipe, open at both ends, is 12 m long and has a radius of 80 cm. Find its external surface area.

____

____

____

Question 3 A cylindrical container, open at one end, is to be made from metal. Find the area of metal needed for the container if it will have a radius of 0.6 m and be 0.7 m high.

____

____

____

____

____

Area, surface area and volume

UNIT 9: Surface area of cylindrical objects

QUESTION 1 The following solids were formed from cylinders. Find the total surface area of each solid correct to two significant figures.

a

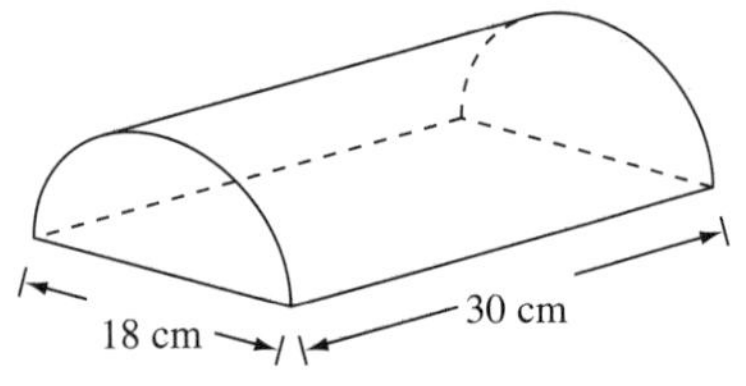

b

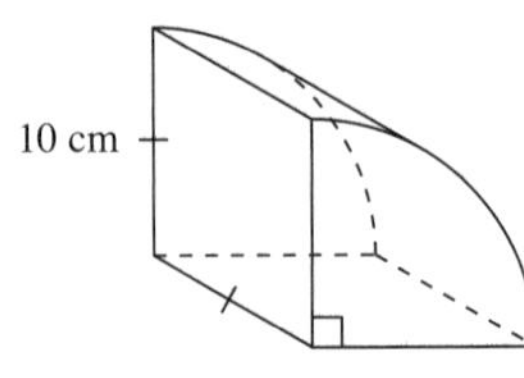

c

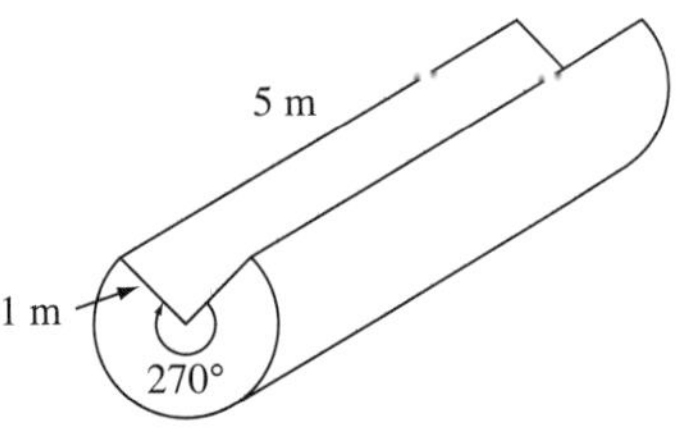

d

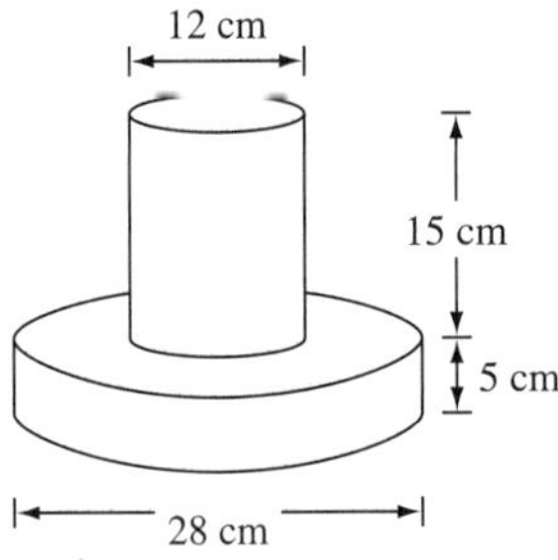

Area, surface area and volume

UNIT 10: Volume of right prisms

QUESTION 1 Calculate the volume of each rectangular prism (correct to one decimal place if necessary).

a

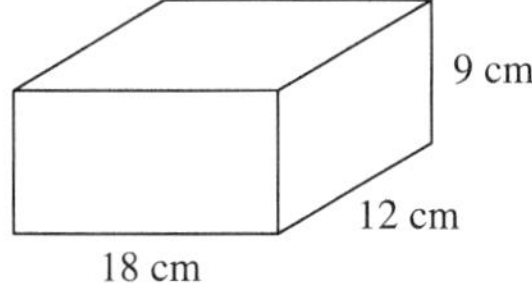

b

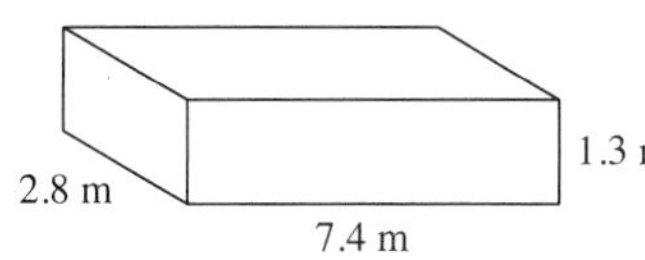

c

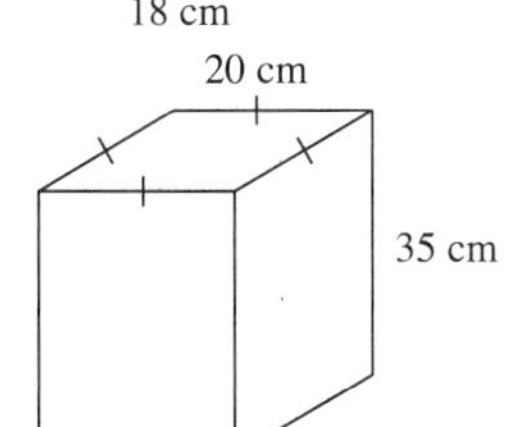

d

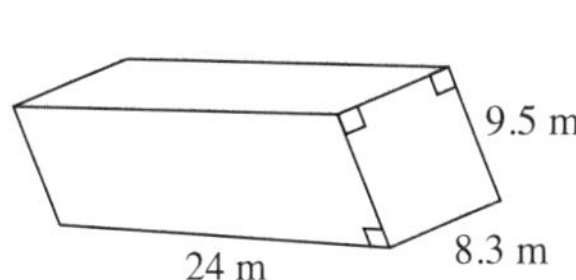

QUESTION 2 Calculate the volume of each triangular prism, giving your answers correct to two significant figures.

a

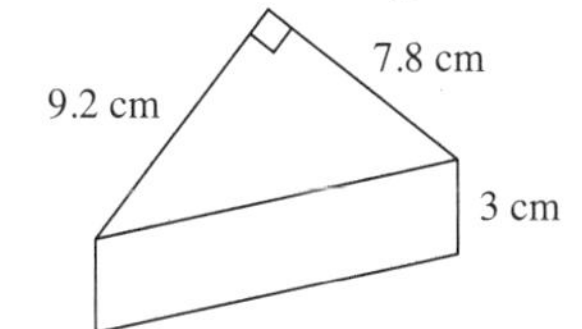

b

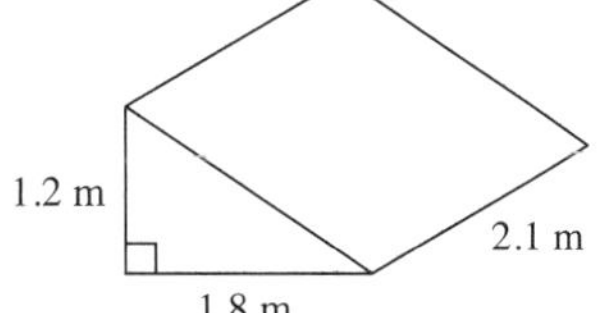

c

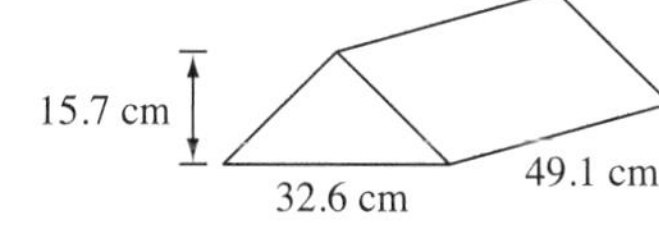

QUESTION 3 Use Pythagoras' theorem to find the height of each triangle, then calculate the volume of each triangular prism to the nearest cubic centimetre.

a

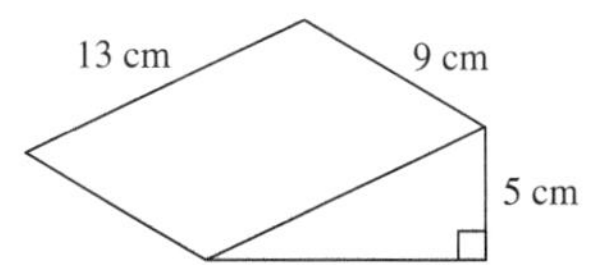

b

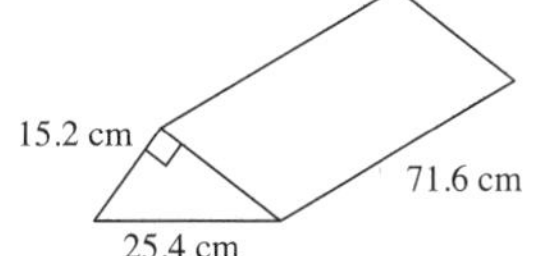

c

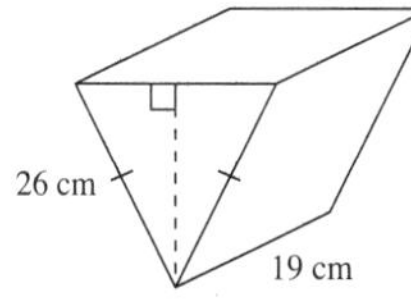

QUESTION 4 Find the volume of each trapezoidal prism to the nearest cubic unit.

a

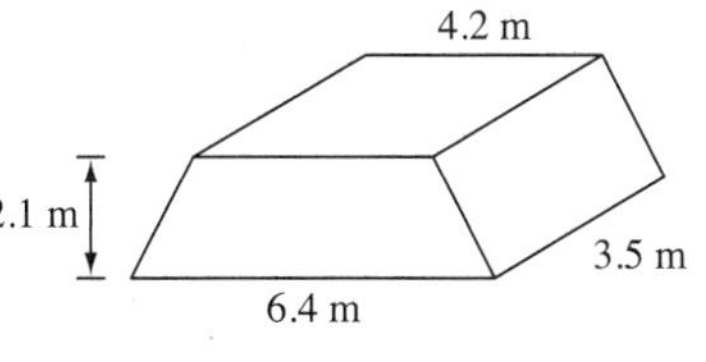

b

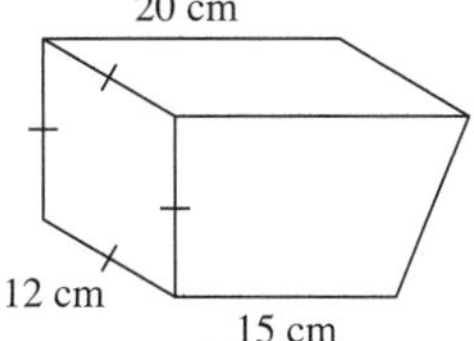

c

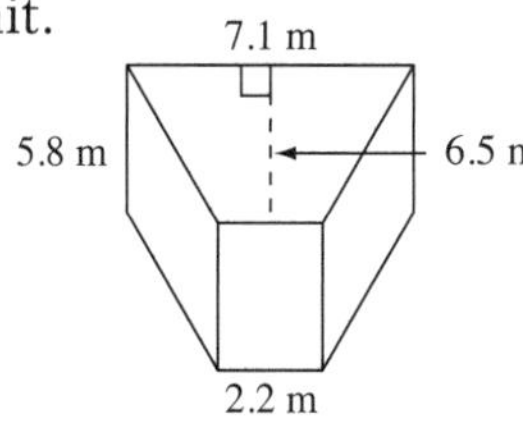

Area, surface area and volume

UNIT 11: Volume of right prisms and composite solids

QUESTION **1** Find the volume, correct to two significant figures.

a

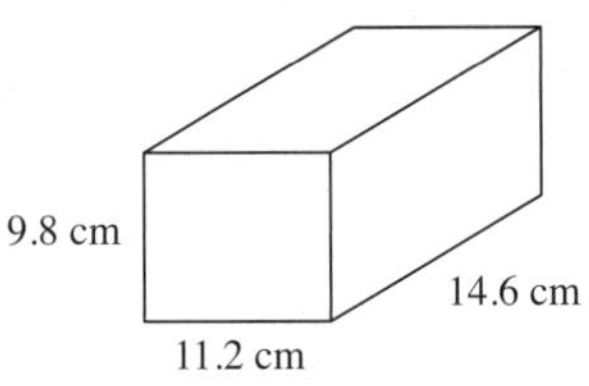

b

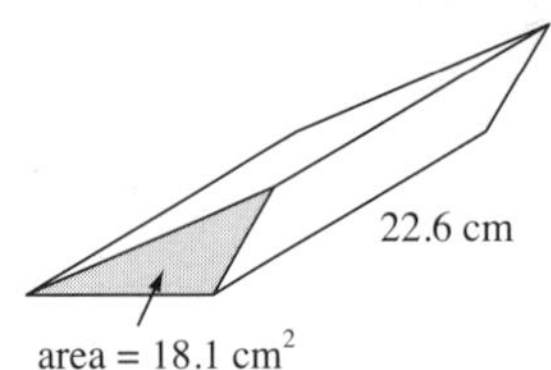

c

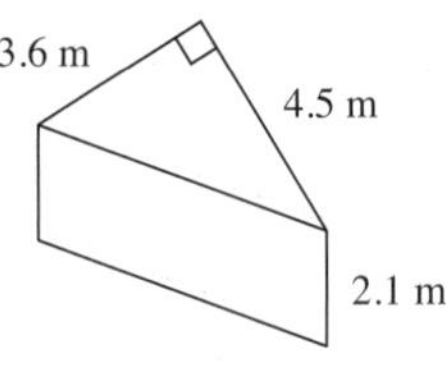

QUESTION **2** Find the volume of each prism (correct to one decimal place if necessary).

a

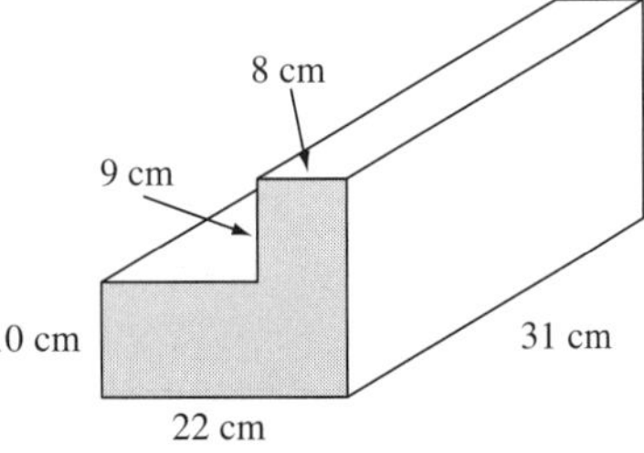

b

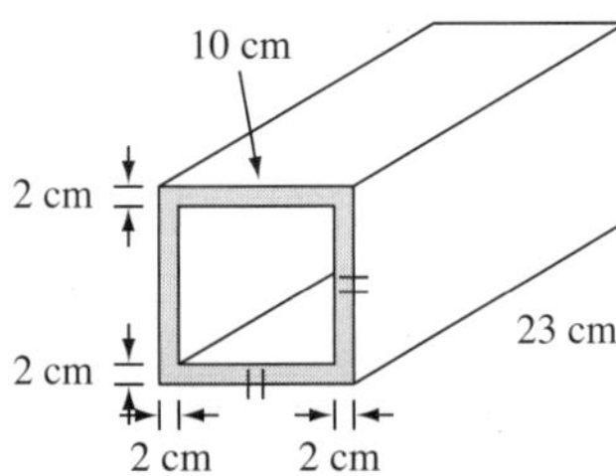

c

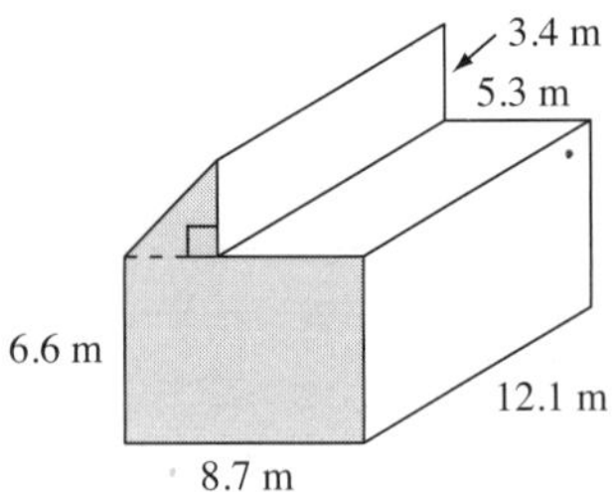

d

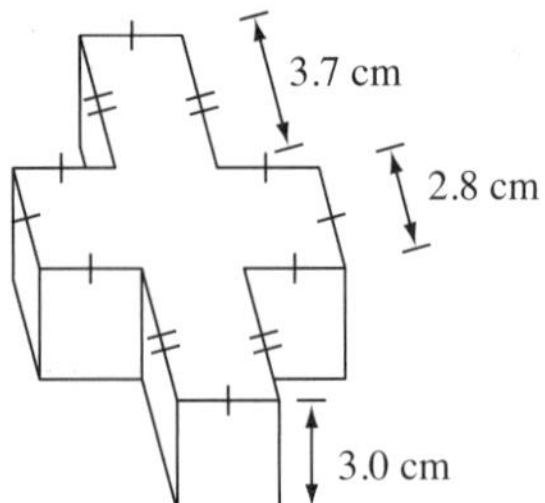

Area, surface area and volume

UNIT 12: Volume of right cylinders

Question 1 Find the volume of each cylinder correct to two significant figures.

a radius 4 cm and height 15 cm

b radius 7.8 cm and height 6.5 cm

c radius 1.4 m and height 1.5 m

d radius 95 cm and height 4.7 m

e radius 0.5 m and height 136 cm

f radius 2.5 m and height 250 cm

Question 2 Find the volume of each shape, correct to two decimal places if necessary.

a

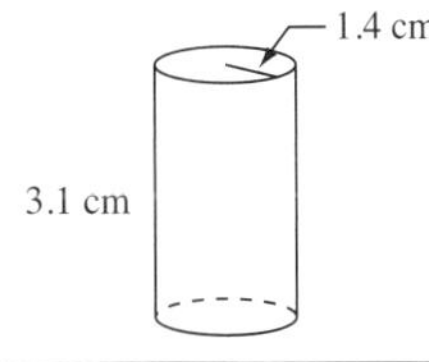

b

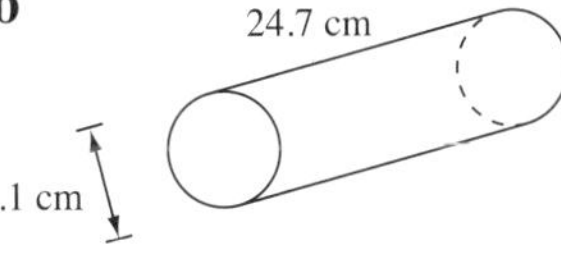

c

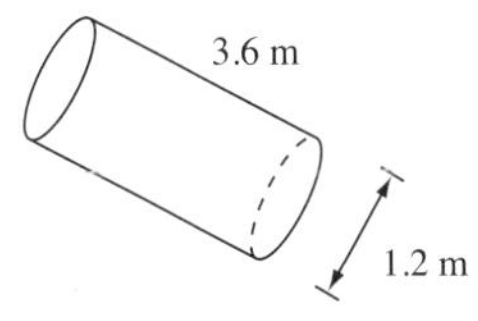

d

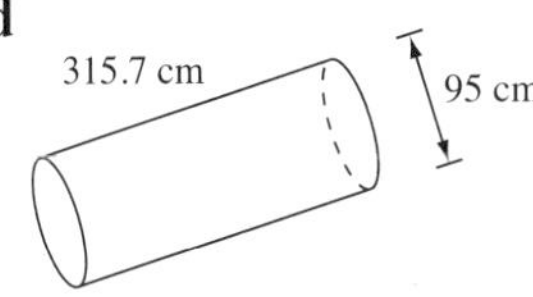

Question 3

a Which of the following cylinders has the larger volume?

i

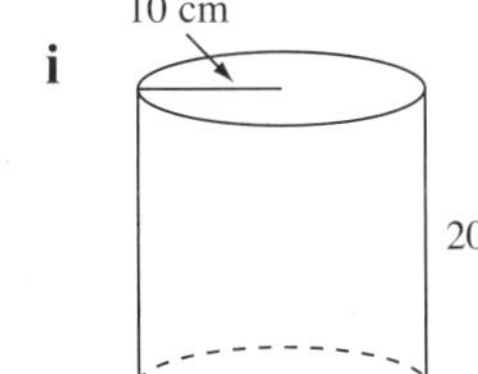

ii

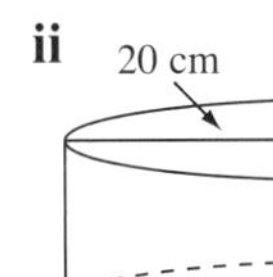

b Are the surface areas of the cylinders the same? Explain.

Question 4 How many times larger is the volume of cylinder **ii** compared to the volume of cylinder **i**?

a **i**

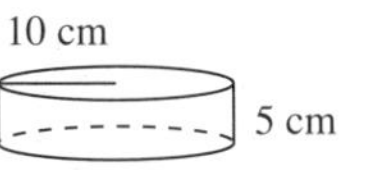

ii

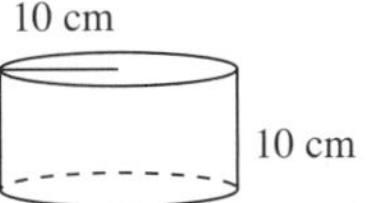

b **i**

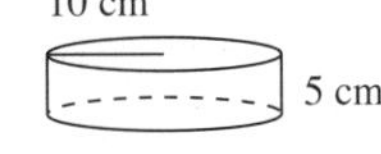

ii 20 cm 5 cm

Area, surface area and volume

UNIT 13: Volume of right cylinders and composite solids

QUESTION 1 Find the volume of each cylindrical can, leaving your answers in terms of π.

a

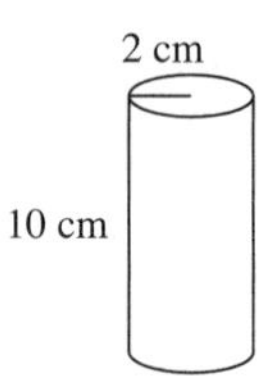

b

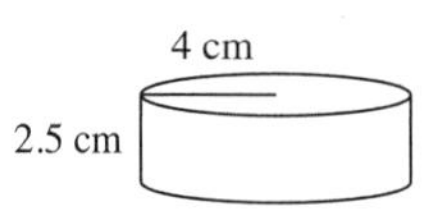

c

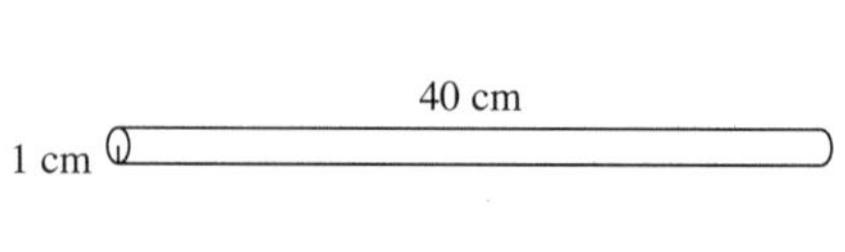

d

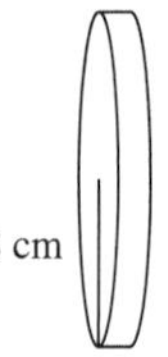

QUESTION 2 Calculate the volume of each solid correct to three significant figures.

a

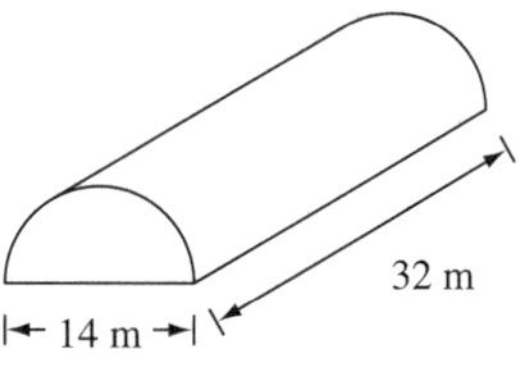

b

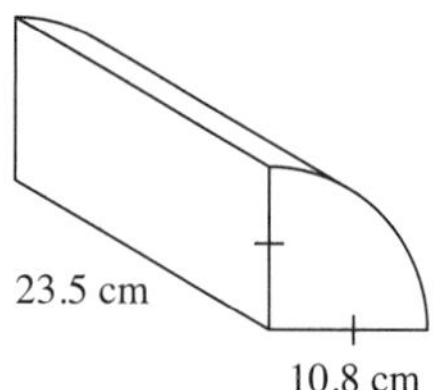

c

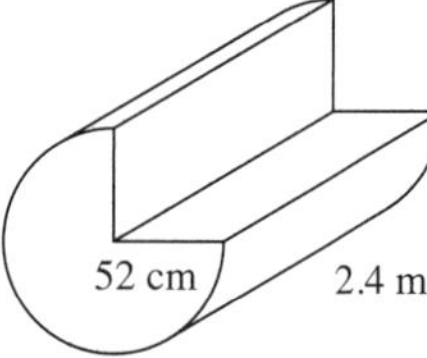

d

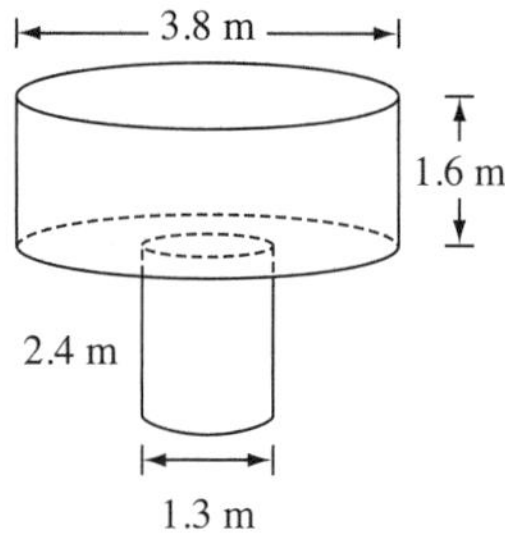

e

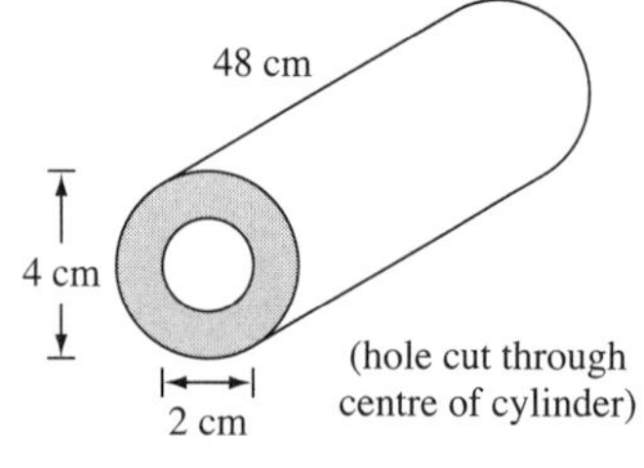

(hole cut through centre of cylinder)

f

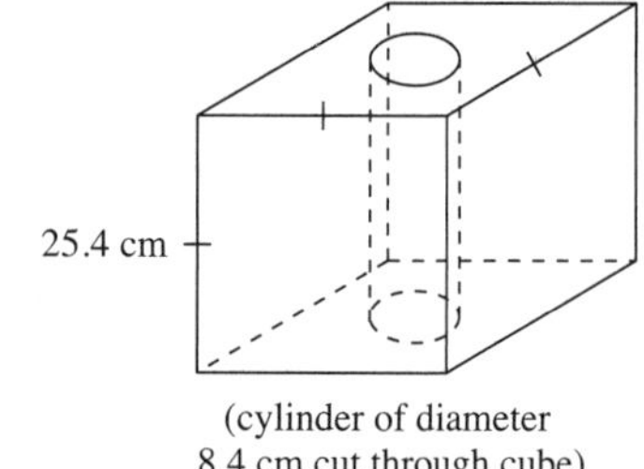

(cylinder of diameter 8.4 cm cut through cube)

Area, surface area and volume

UNIT 14: Problems involving volume and surface area

QUESTION **1** A flat rectangular roof is 18 m long and 11 m wide.

a If 10 mm of rain falls on the roof, find the total volume of water in cubic metres.

b How many litres of water is this? (1 m^3 = 1000 L)

c The water flows into a cylindrical tank of radius 1.5 m. How much will the height of water in the tank increase? Give the answer to the nearest centimetre.

QUESTION **2** A building has two walls that are pentagonal in shape and two other rectangular walls.

a Find the area of a pentagonal wall.

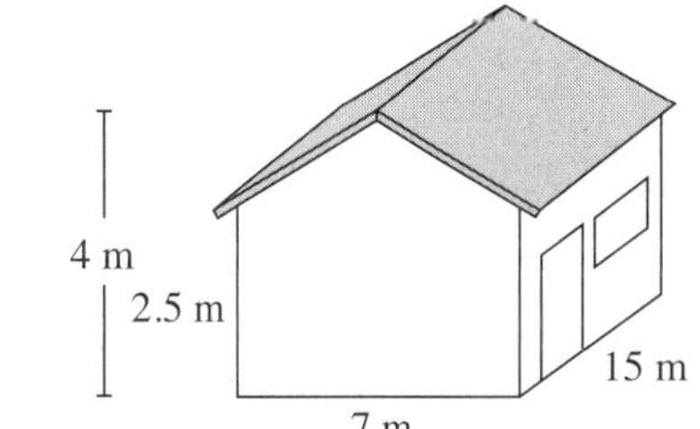

b Find the total area of all four walls.

c Find the area to be painted if a door 1.8 m wide and 2 m tall and a window 1.5 m wide and 1.2 m tall are not painted.

d Find the total amount of paint required, to the nearest litre, if the walls require two coats and one litre of paint covers 13 m^2.

QUESTION **3** A concrete bollard will be cylindrical in shape. It has height 1.2 m and radius 15 cm.

a Find the amount of concrete needed to make the bollard.

b How many of the bollards could be made from 8 m^3 of concrete?

Area, surface area and volume

TOPIC TEST — PART A

Instructions
- This part consists of 10 multiple-choice questions.
- Fill in only ONE CIRCLE for each question.
- Each question is worth 1 mark.

Time allowed: 15 minutes **Total marks: 10**

Marks

1 What is the surface area of a cube of side length 5 m?
(A) 100 m^2 (B) 125 m^2 (C) 150 m^2 (D) 225 m^2 1

2 What is the volume of a pentagonal prism if the area of the cross-section is 87 m^2 and the perpendicular height is 11 m?
(A) 191.4 m^3 (B) 696 m^3 (C) 957 m^3 (D) 4785 m^3 1

3 A cylinder has height 5 m and diameter 3.2 m. Its volume is closest to:
(A) 20.1 m^3 (B) 40.2 m^3 (C) 80.4 m^3 (D) 160.8 m^3 1

4 The shaded area is closest to:
(A) 38 cm^2 (B) 93 cm^2
(C) 154 cm^2 (D) 374 cm^2 1

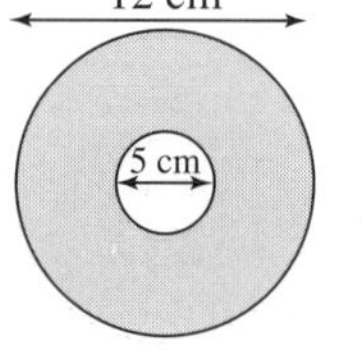

5 What is the volume of a cube of side length 8 cm?
(A) 256 cm^3 (B) 384 cm^3 (C) 448 cm^3 (D) 512 cm^3 1

6 What is the volume of the prism at right?
(A) 848 cm^3 (B) 540 cm^3
(C) 462 cm^3 (D) 231 cm^3 1

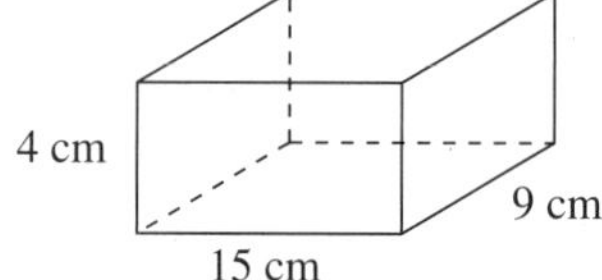

7 What is the surface area of the prism above?
(A) 848 cm^2 (B) 540 cm^2 (C) 462 cm^2 (D) 231 cm^2 1

8 Which is closest to the curved surface area of a cylinder of radius 14 cm and height 20 cm?
(A) 1230 cm^2 (B) 1760 cm^2 (C) 2990 cm^2 (D) 3520 cm^2 1

9 What is the surface area of the prism on the right?
(A) 1428 cm^2 (B) 1470 cm^2
(C) 2940 cm^2 (D) 4900 cm^2 1

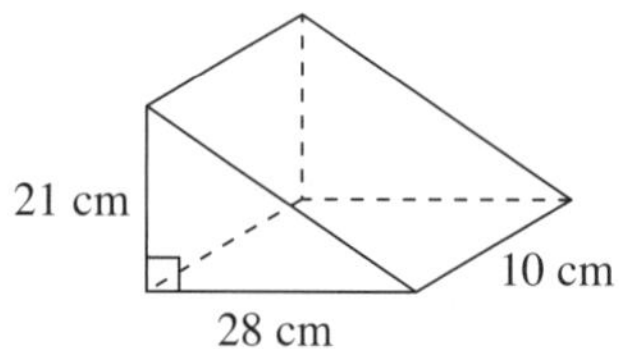

10 What is the volume of the prism on the right?
(A) 1428 cm^3 (B) 1470 cm^3
(C) 2940 cm^3 (D) 4900 cm^3 1

Total marks achieved for PART A 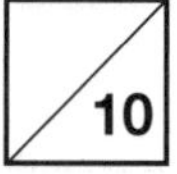 /10

Area, surface area and volume

TOPIC TEST PART B

Instructions
- This part consists of 6 questions.
- Write only the answer in the answer column.
- For any working use the question column.

Time allowed: 20 minutes **Total marks: 15**

Questions | **Answers** | **Marks**

1 For this prism, find the:

a volume 1

b surface area 1

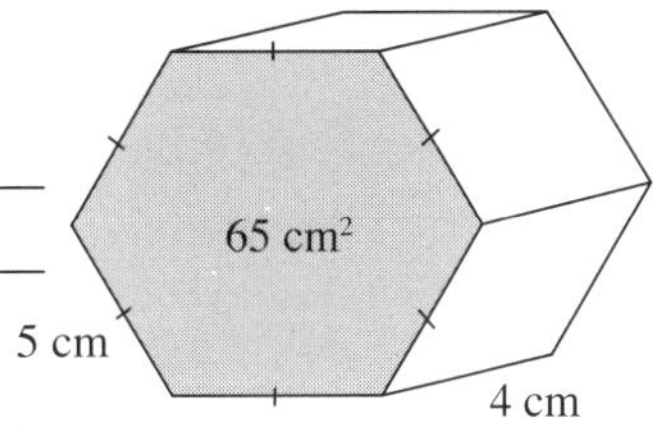

2 For the closed cylinder on the right, find the:

a volume (to nearest cubic centimetre) 1

b capacity in litres (1 cm^3 = 1 mL) 1

c surface area (to nearest square centimetre) 1

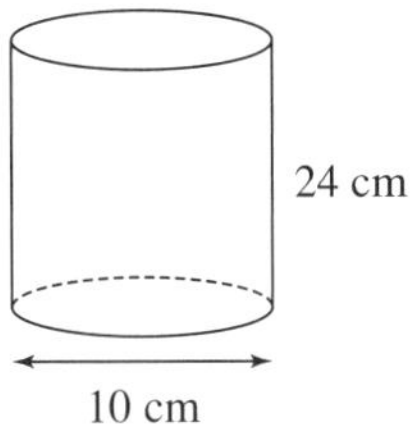

3 **a** Find the perpendicular height of the triangular face of this prism. 1

b Find the area of the triangular face. 1

c Find the volume of the prism. 1

d Find the total surface area of the prism. 1

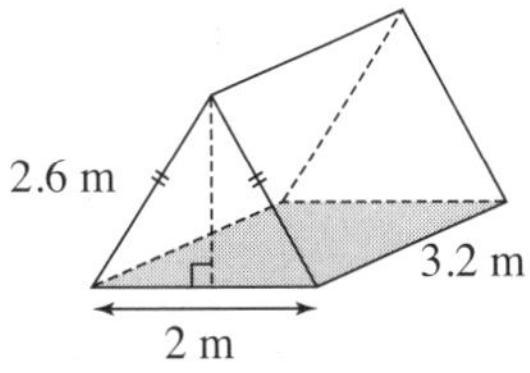

Continued on the next page

Area, surface area and volume

TOPIC TEST — PART B

Questions	Answers	Marks
4 a Find the shaded area (to the nearest square centimetre).		1
b The shaded area shown is the cross-section of a prism. The perpendicular height of the prism is 48 cm. Find the volume of the prism.		1
5 Find the surface area of this prism in square metres.		1
6 The machinery part is made up of a right-angled triangle and semicircle. a What is the diameter of the semicircle?		1
b What is the shaded area in square centimetres to one decimal place?		1
c What is the volume if the part is 3.6 cm thick? Give the answer to the nearest cubic centimetre.		1

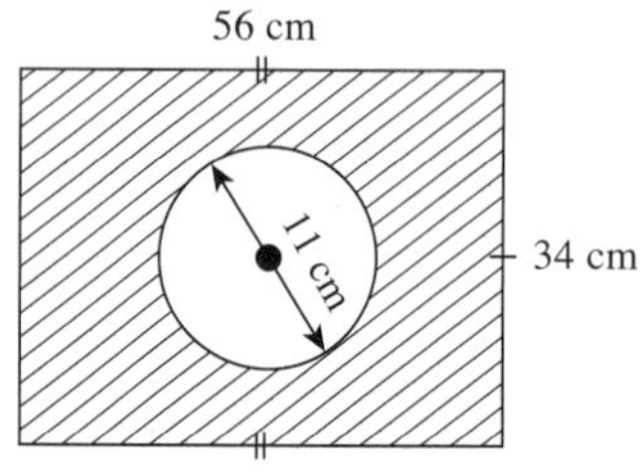

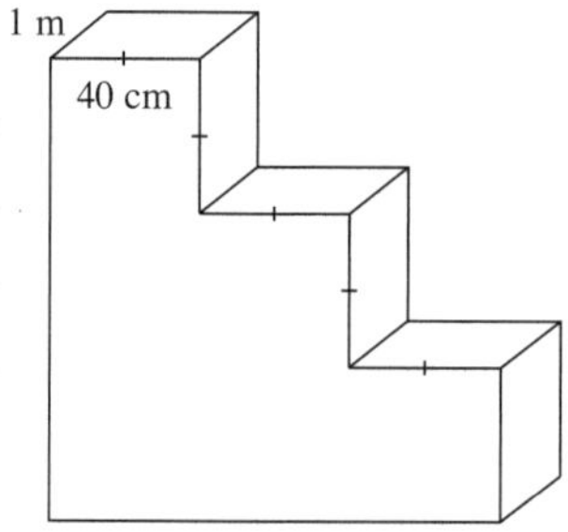

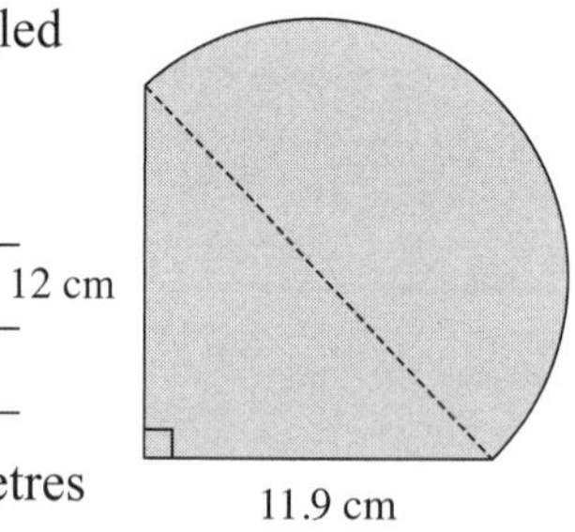

Total marks achieved for PART B ___ /15

CHAPTER 9
Further surface area and volume

UNIT 1: Surface area of different solids

QUESTION 1 Find the surface area of the following rectangular prisms.

a

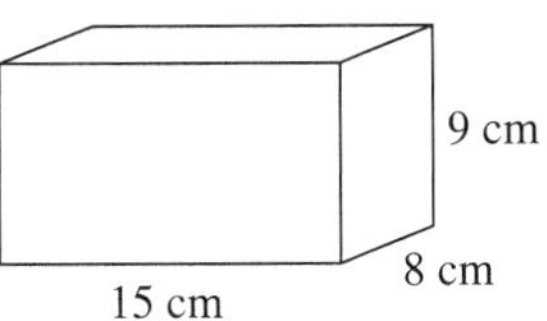

b

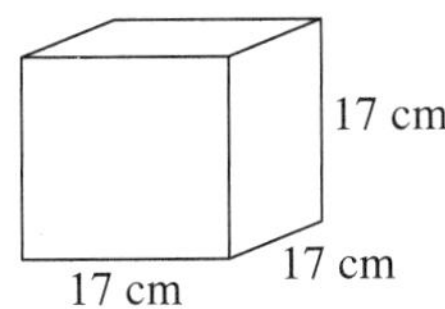

c

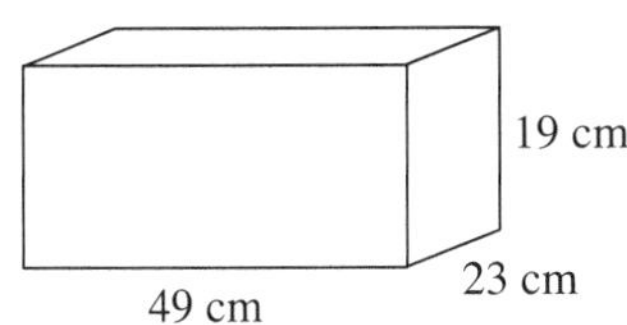

QUESTION 2 Find the surface area of the following triangular prisms.

a

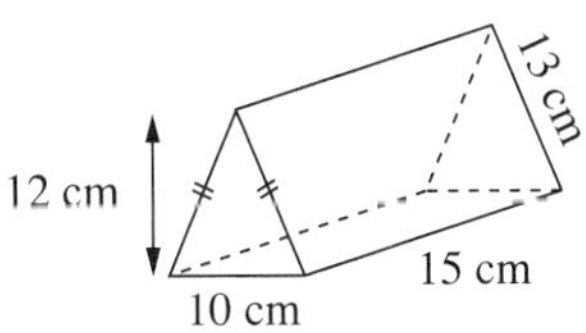

b

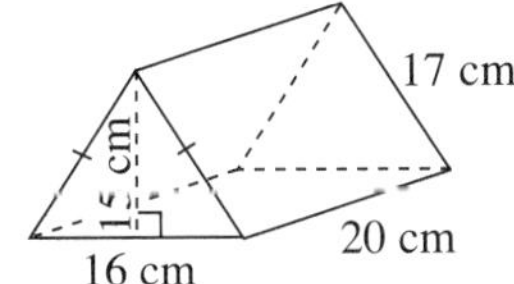

c

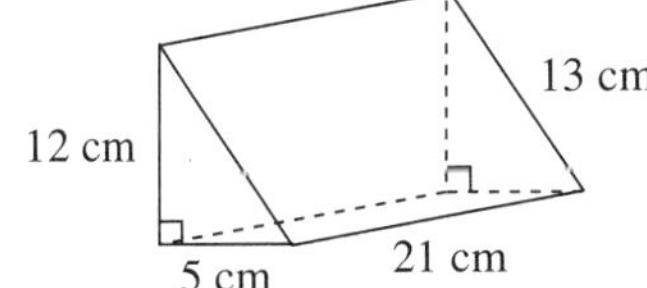

QUESTION 3 Find the surface area of the following trapezoidal prisms.

a

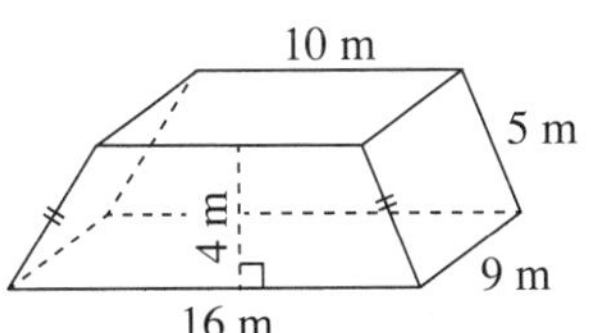

b

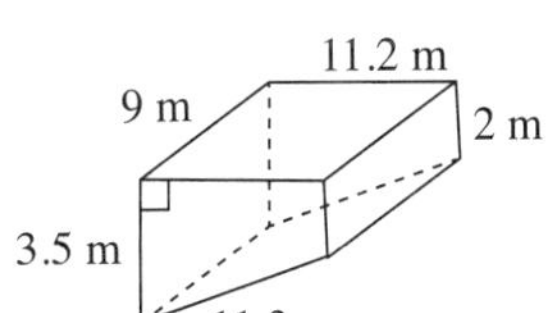

c

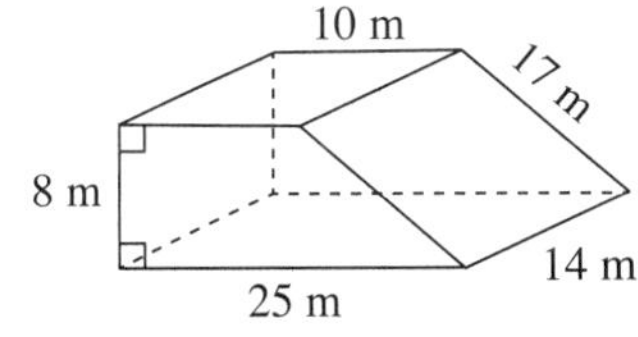

QUESTION 4 Find the surface area of the following closed cylinders correct to one decimal place.

a

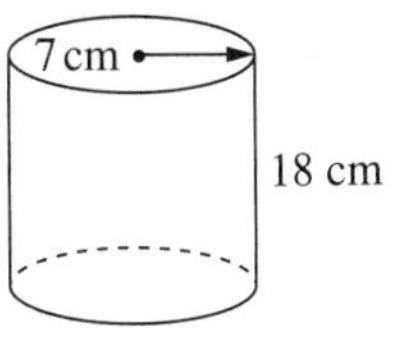

b

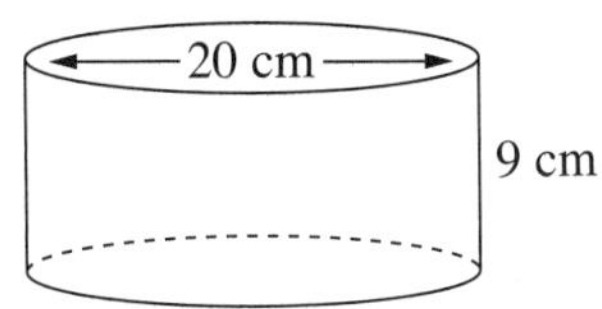

c

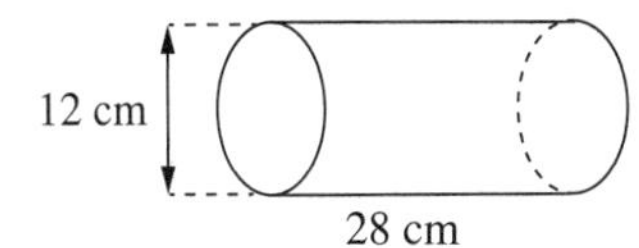

Further surface area and volume

UNIT 2: Perpendicular height and slant height

QUESTION 1 This cone has radius 39 mm and perpendicular height 80 mm. Find the slant height of the cone.

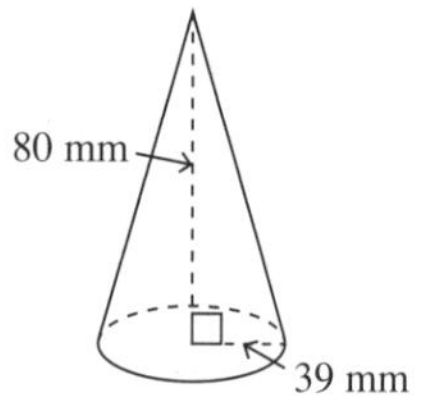

80 mm

39 mm

QUESTION 2 This cone has diameter 32 cm and slant height 28.1 cm. Find the perpendicular height of the cone.

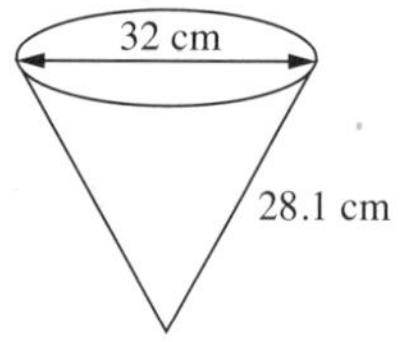

32 cm

28.1 cm

QUESTION 3 The perpendicular height, OP, of this rectangular pyramid is 12 cm. M is the midpoint of AB and N is the midpoint of BC. $AB = 32$ cm and $BC = 10$ cm. Find the length of:

a PM **b** PN

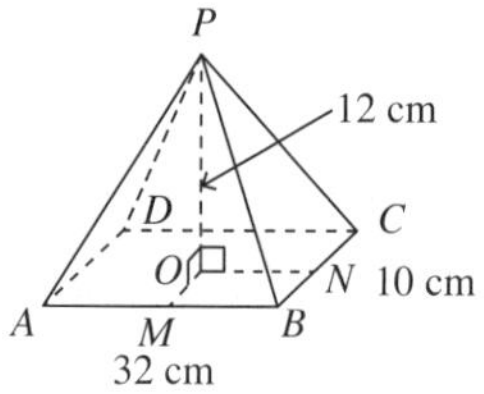

P, 12 cm, D, C, O, N, 10 cm, A, M, B, 32 cm

QUESTION 4 In this rectangular pyramid $AB = 30$ m and $BC = 12$ m. M is the midpoint of AB and N is the midpoint of BC. $PM = 10$ m. Find:

a the perpendicular height OP **b** the slant height PN

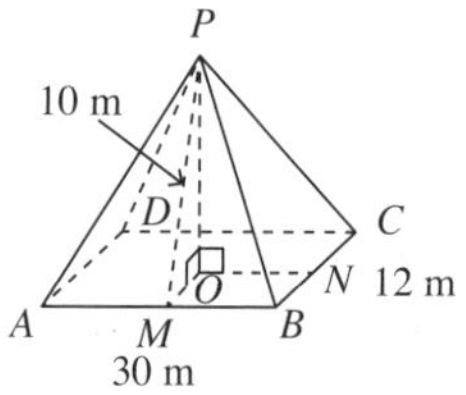

P, 10 m, D, C, O, N, 12 m, A, M, B, 30 m

QUESTION 5 A square pyramid has perpendicular height, OP, of 7 cm and slant height, AP, of 21 cm. Find the length of the base of the pyramid.

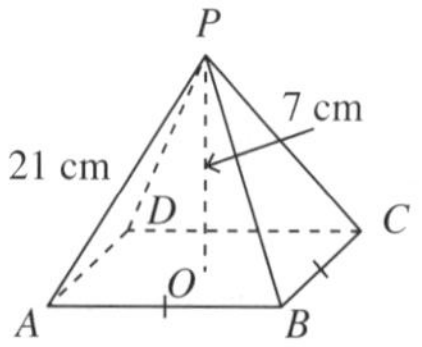

P, 7 cm, 21 cm, D, C, O, A, B

QUESTION 6 A rectangular pyramid has perpendicular height OP of 3.6 m. M is the midpoint of AB and N is the midpoint of BC. $PM = 6$ m and $PN = 3.9$ m. Find the length and width of the base.

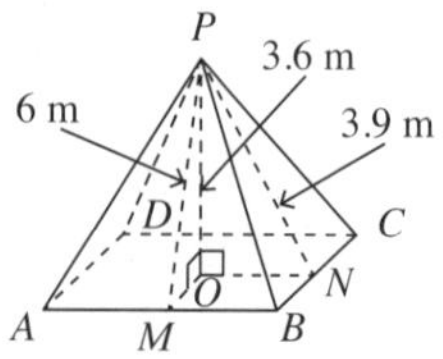

P, 3.6 m, 6 m, 3.9 m, D, C, O, N, A, M, B

Further surface area and volume

UNIT 3: Surface area of pyramids

QUESTION 1 Calculate the surface area of the following square pyramids correct to four significant figures.

a

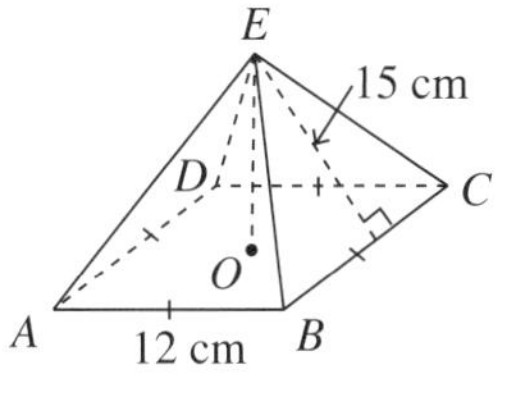

b

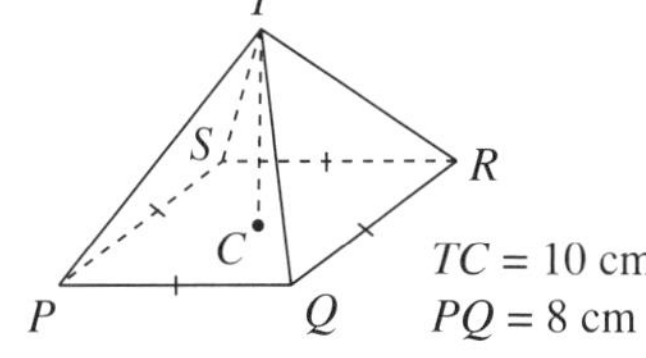

c

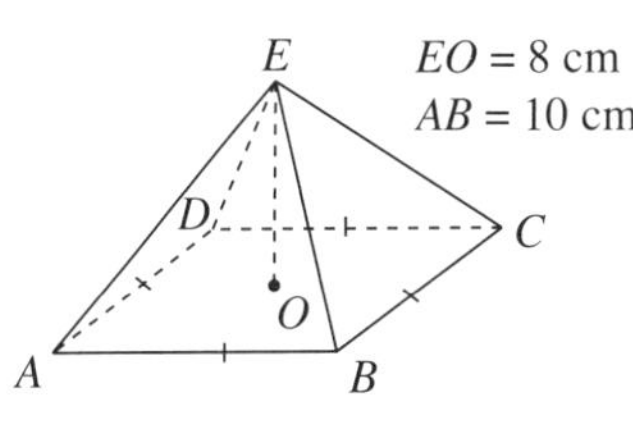

QUESTION 2 Calculate the surface area of the following rectangular pyramids correct to four significant figures.

a

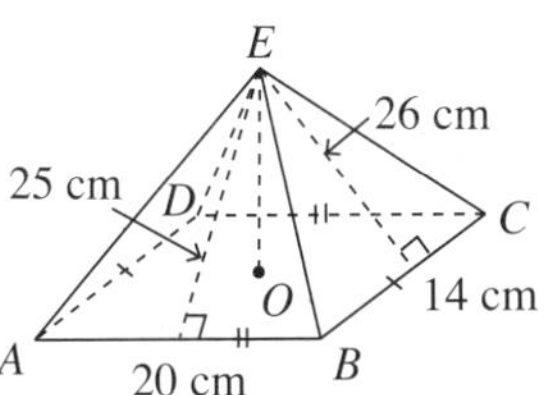

b

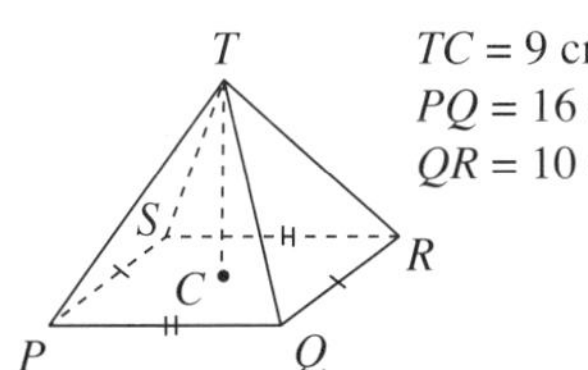

c

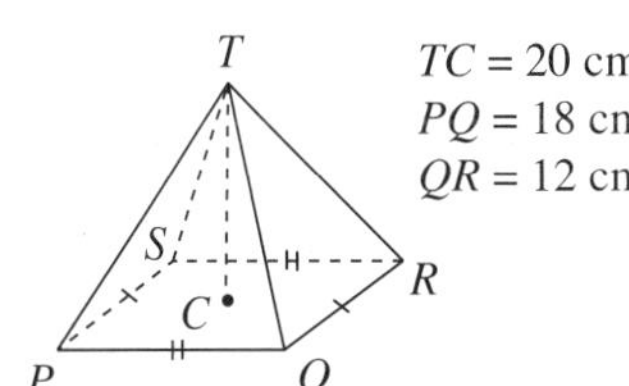

Further surface area and volume

UNIT 4: Surface area of cones

QUESTION **1** Find the **curved** surface area of the following cones correct to one decimal place.

a

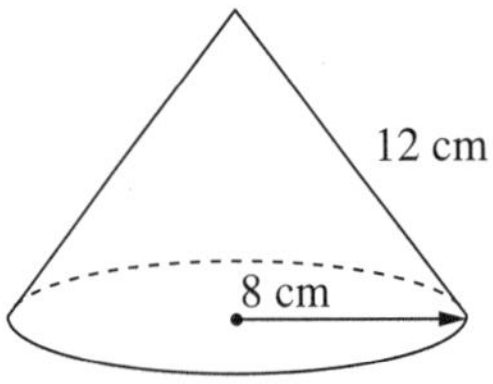

12 cm
8 cm

b

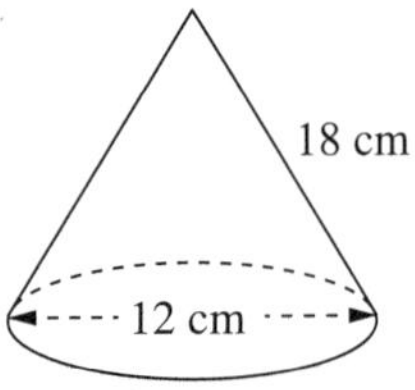

18 cm
12 cm

c

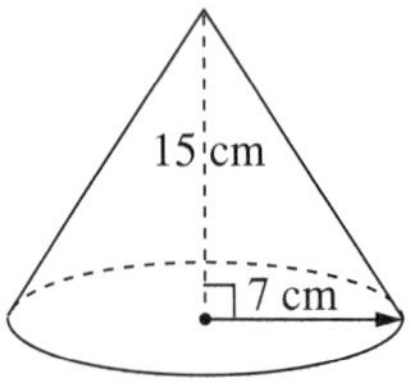

15 cm
7 cm

d

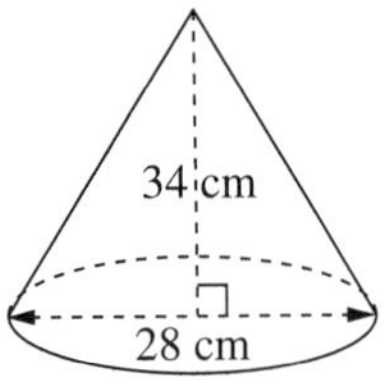

34 cm
28 cm

QUESTION **2** Find the total surface area of a cone with the following measurements. Give answers in terms of π.

a diameter 24 cm, slant height 10 cm

b radius 16 cm, perpendicular height 30 cm

QUESTION **3** The **curved** surface area of a cone is 795 cm^2. The radius of the cone is 11 cm. Find the slant height.

QUESTION **4** The total surface area of a cone is 36π cm^2. The slant height is 5 cm. Find the diameter of the cone.

Further surface area and volume

UNIT 5: Surface area of spheres

QUESTION 1 Find the surface area of the following spheres, giving your answers in terms of π, with:

a radius = 7 cm

b diameter = 18 cm

c radius = 28 cm

d diameter = 42 cm

e radius = 8.3 cm

f diameter = 23.9 cm

QUESTION 2 Calculate the surface area of the following spheres. Leave your answer in terms of π.

a

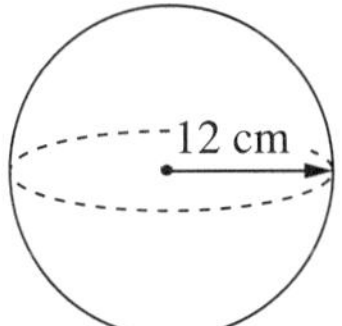

b

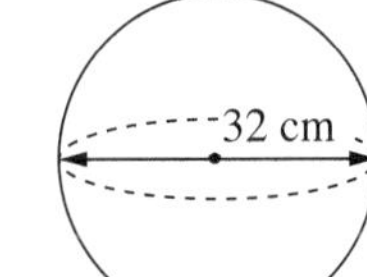

QUESTION 3 Calculate the surface area of the following hemispheres correct to two decimal places.

a

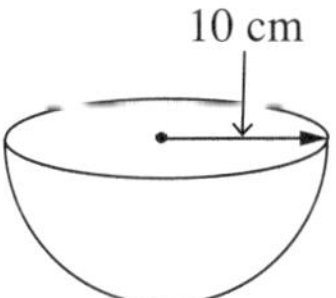

b

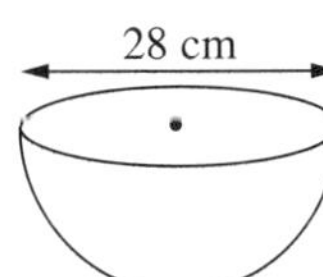

QUESTION 4 Find the external surface area of the following solids correct to three significant figures.

a

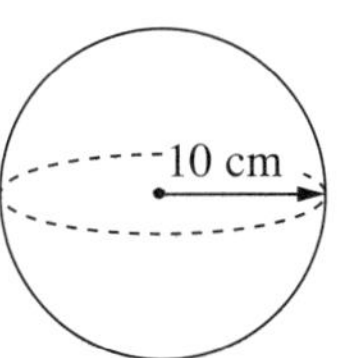

b

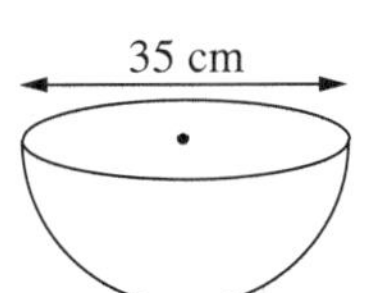

c

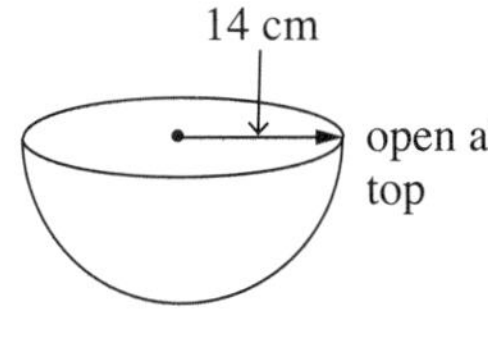

QUESTION 5 A sphere has a surface area of 360 cm^2. Find its radius correct to two decimal places.

Further surface area and volume

UNIT 6: Volume of different solids

QUESTION 1 Find the volume of the following rectangular prisms. Give answers correct to one decimal place.

a

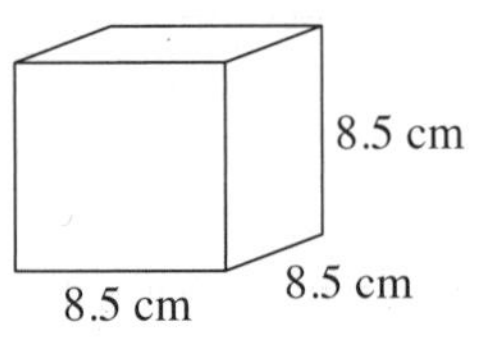

b

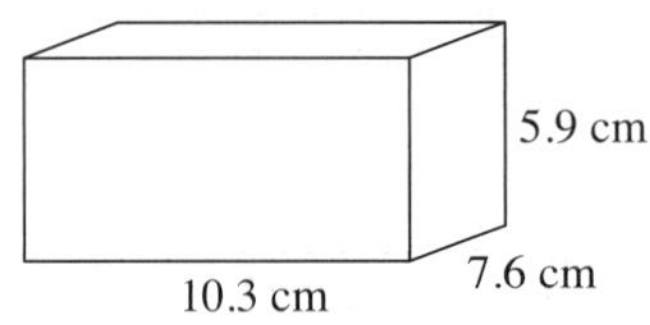

c

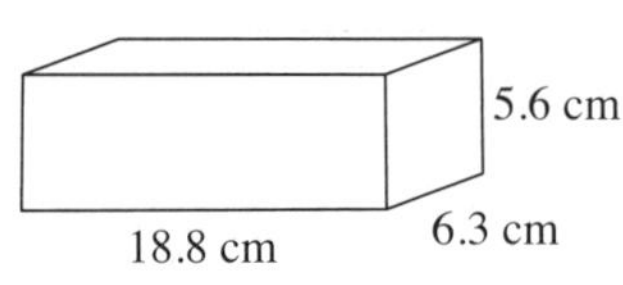

QUESTION 2 Find the volume of the following triangular prisms. Give answers correct to four significant figures.

a

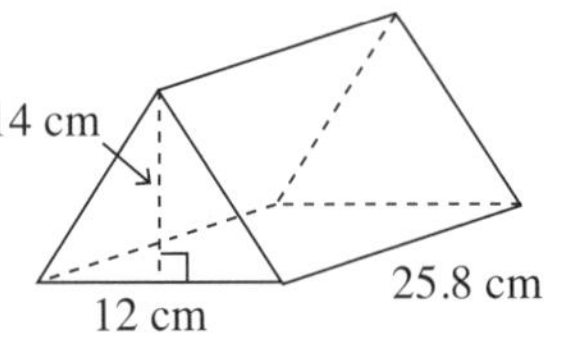

b

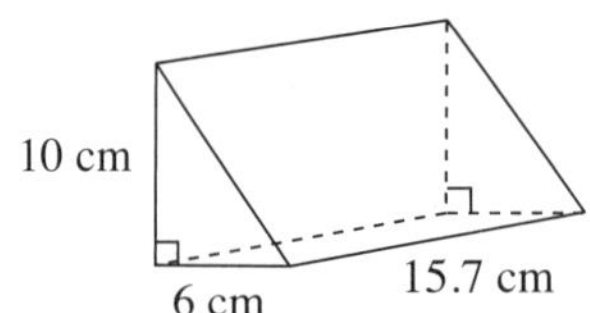

c

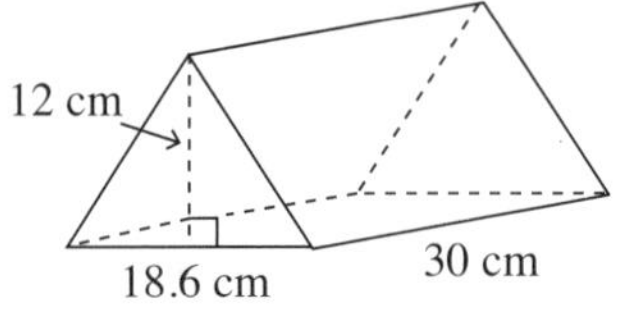

QUESTION 3 Find the volume of the following trapezoidal prisms. Give answers correct to two decimal places if necessary.

a

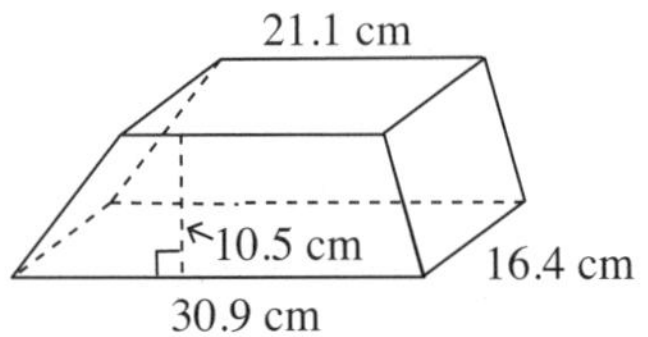

b

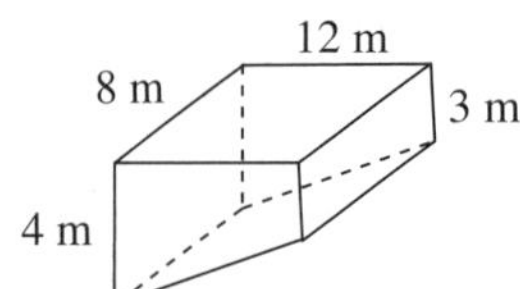

c

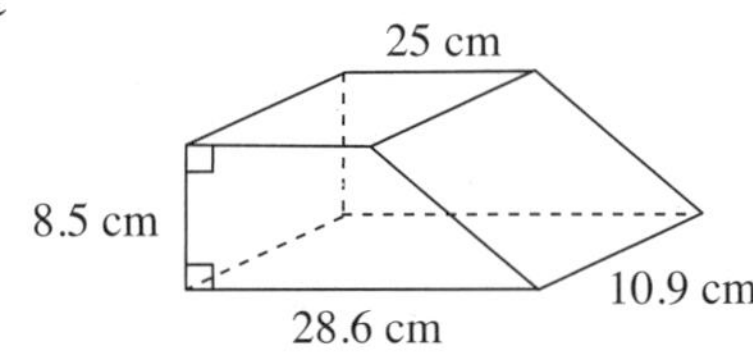

QUESTION 4 Find the volume of the following solids. Give answers correct to one decimal place.

a

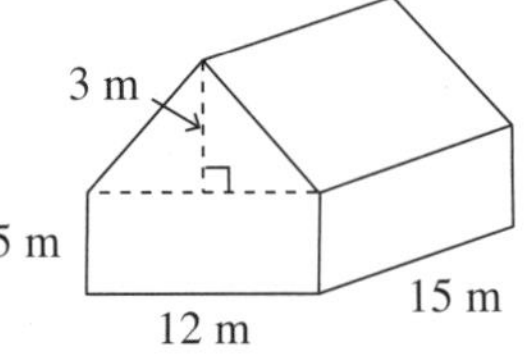

b

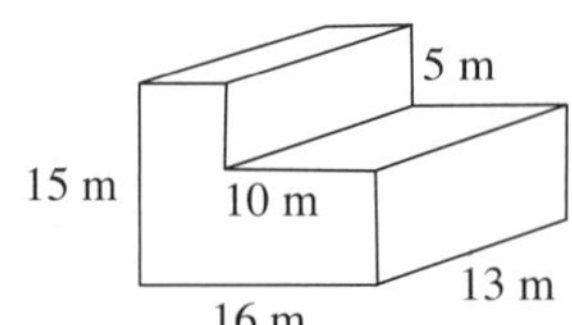

c

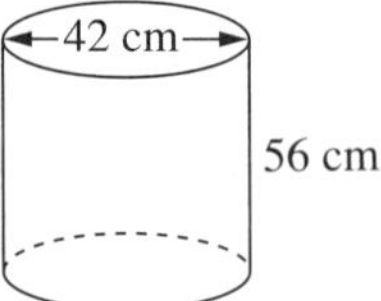

Further surface area and volume

UNIT 7: Volume of right pyramids

QUESTION 1 Calculate the volume of the following square pyramids correct to one decimal place.

a

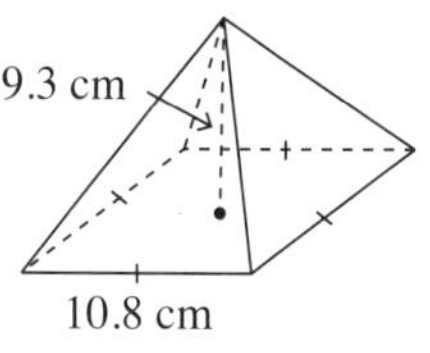

b

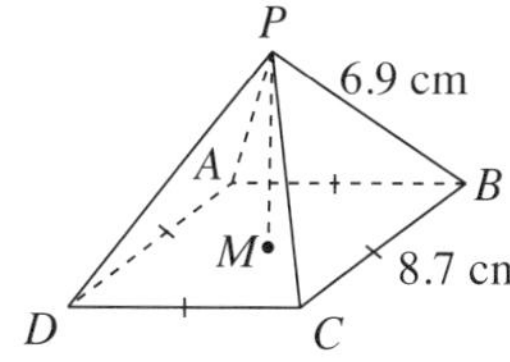

c

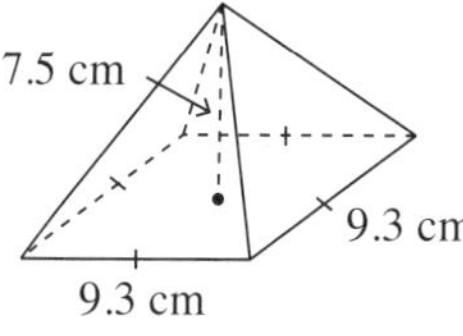

QUESTION 2 Calculate the volume of the following rectangular pyramids correct to two decimal places.

a

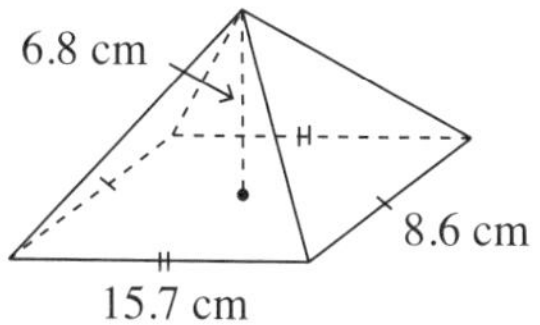

b

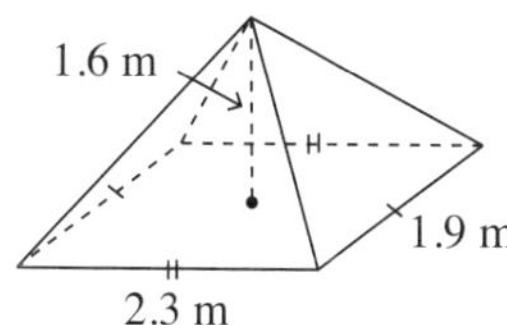

c

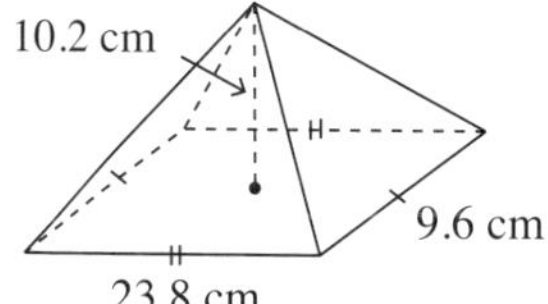

QUESTION 3 Find the volume of the following pyramids, given base area A cm^2 and perpendicular height h cm.

a Octagonal pyramid; $A = 225, h = 16.4$

b Hexagonal pyramid; $A = 98, h = 12$

QUESTION 4 Find the volume of these pyramids correct to one decimal place.

a

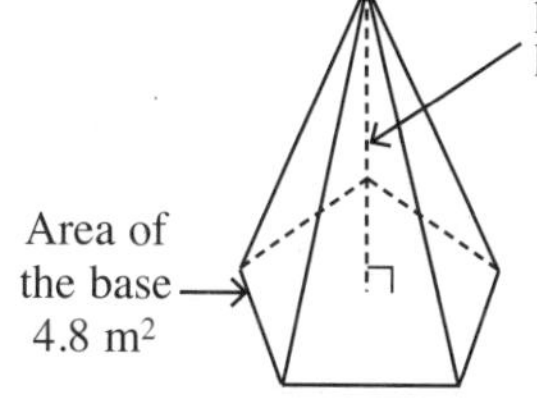

b

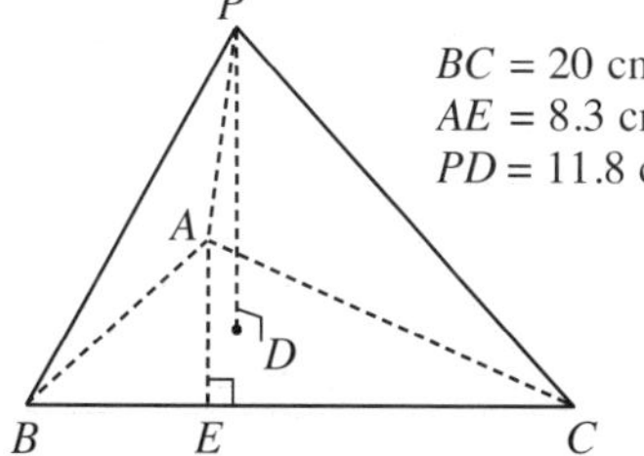

c

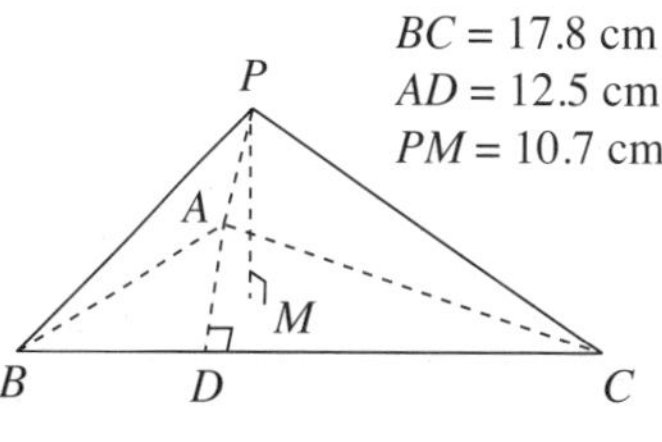

Further surface area and volume

UNIT 8: Volume of cones

QUESTION 1 Find the volume, correct to one decimal place, of these cones.

a

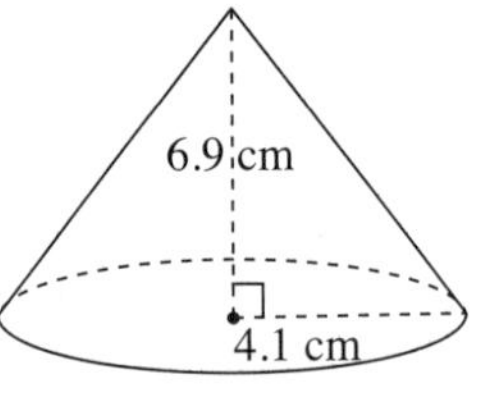

b

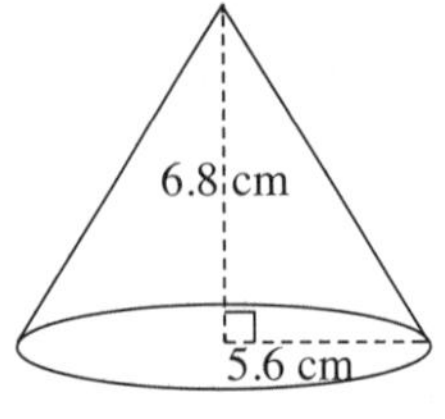

c

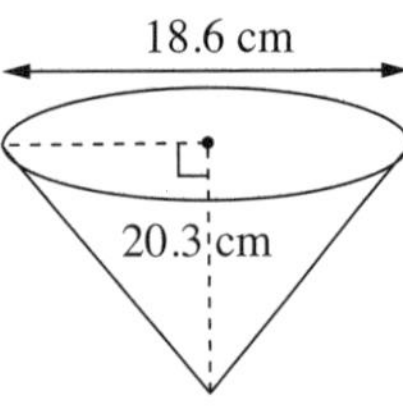

d

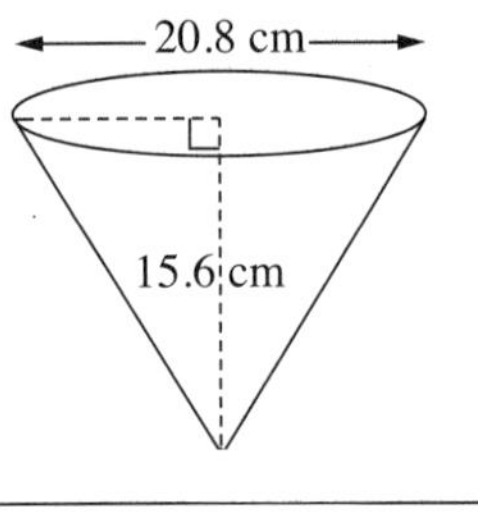

e

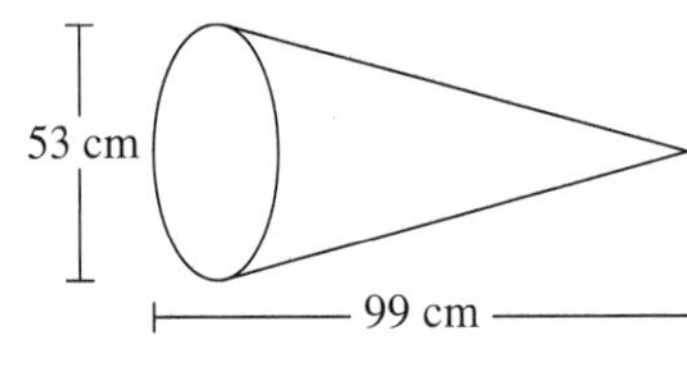

f

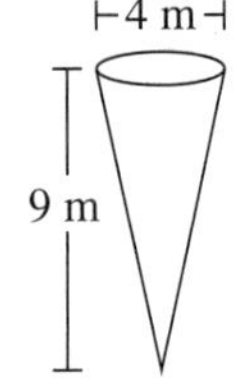

QUESTION 2 Find the volume of a cone with the following dimensions correct to one decimal place.

a radius 8 cm, perpendicular height 15 cm

b diameter 5.2 cm, perpendicular height 7.9 cm

QUESTION 3 Find the volume of these cones correct to two decimal places.

a

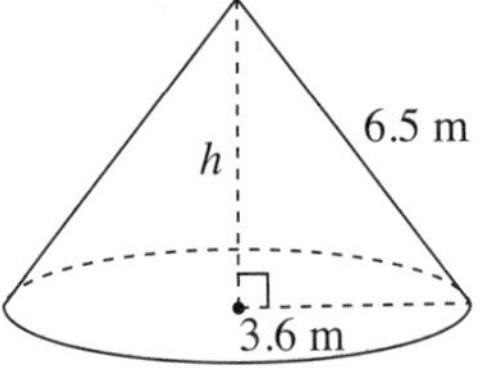

b

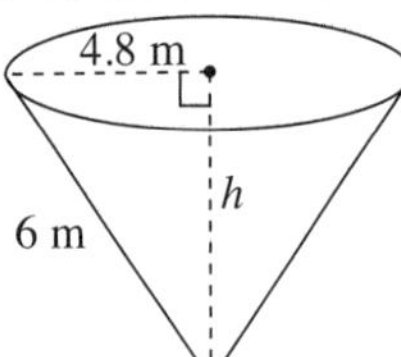

c

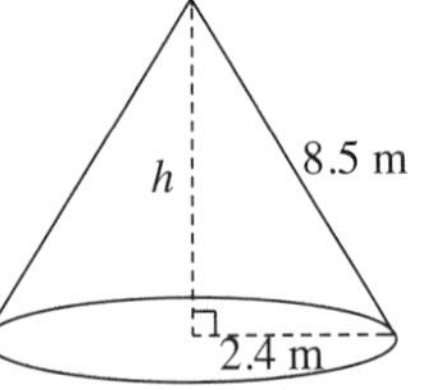

Further surface area and volume

UNIT 9: Volume of spheres

QUESTION **1** Find the volume of the following spheres (correct to one decimal place) with:

a radius = 9 cm

b diameter = 20 cm

c radius = 30 cm

d diameter = 35 cm

e radius = 15.3 cm

f diameter = 56 cm

QUESTION **2** Calculate the volume of the following spheres correct to one decimal place.

a

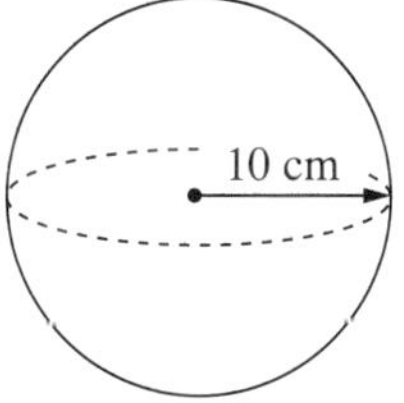

b

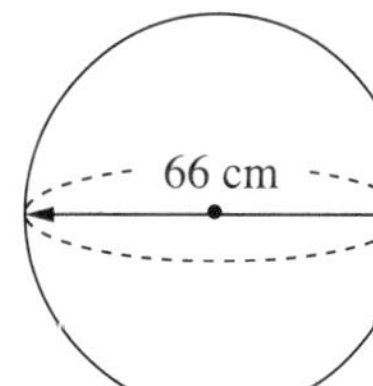

c

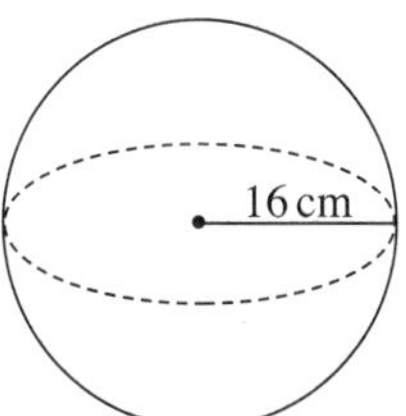

d

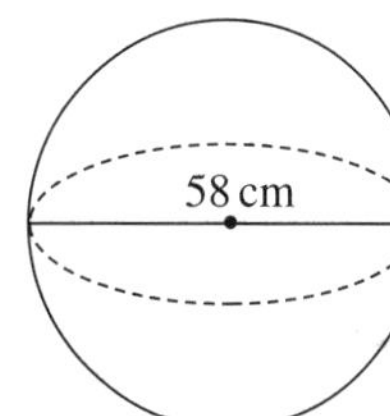

QUESTION **3** Calculate the volume of the following hemispheres correct to one decimal place.

a

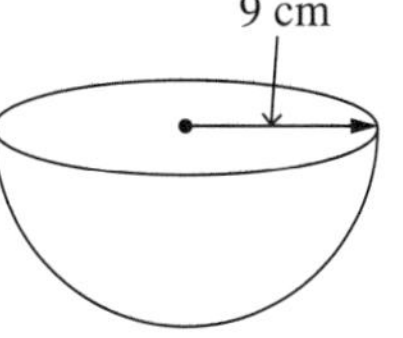

b

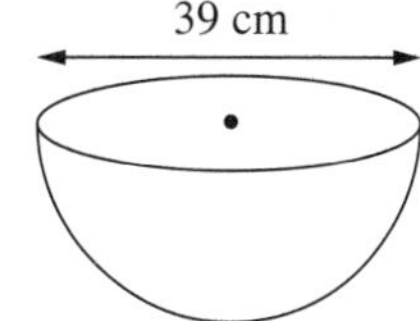

Further surface area and volume

TOPIC TEST PART A

Instructions
- This part consists of 10 multiple-choice questions.
- Fill in only ONE CIRCLE for each question.
- Each question is worth 1 mark.

Time allowed: 15 minutes **Total marks: 10**

Marks

1 A cone has a base diameter of 12 cm and a vertical height of 8 cm. Its volume is: 1

(A) 8π cm^3 (B) 24π cm^3 (C) 72π cm^3 (D) 96π cm^3

2 The volume of a sphere of radius 5 cm is closest to: 1

(A) 515 cm^3 (B) 524 cm^3 (C) 864 cm^3 (D) 1765 cm^3

3 Approximately how many spherical balls of diameter 0.5 cm could be made from a melted down cube of side length 5 cm? 1

(A) 19 (B) 190 (C) 1900 (D) 19 000

4 The volume of a cone with diameter 7 cm and height 8 cm is closest to: 1

(A) 56 cm^3 (B) 103 cm^3 (C) 392 cm^3 (D) 448 cm^3

5 A cone has a perpendicular height of 65.1 cm and slant height of 70.1 cm. What is the diameter of the cone? 1

(A) 26 cm (B) 48 cm (C) 52 cm (D) 96 cm

6 The surface area of a sphere of diameter 28 cm is closest to: 1

(A) 2463 cm^2 (B) 3284 cm^2 (C) 9852 cm^2 (D) 11 494 cm^2

7 The curved surface area of a cone with radius 9 cm and slant height 15 cm is closest to: 1

(A) 339 cm^2 (B) 424 cm^2 (C) 679 cm^2 (D) 1018 cm^2

8 What is the surface area of this square-based pyramid? 1

20 cm
24 cm

(A) 1344 cm^2 (B) 1536 cm^2
(C) 2112 cm^2 (D) 2496 cm^2

9 A triangular pyramid has base area 72 cm^2 and perpendicular height 16 cm. What is the volume of the pyramid? 1

(A) 1152 cm^3 (B) 384 cm^3 (C) 576 cm^3 (D) 128 cm^3

10 The total surface area of a cone with radius 12 cm is 980 cm^2. What is the slant height? 1

(A) 26 cm (B) 28 cm (C) 13 cm (D) 14 cm

Total marks achieved for PART A /10

Further surface area and volume

TOPIC TEST — PART B

Instructions
- This part consists of 6 questions.
- Write only the answer in the answer column.
- For any working use the question column.

Time allowed: 20 minutes **Total marks: 15**

Questions **Marks**

1 Find the following for this cone (to one decimal place).

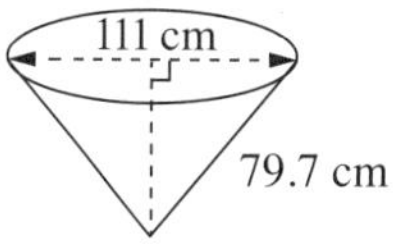

111 cm
79.7 cm

a perpendicular height **b** curved surface area **c** volume

3

Find the surface area and the volume of the following to the nearest whole number.
All measurements are in centimetres.

2 9

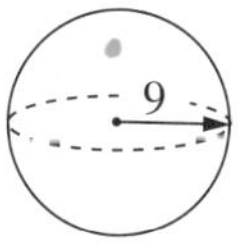

a surface area **b** volume

2

3 12 10

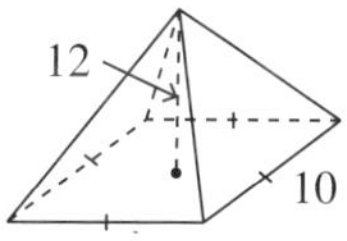

a surface area **b** volume

2

4 8 10 6

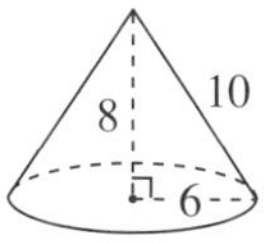

a surface area **b** volume

2

5 70 30

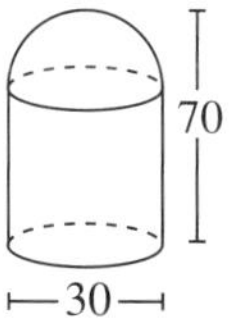

a surface area **b** volume

2

6 Consider this rectangular pyramid. Find the:

a length *AB* **b** width *BC*

c surface area **d** volume

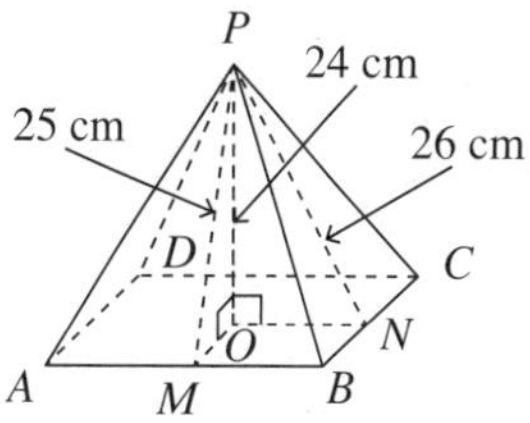

P 24 cm 25 cm 26 cm D C O N A M B

4

Total marks achieved for PART B /15

CHAPTER 10
Properties of geometrical figures

UNIT 1: Angle properties

QUESTION **1** Find the value of the pronumeral in each of the following. Give reasons to justify your answer.

a

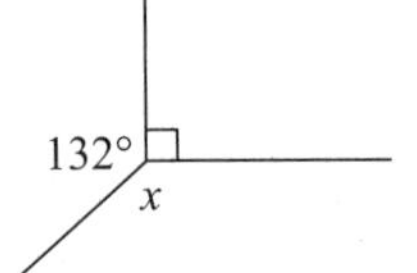

b

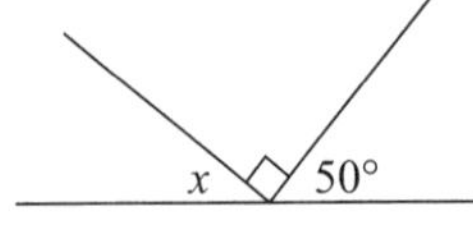

c

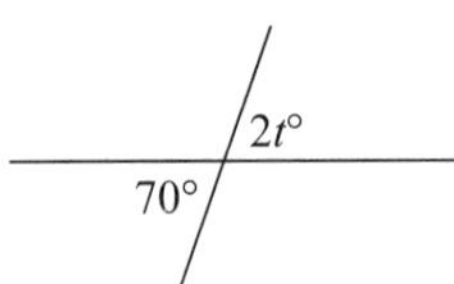

d

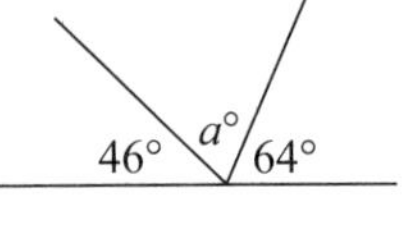

e

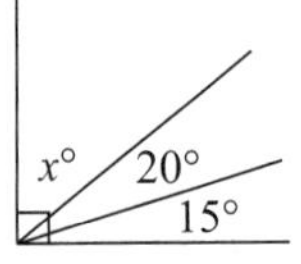

f

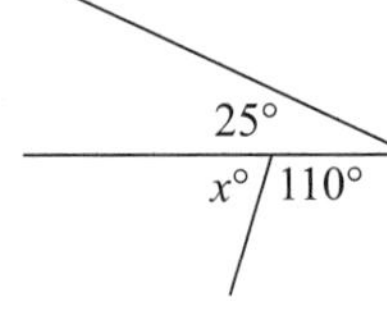

g

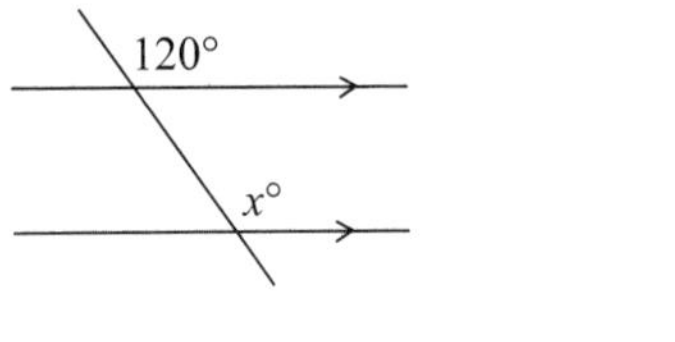

h

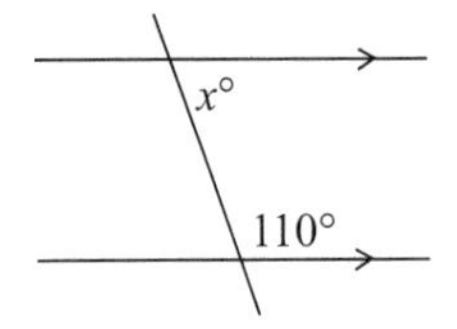

i

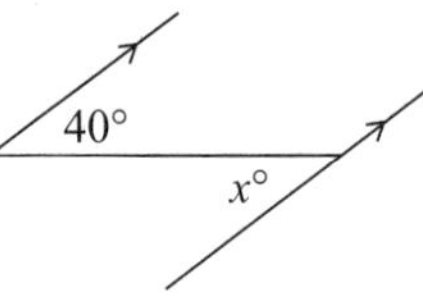

j

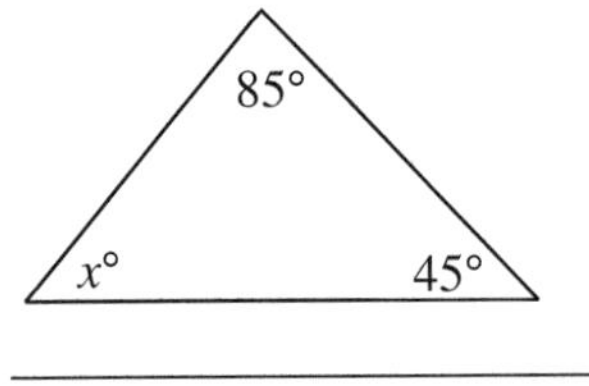

k

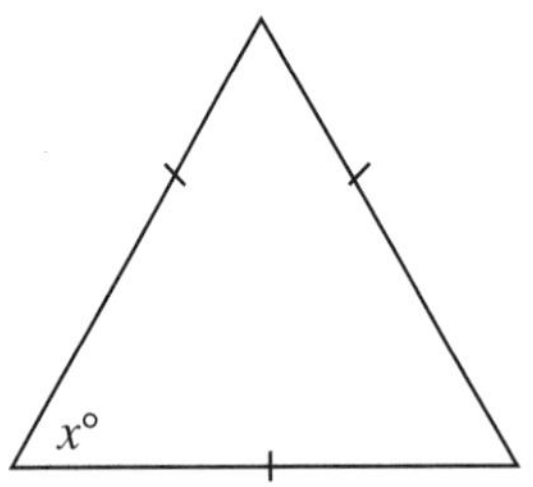

l

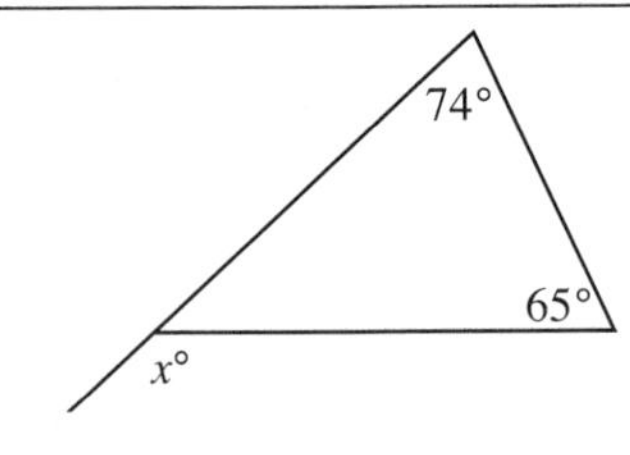

m

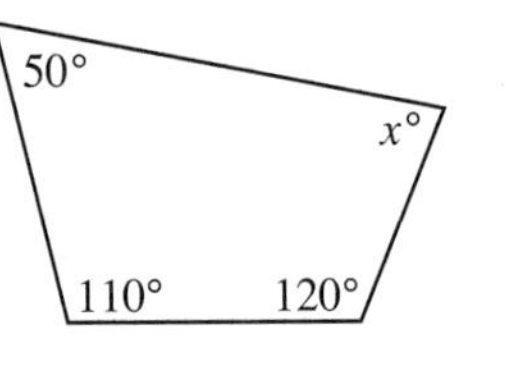

n

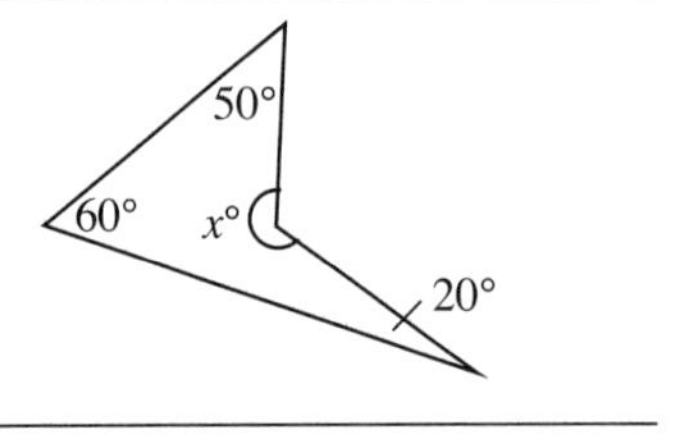

o

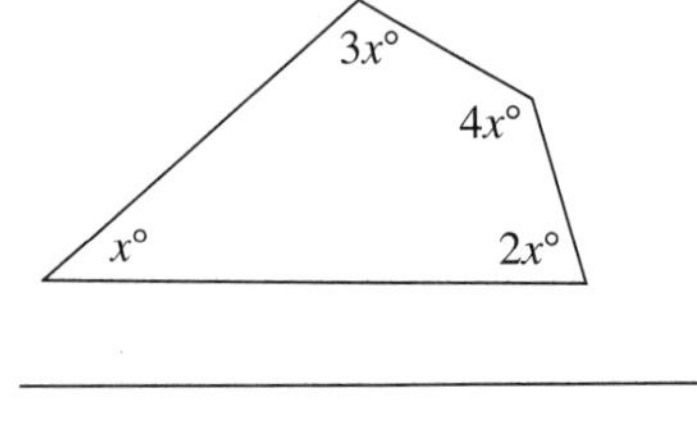

Properties of geometrical figures

UNIT 2: Using equations in geometry

QUESTION 1 Find the value of x in each diagram:

a

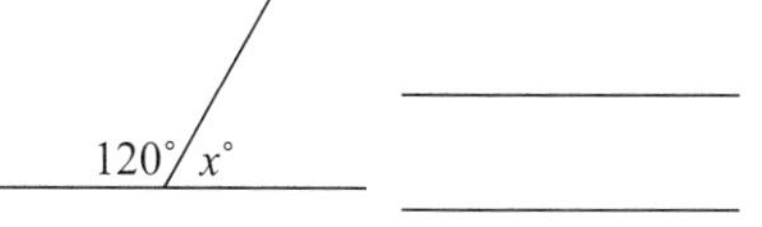

b

c

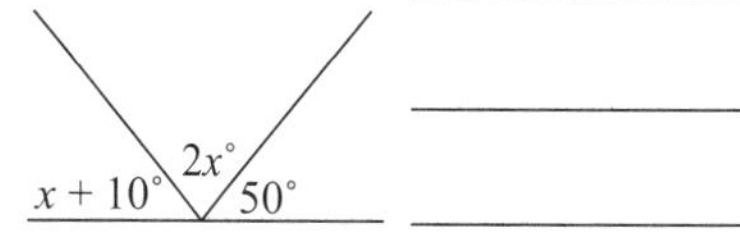

d

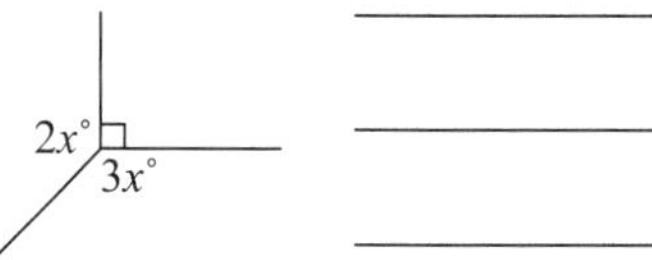

e

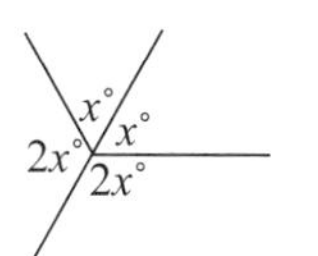

f

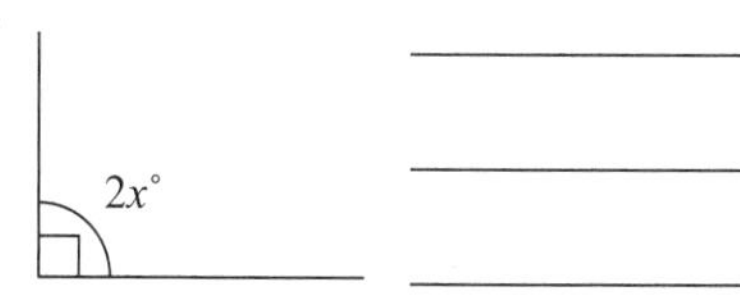

g

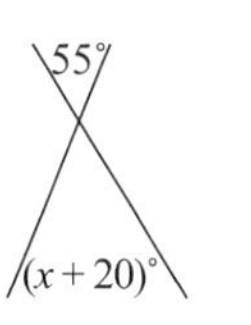

h

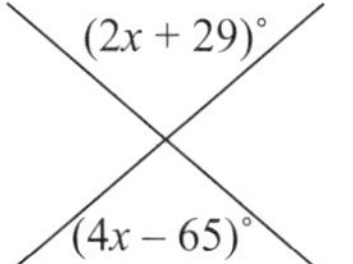

i

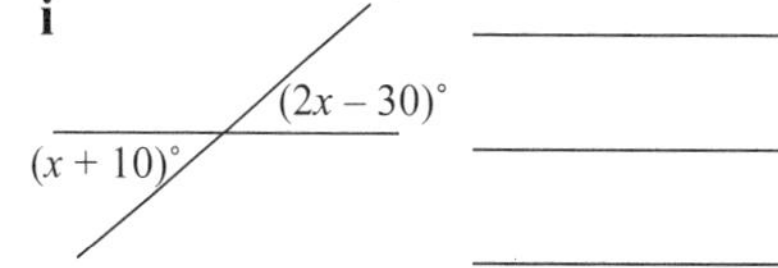

QUESTION 2 Find the value of the pronumeral in each diagram. All length measurements are in centimetres.

a Perimeter = 40 cm

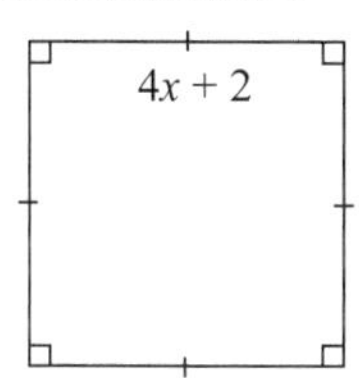

b Perimeter = 60 cm

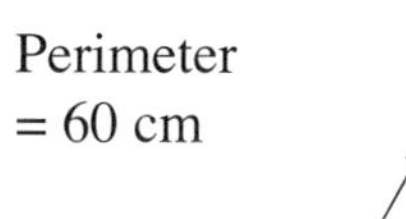

c Perimeter = 128 cm

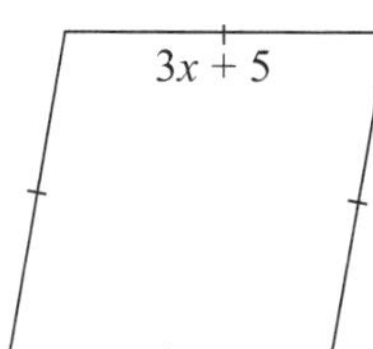

d

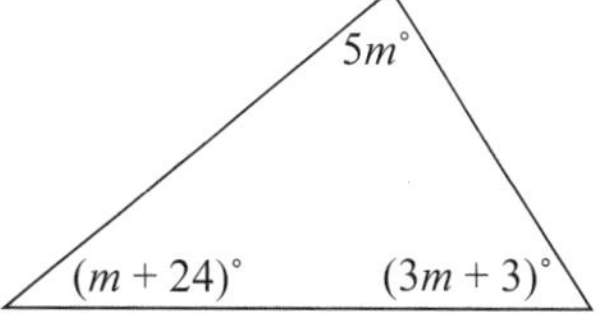

e

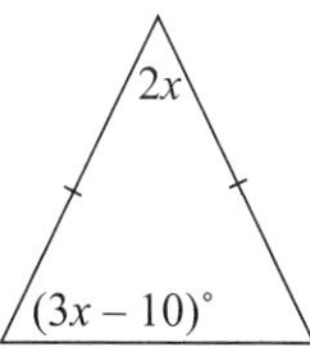

f

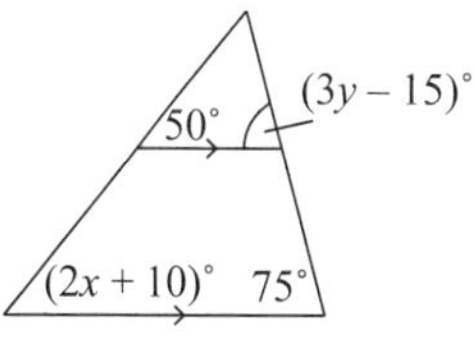

g

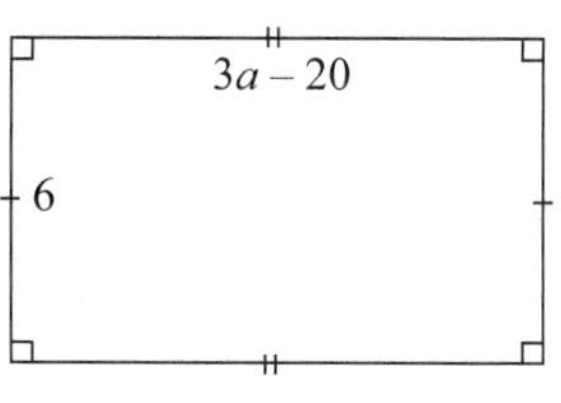

Area = 96 cm²

h

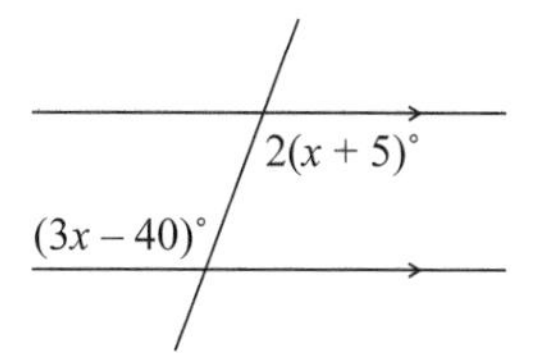

i

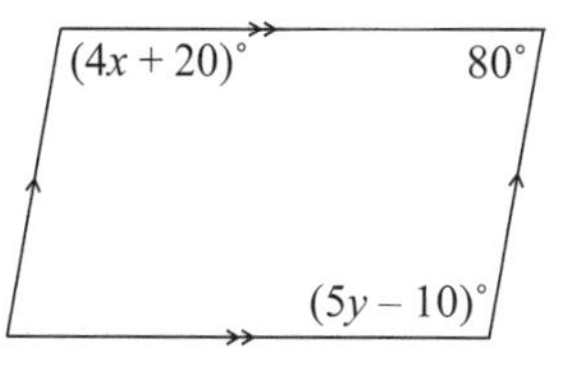

Properties of geometrical figures

UNIT 3: Polygons

QUESTION **1** The diagram shows a hexagon divided into triangles.

a How many triangles is the hexagon divided into? __________

b What is the angle sum of a hexagon?

c What is the size of each angle of a regular hexagon?

QUESTION **2** Find the angle sum of:

a a pentagon

b an octagon

c a dodecagon

QUESTION **3** What is the size of each angle of a regular:

a pentagon?

b octagon?

c dodecagon?

QUESTION **4** Complete:

The sum of the exterior angles of any polygon is ____________________.

QUESTION **5** For a regular decagon, what is the size of each

a exterior angle?

b interior angle?

QUESTION **6** Find the value of x.

a $100°$, $x°$, $x°$

b $x°$

c $x°$

Properties of geometrical figures

UNIT 4: Problem solving in geometry

1 In a right-angled triangle, if one angle is 55°, find the other acute angle.

2 In a right-angled triangle the two shorter sides are equal. What is the size of each acute angle?

3 In a right-angled triangle, one acute angle is twice the size of the other. What is the size of each angle?

4 The angles of a triangle are $x°$, $2x°$ and $3x°$. Find the size of each angle.

5 The sides of a rectangle are 5 cm and 12 cm. How long is the diagonal?

6 *ABCD* is a rectangle. If $\angle BDC = 35°$, find $\angle DBC$.

A B D C 35°

7 Three angles of a quadrilateral are 120°, 70° and 110°. Find the fourth angle.

8 If one of the base angles of an isosceles triangle is 68°, find the size of the vertical angle.

9 If the vertical angle of an isosceles triangle is 86°, find the size of each of the base angles.

10 In ΔABC, $AB = BC$ and $BC = AC$. What is the size of $\angle A$?

Properties of geometrical figures

UNIT 5: Congruence

QUESTION 1 Complete.

a If two figures are congruent, they are exactly the same __________ and exactly the same __________ .

b If two figures are congruent, the corresponding angles are __________ and the corresponding sides are the same __________ .

c The symbol __________ means 'is congruent to'.

QUESTION 2 Given that $ABCD$ is congruent to $SPRQ$, write down the corresponding sides and angles of $SPRQ$ that match these of $ABCD$.

a	AB	__________	**b**	AC	__________	**c**	CD	__________
d	BD	__________	**e**	DA	__________	**f**	BC	__________
g	$\angle ABC$	__________	**h**	$\angle ADB$	__________	**i**	$\angle BCD$	__________
j	$\angle CAB$	__________	**k**	$\angle DBC$	__________	**l**	$\angle CDA$	__________

QUESTION 3 It is known that ΔJLK is congruent to ΔXYZ.

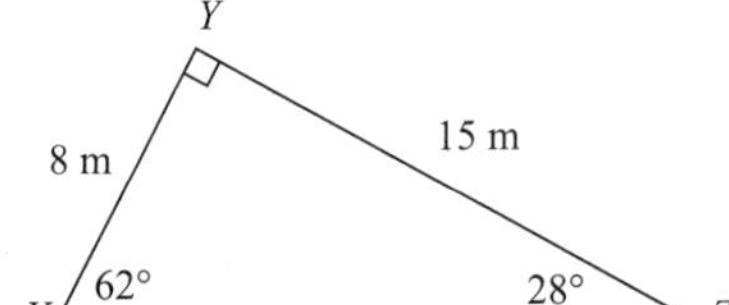

a What is the length of side:

i JL? __________ **ii** LK? __________ **iii** JK? __________

b What is the size of:

i $\angle JKL$? __________ **ii** $\angle LJK$? __________ **iii** $\angle KLJ$? __________

QUESTION 4 By measuring, find all pairs of congruent triangles.

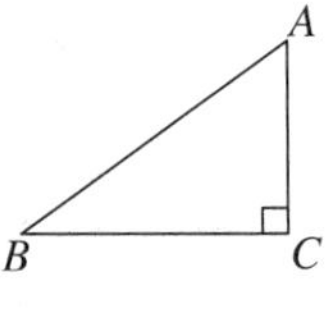

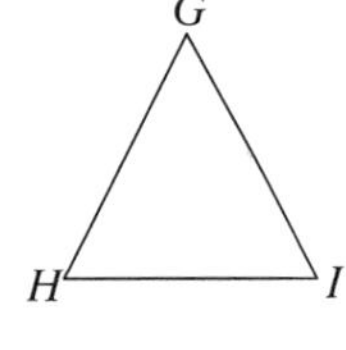

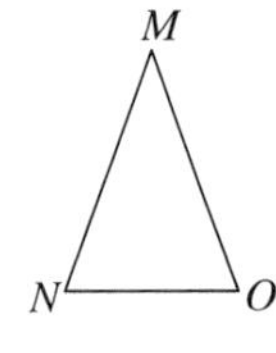

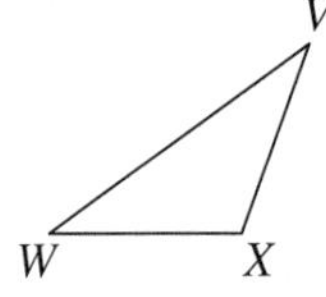

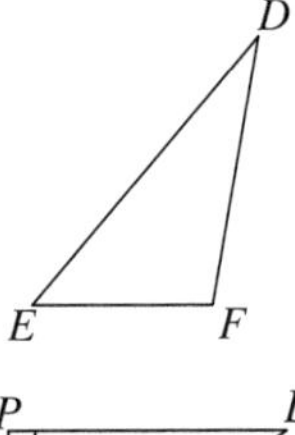

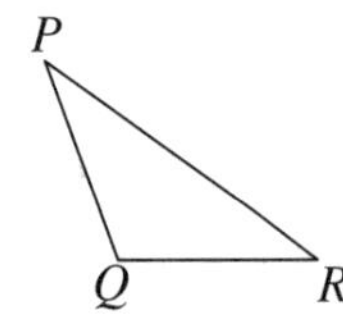

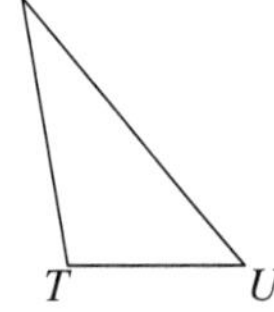

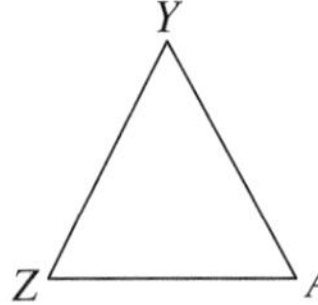

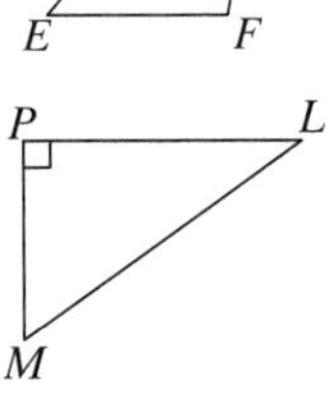

P Q R

____

____

____

____

Properties of geometrical figures

UNIT 6: Congruent triangles

QUESTION 1 Complete:

a If two triangles are congruent, they are exactly the same __________ and __________ .

b If ΔPQR is congruent to ΔABC, then $\angle P$ corresponds to __________, $\angle Q$ corresponds to __________ and $\angle R$ corresponds to __________ .

QUESTION 2 In the following pairs of congruent triangles:
i find all pairs of corresponding angles **ii** find all pairs of corresponding sides

a

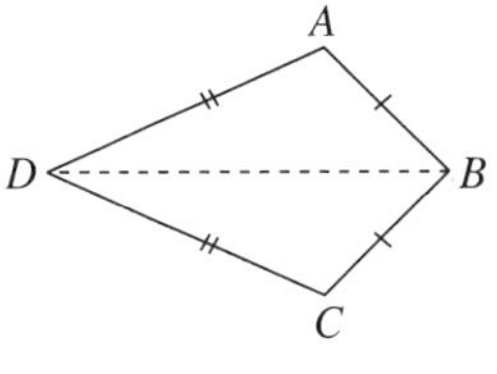

b

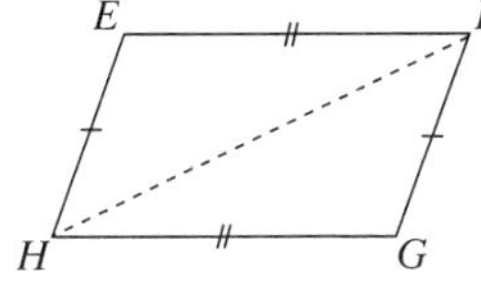

c

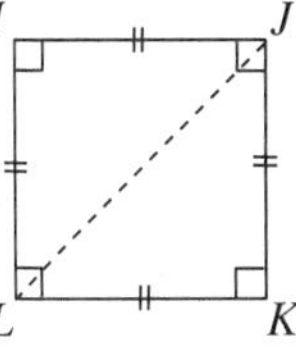

d

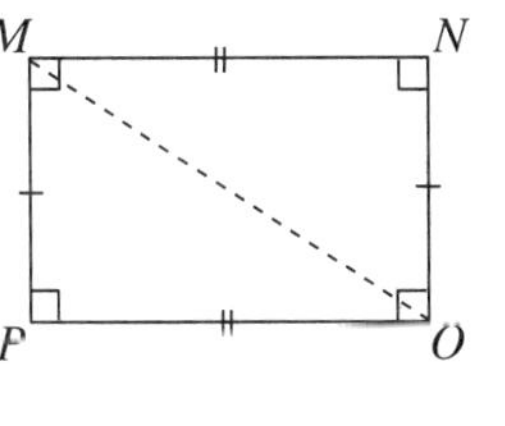

e

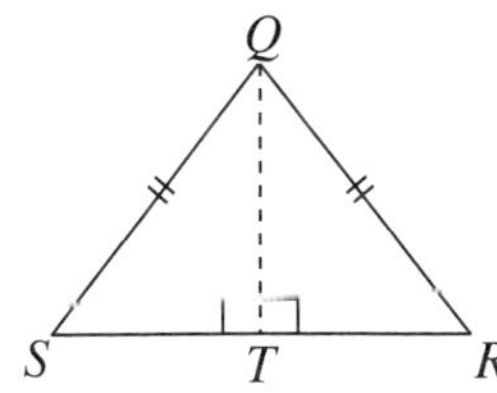

f

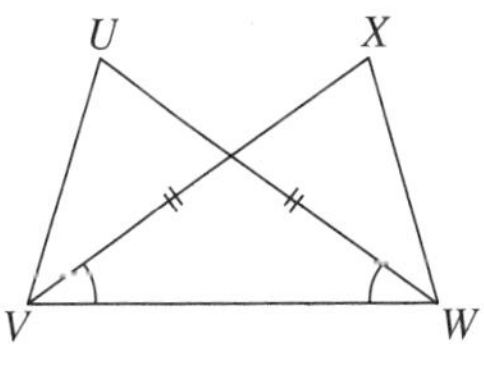

QUESTION 3 ΔEFG and ΔGHE are congruent.

a Find all pairs of corresponding angles.

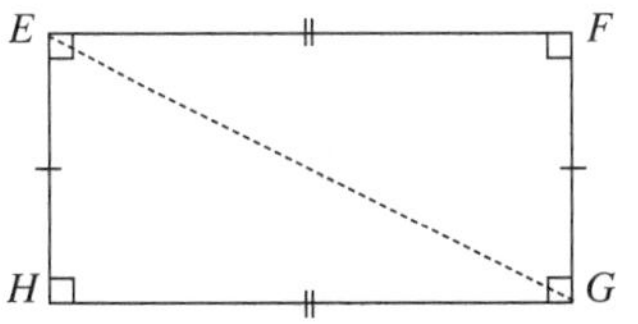

b Find all pairs of corresponding sides.

QUESTION 4 Complete.

a ΔJIM ≡ Δ __________

b ΔIKJ ≡ Δ __________

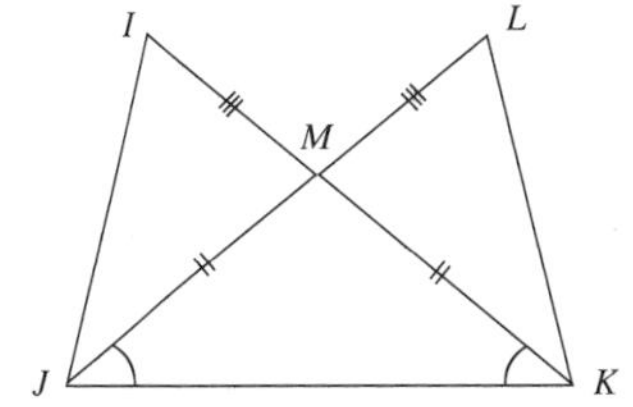

Properties of geometrical figures

UNIT 7: Tests for congruent triangles

Question 1 Complete.

a Two triangles are congruent if three sides of one triangle are equal to ______________________ of the other triangle.

b Two triangles are congruent if two angles and a side of one triangle are equal to

____ of the other triangle.

c Two triangles are congruent if two sides and the included angle of one triangle are equal to

____ of the other triangle.

d Two right-angled triangles are congruent if the hypotenuse and one side of one triangle are equal to

____ of the other triangle.

Question 2 In each pair of triangles, write the congruency test that would be used to prove that the triangles are congruent.

a ______________

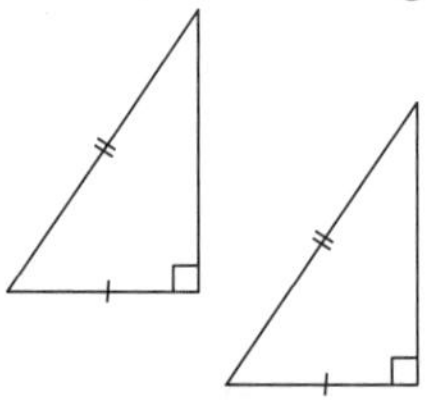

b ______________

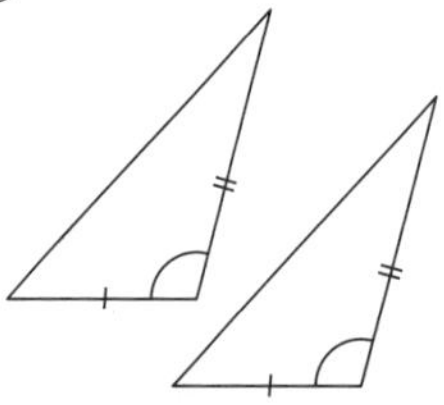

c ______________

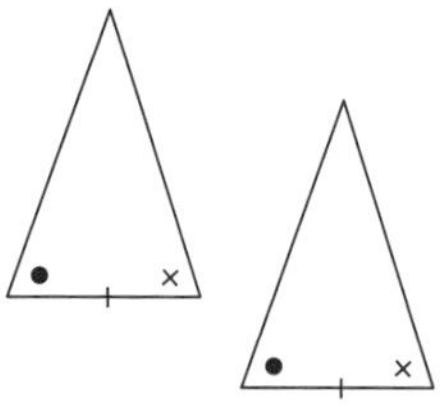

d ______________

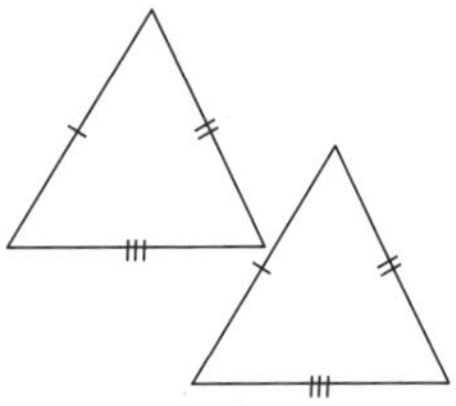

e ______________

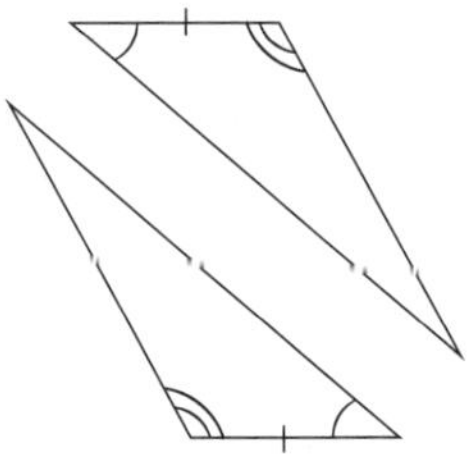

f ______________

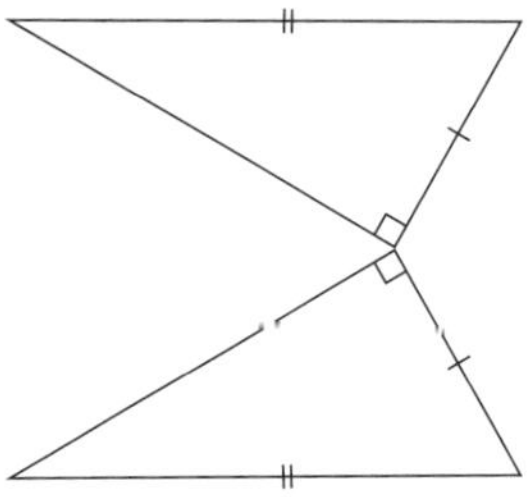

g ______________

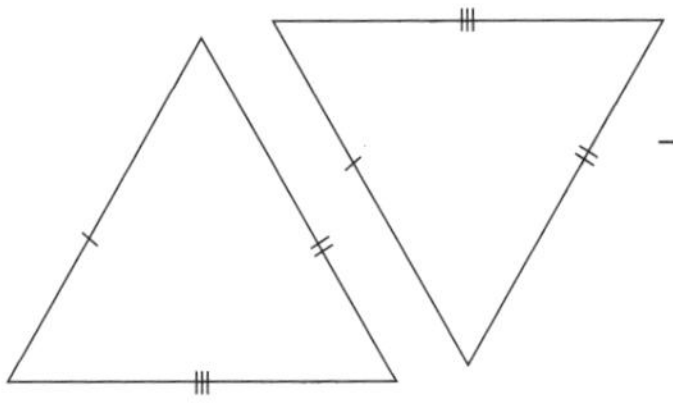

h ______________

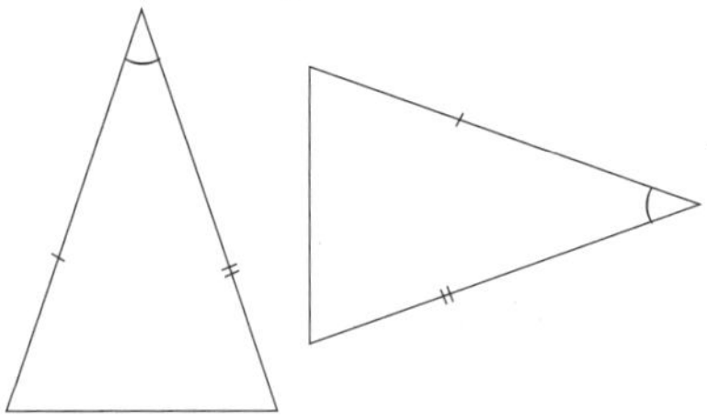

Question 3 In the diagram, O is the centre of the circle. OC is drawn perpendicular to AB.

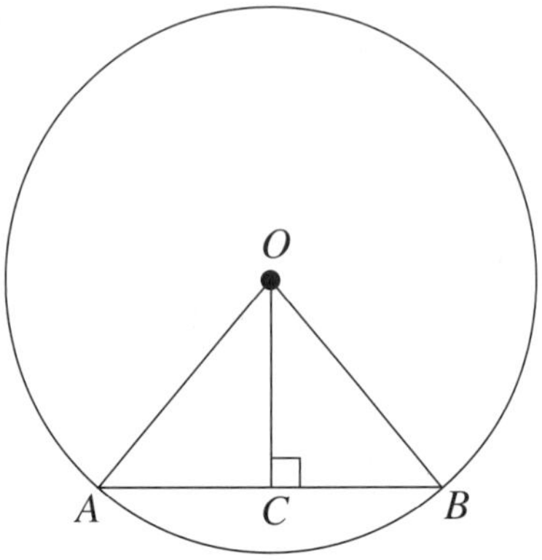

a Name the common side in ΔOAC and ΔOBC.

____

b Name the pair of sides that are equal.

____

c Are the triangles congruent?

____

d If they are congruent, name the test you can use to prove it.

____

O

A C B

Properties of geometrical figures

UNIT 8: Triangle congruence tests and numerical problems

QUESTION 1 Find the value of each pronumeral and justify your answer.

a

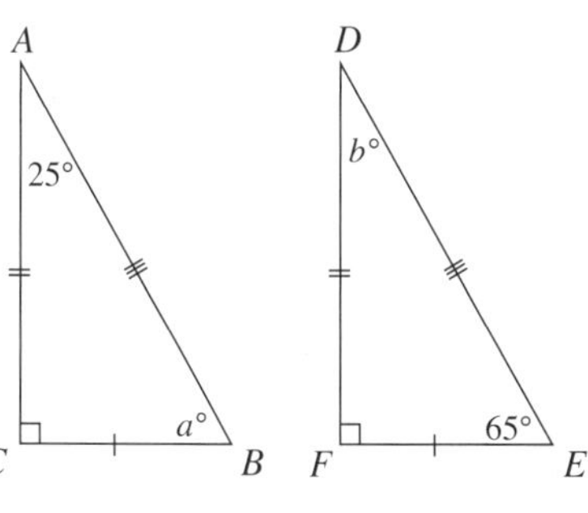

b

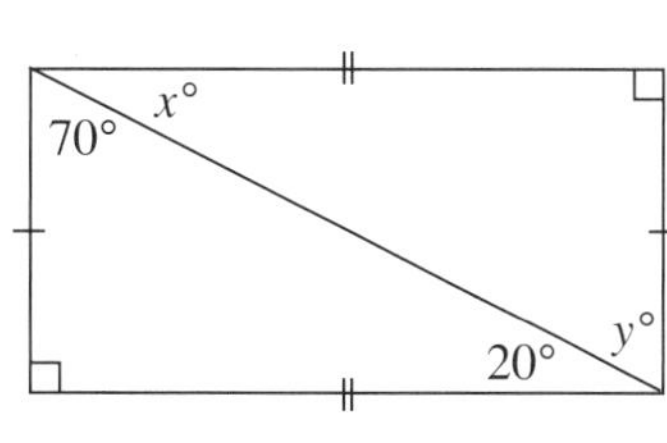

c

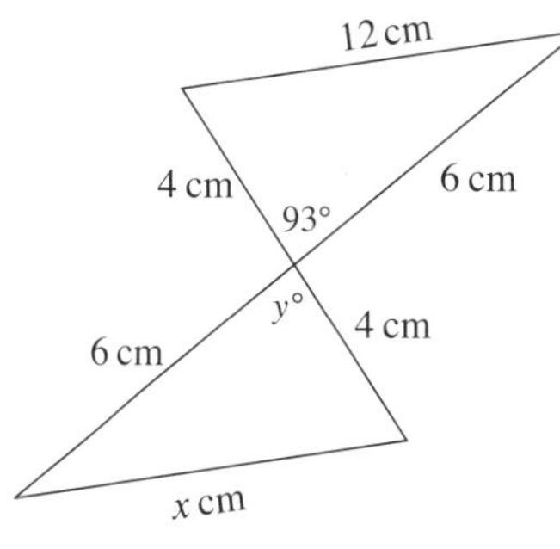

QUESTION 2 Find the value of each pronumeral.

a

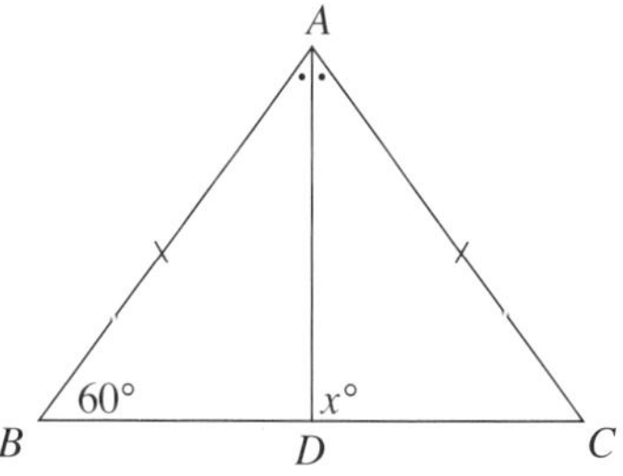

b

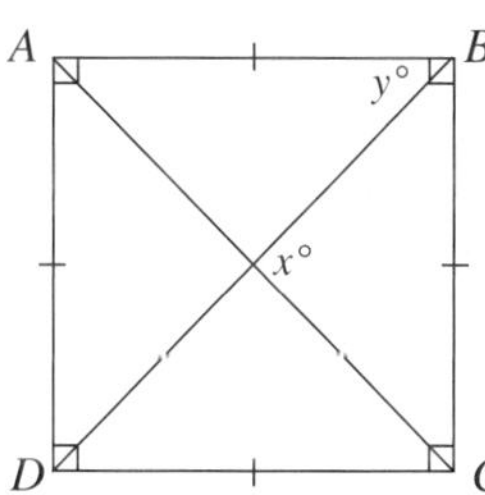

c

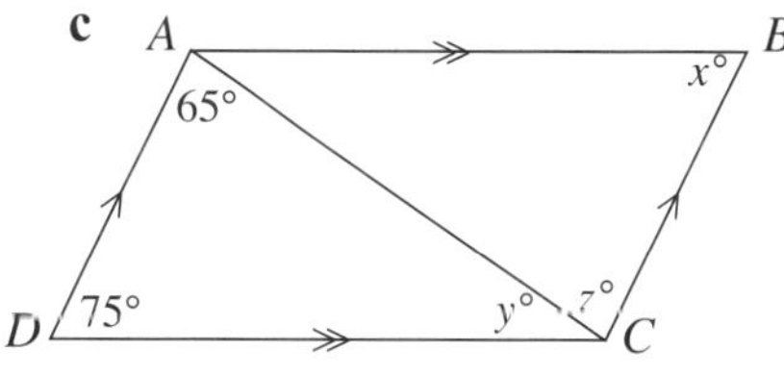

QUESTION 3 Show the congruence relationship between two triangles, stating the relevant test, and then find the value of each pronumeral.

a

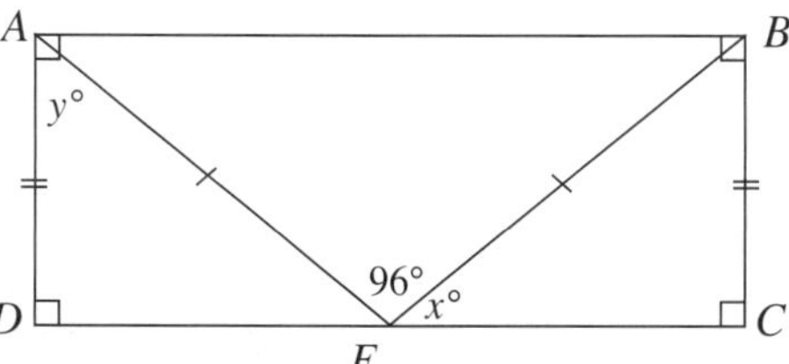

b

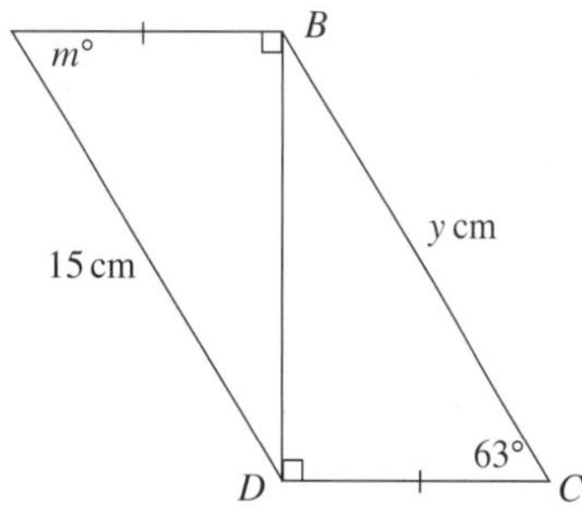

Properties of geometrical figures

UNIT 9: Similar triangles

QUESTION 1 Complete the following statements.

a The symbol for similar triangles is ______________________.

b Two triangles are similar if two angles of one triangle are equal to ______________________ of the other triangle.

c Two triangles are similar if their corresponding sides are in the ______________________.

d Two triangles are similar if one angle of one triangle is equal to ______________________ of the other and the lengths of the sides that form the angle are in the ______________________.

e Two triangles are similar if the hypotenuse and another side of one right-angled triangle are proportional to the ______________________ and another ______________________ of a second ______________________ triangle.

QUESTION 2

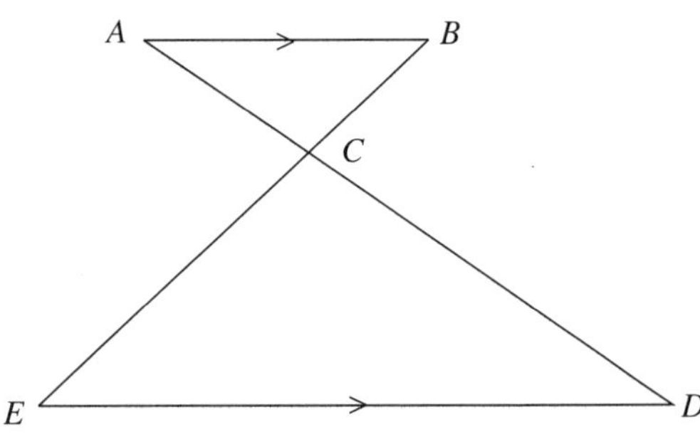

a Why does $\angle ABC = \angle DEC$? ______________________

b Why does $\angle ACB = \angle DCE$? ______________________

c Complete $\Delta ABC \mid\mid\mid \Delta$ ______________________

d Why are the triangles similar?

____

e If $AB = 3$ cm and $DE = 6$ cm, what is the enlargement factor? ______________________

QUESTION 3

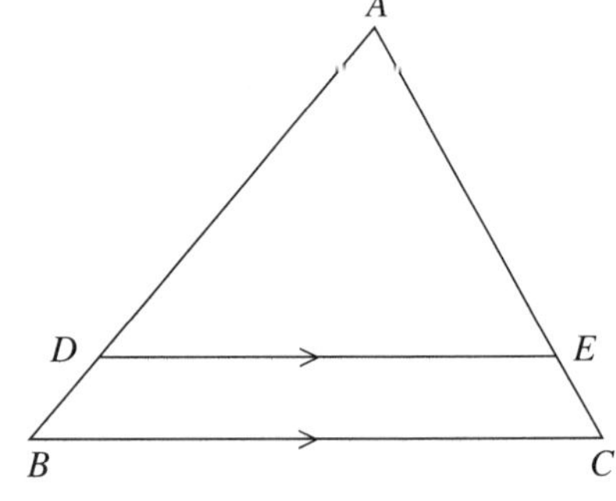

a Why does $\angle DAE = \angle BAC$? ______________________

b Why does $\angle ADE = \angle ABC$? ______________________

c Complete $\Delta ADE \mid\mid\mid \Delta$ ______________________

d Why are the triangles similar? ______________________

QUESTION 4 The triangles drawn below are similar triangles.

a Find $\frac{AB}{DE}$ in simplest form. ______________________

b Find $\frac{AC}{DF}$ in simplest form. ______________________

c Which test shows the triangles are similar? ______________________

____

d What is the enlargement factor? ______________________

e List the pairs of corresponding angles. ______________________

____

f List the pairs of corresponding sides. ______________________

____

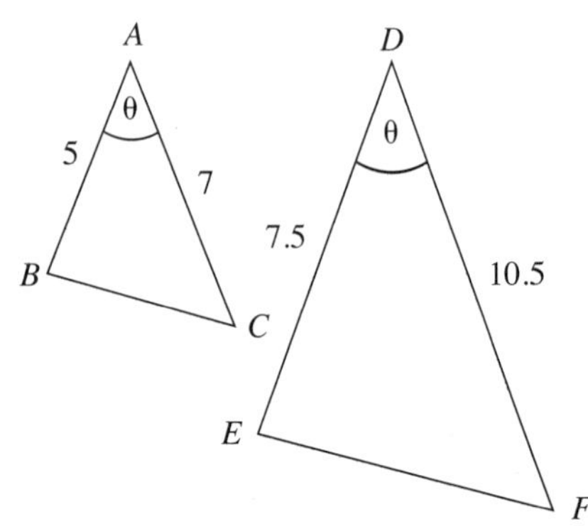

Properties of geometrical figures

UNIT 10: Using similar triangles to find the value of pronumerals

QUESTION **1** In each diagram, use a test of similarity to find the value of the pronumeral. All lengths are in centimetres.

a

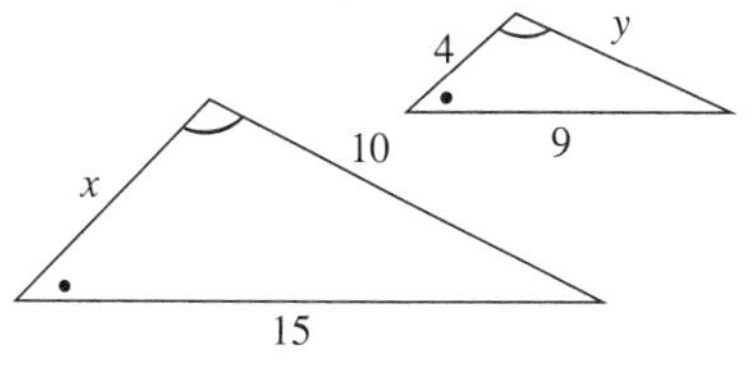

b

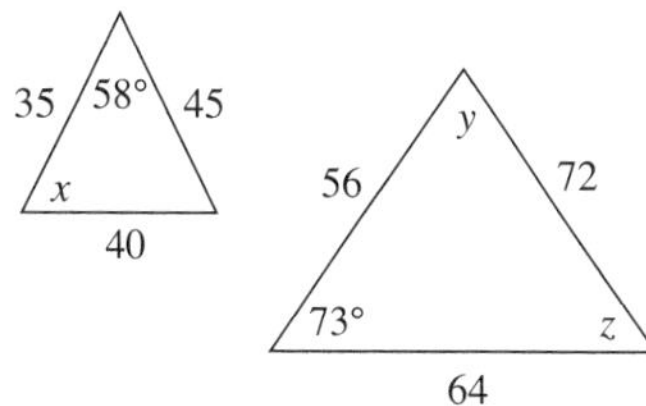

c

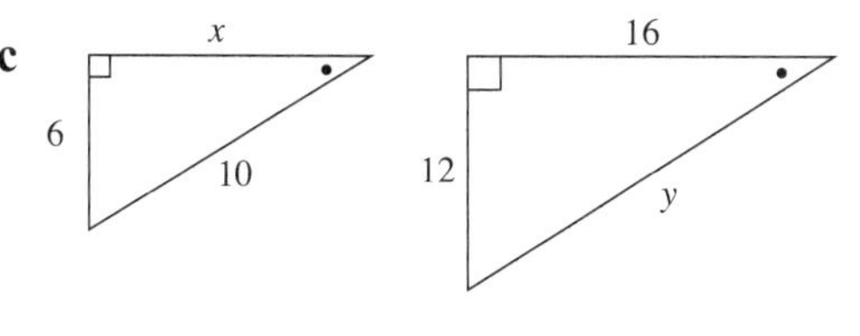

d

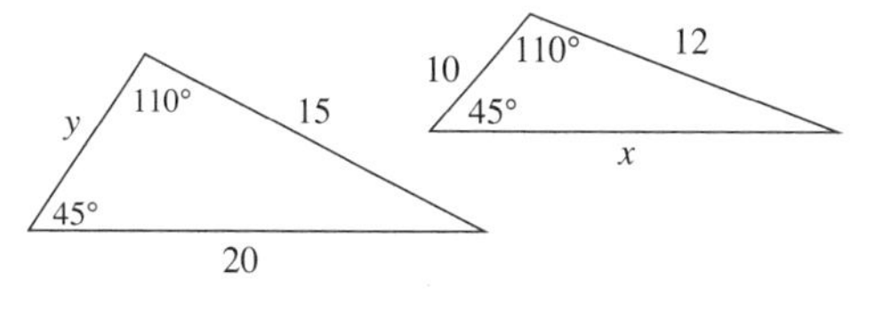

QUESTION **2** For each pair of similar triangles, find the values of the pronumerals. All lengths are in centimetres.

a

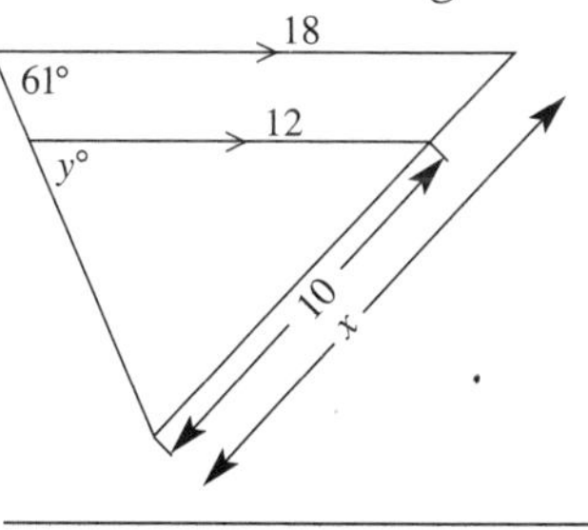

b

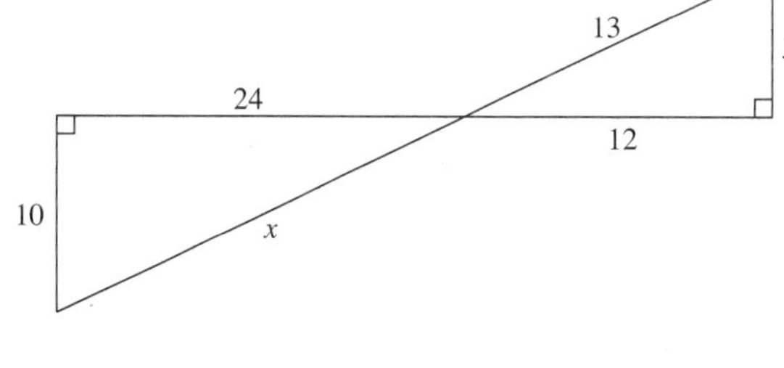

c

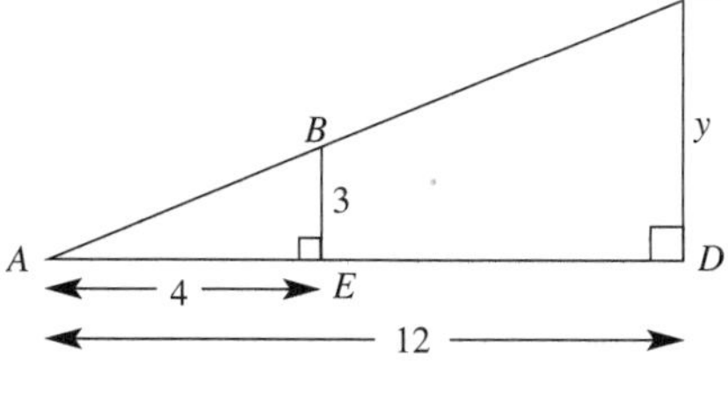

d

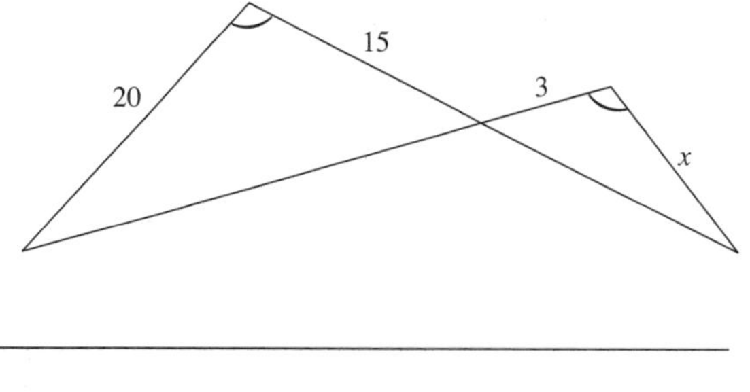

Properties of geometrical figures

TOPIC TEST — PART A

Instructions
- This part consists of 10 multiple-choice questions.
- Fill in only ONE CIRCLE for each question.
- Each question is worth 1 mark.

Time allowed: 15 minutes — **Total marks: 10**

Marks

1 $x =$

(A) 20 (B) 50 (C) 60 (D) 70 — 1

2 The sum of the exterior angles of any polygon is equal to:

(A) 90° (B) 180° (C) 270° (D) 360° — 1

3 $x =$

(A) 50 (B) 60 (C) 70 (D) 80 — 1

4 Which test would be used to prove the two triangles are congruent?

(A) SAS (B) SSS
(C) AAS (D) RHS — 1

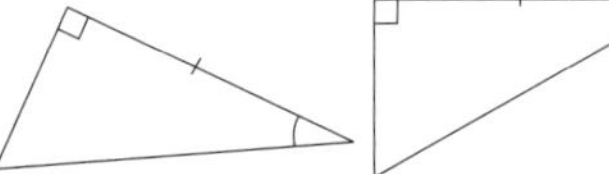

5 All similar triangles are:

(A) equilateral (B) equiangular (C) different (D) congruent — 1

6 $\Delta GHI \equiv \Delta ABC$. What is the size of $\angle HIG$?

(A) 40° (B) 50°
(C) 90° (D) not enough information — 1

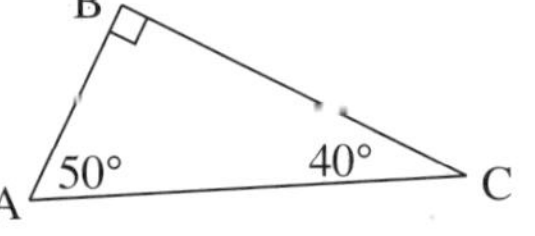

7 A diagonal of a parallelogram divides the parallelogram into two triangles that are:

(A) equilateral (B) isosceles (C) congruent (D) none of these — 1

8 $x =$

(A) 30 (B) 40 (C) 70 (D) 80 — 1

9 If the corresponding angles of two triangles are equal, the triangles are definitely:

(A) congruent (B) similar (C) isosceles (D) equilateral — 1

10 These two triangles are:

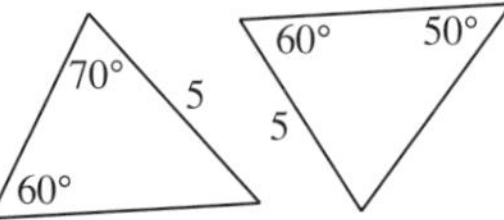

(A) similar but not congruent (B) congruent but not similar
(C) both similar and congruent (D) neither similar nor congruent — 1

Total marks achieved for PART A

Properties of geometrical figures

TOPIC TEST — PART B

Instructions
- This part consists of 5 questions.
- Write only the answer in the answer column.
- For any working use the question column.

Time allowed: 20 minutes — **Total marks: 15**

Questions | **Marks**

1 In the diagram,
O is the centre of the circle.
OC is drawn perpendicular to *AB*.

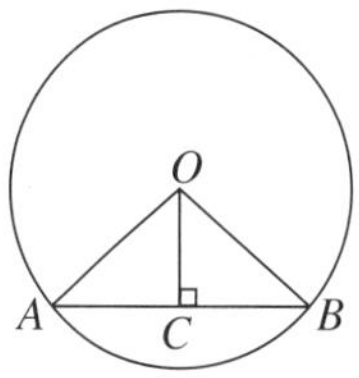

a Name triangles that are congruent. ____________ 1

b State the congruency test. ____________ 1

c Name the pairs of equal sides. ____________ 1

d Name the pairs of equal angles. ____________ 1

2 *ABCD* is a rectangle.

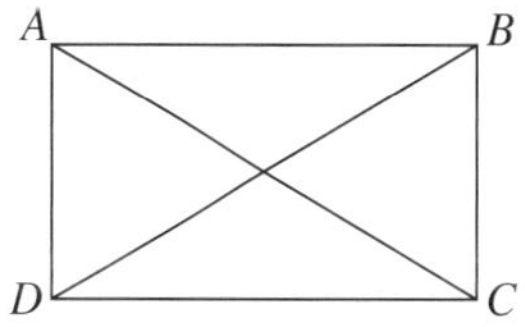

a Why does $AB = CD$? ____________ 1

b Why does $\angle ABC = \angle CDA$? ____________ 1

c Why does $BC = DA$? ____________ 1

d Which test shows that $\Delta ABC \equiv \Delta CDA$? ____________ 1

e Explain why this proves that the diagonals of a rectangle are equal.

____________ 1

3 In the quadrilateral $PS = PQ$ and $SR = QR$.

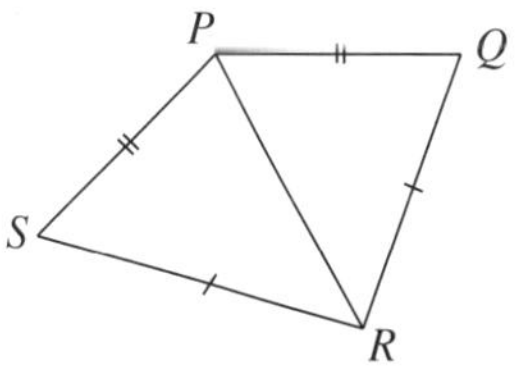

a Complete $\Delta PSR \equiv \Delta$ ____________ 1

b Which test is used to show the triangles are congruent? ____________ 1

4 *ABCDE* is a regular pentagon.

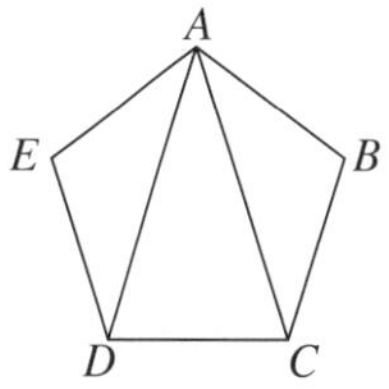

a Which test is used to show $\Delta AED \equiv \Delta ABC$? ____________ 1

b What is the size of $\angle AED$?

____

____________ 1

c What is the size of $\angle ADC$? ____________ 1

5 What is the value of *x*?

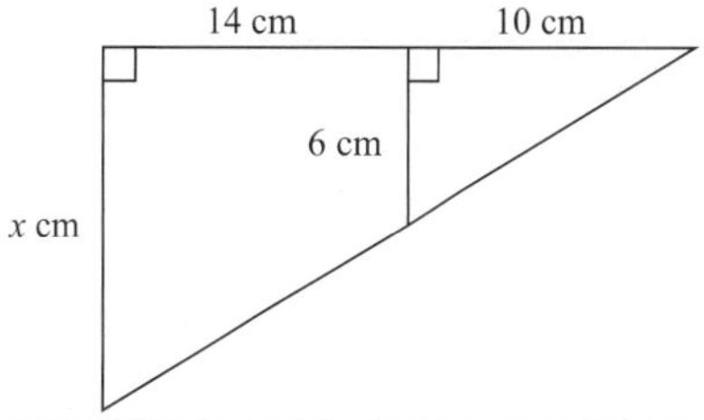

____

____________ 1

Total marks achieved for PART B ___ /15

CHAPTER 11

Data analysis

UNIT 1: Review of basic statistics

QUESTION 1 For the scores 3, 5, 7, 7, 7, 8, 8, 9, 9, 10, 10, find the following.

a mode ______________ **b** median ______________

c mean ______________ **d** range ______________

QUESTION 2 For the scores 9, 2, 7, 6, 2, 5, 4, 8, 2, 5, find the following.

a mode ______________ **b** median ______________

c mean ______________ **d** range ______________

QUESTION 3 The mean mass of 3 students is 53 kg. A fourth student of mass 61 kg joins the group. What is the mean mass of the 4 students?

____

QUESTION 4

a Complete the frequency distribution table.

Score (x)	Frequency (f)	$f \times x$	Cumulative frequency
5	2		
6	6		
7	8		
8	9		
9	7		
10	5		
Total			

Find the:

b mean (to one decimal place)

____

c mode ______________

d median ______________

e range ______________

QUESTION 5 For the scores in the stem-and-leaf plot find the:

Stem	Leaf
1	7
2	0 3
3	1 4 5 8
4	0 2 2 5 6 7 9
5	1 3 4 6 8 8 8 9

a mean ______________

b mode ______________

c median ______________

d range ______________

e Describe the shape of the stem-and-leaf plot. ______________

QUESTION 6 Use your calculator to find the mean ($\bar{x}$) and standard deviation (σ) correct to one decimal place for each set of scores.

a $3, 5, 6, 7, 8, 9, 10$ $\bar{x} =$ ______________ $\sigma =$ ______________

b $32, 43, 45, 37, 33, 38, 39, 34$ $\bar{x} =$ ______________ $\sigma =$ ______________

c $6, 9, 11, 16, 16, 11, 9, 10, 19, 21, 19, 16, 11, 16$ $\bar{x} =$ ______________ $\sigma =$ ______________

d

Score	5	10	15	20	25	30	35
Frequency	2	3	2	4	6	4	1

$\bar{x} =$ ______________ $\sigma =$ ______________

Data analysis

UNIT 2: Quartiles and interquartile range

QUESTION 1 For the set of scores 19, 22, 22, 24, 25, 27, 28, 28, 29, 29, 29, 30, find the:

a median

b lower quartile

c upper quartile

d interquartile range

QUESTION 2 For a particular set of scores the lower extreme is 5, the lower quartile is 9, the median is 12, the upper quartile is 17 and the upper extreme is 20. What percentage of scores are between:

a 5 and 20? ________

b 9 and 17? ________

c 17 and 20? ________

d 5 and 17? ________

e 9 and 12? ________

f 12 and 20? ________

QUESTION 3 For the scores 11, 13, 14, 14, 15, 16, 17, 17, 17, 19, 20, 20, 21, 23, 24, 25 what is the:

a lower extreme? ________

b upper extreme? ________

c median? ________

d lower quartile? ________

e upper quartile? ________

f interquartile range? ________

QUESTION 4 For each of the following distribution displays, find:
i the median **ii** the lower quartile **iii** the upper quartile **iv** the interquartile range

a

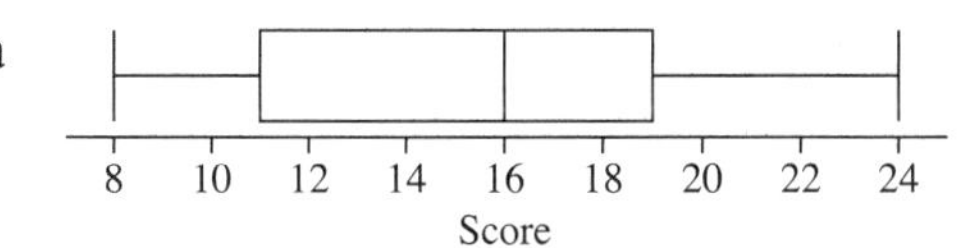

i ________ **ii** ________

iii ________ **iv** ________

b

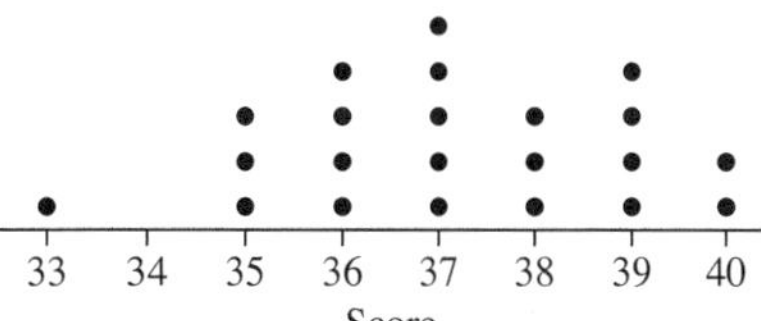

i ________ **ii** ________

iii ________ **iv** ________

c

Stem	Leaf
4	4 6 8
5	0 1 1 2 3 5 9
6	2 3 3 4 5 6 7 8 8
7	0 1 2 2 3 7
8	3 7 9
9	2

i ________ **ii** ________

iii ________ **iv** ________

d

Score	Frequency
7	2
8	3
9	5
10	8
11	12
12	10
13	8
14	6
15	1

i ________

ii ________

iii ________

iv ________

Data analysis

UNIT 3: Standard deviation

QUESTION **1** Consider these scores: 12, 18, 22, 17, 19, 20, 18, 20, 16, 21, 18, 15. Use a calculator to find (to one decimal place where necessary) the:

a mean ______________ **b** standard deviation ______________

QUESTION **2** If another score of 18 was added to the scores in question 1:

a briefly explain what will happen to the mean. Justify your answer. ______________

____

b briefly explain what will happen to the standard deviation. Justify your answer. ______________

____

QUESTION **3** If a score of 10 was added to the scores in question 1:

a briefly explain what will happen to the mean. Justify your answer. ______________

____

b briefly explain what will happen to the standard deviation. Justify your answer. ______________

____

QUESTION **4** Consider these scores: 17, 23, 27, 22, 24, 25, 23, 25, 21, 26, 23, 20. Use a calculator to find (to one decimal place where necessary) the:

a mean ______________ **b** standard deviation ______________

QUESTION **5** Compare the scores in question 1 and question 4.

a Describe any relationship between the scores. ______________

b Describe the relationship between the means of the two sets of scores and briefly explain why this is so.

____

c Describe the relationship between the standard deviations of the two sets of scores and briefly explain why this is so. ______________

QUESTION **6** Consider these scores: 24, 36, 44, 34, 38, 40, 36, 40, 32, 42, 36, 30. Use a calculator to find (to one decimal place where necessary) the:

a mean ______________ **b** standard deviation ______________

QUESTION **7** Compare the scores in question 1 and question 6.

a Describe any relationship between the scores. ______________

____

b Describe the relationship between the means of the two sets of scores and briefly explain why this is so.

____

c Describe the relationship between the standard deviations of the two sets of scores and briefly explain why this is so. ______________

Data analysis

UNIT 4: Interpreting data displays

QUESTION 1 Find, for each cumulative frequency histogram and polygon:
i the median **ii** the lower quartile **iii** the upper quartile **iv** the interquartile range

a

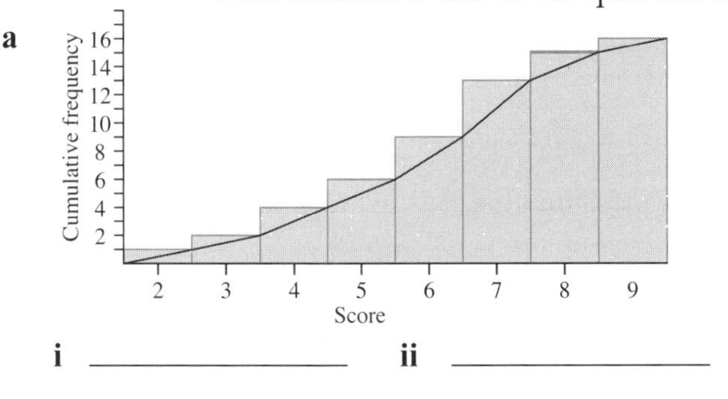

i ______ **ii** ______

iii ______ **iv** ______

b

Cumulative frequency
90 80 70 60 50 40 30 20 10
15 16 17 18 19 20 21
Score

i ______ **ii** ______

iii ______ **iv** ______

QUESTION 2 This dot plot shows the number of mistakes made in a spelling competition.

a Briefly describe the shape of the dot plot.

______

b Without drawing a box plot, briefly comment on the features you might expect to see in one based on the shape of the dot plot.

______

______

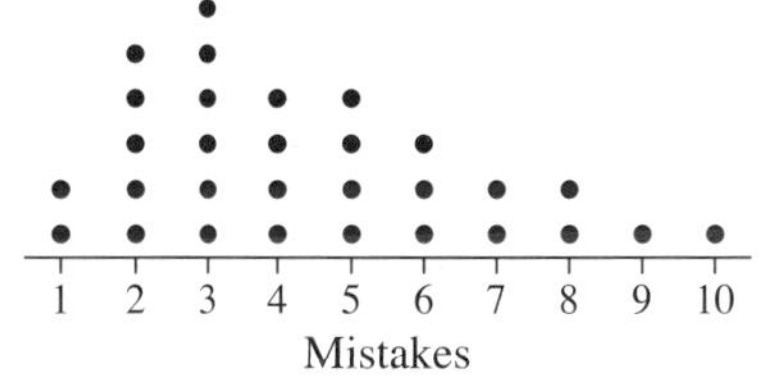

QUESTION 3 The back-to-back stem-and-leaf plot shows marks for two components of a competition.

a Which component has a symmetric display? ______

b How could you describe the skewness of the other component? ______

c Find the five-number summary for each component.

i artistic ______

ii technical ______

d Draw box plots on the same scale.

Artistic		Technical
8	0	5 7 9
5 2	1	0 2 6 8 8
7 3 0	2	1 3 4 7
6 5 4 1 1	3	0 1 5 6
8 7 7 5 2 1	4	1 2 3 3 7
6 4 4 4 3 1 0	5	0 3 4

e Compare the relative merits of each type of display.

______

______

Data analysis

UNIT 5: Comparing data sets

QUESTION **1** Consider these scores: 9, 12, 7, 16, 8, 10, 11, 9, 15, 13, 12, 9, 15.

a What is the range? ______

b What is the interquartile range? ______

c What is the standard deviation? ______

d Briefly comment on the advantages or disadvantages of each measure of spread. ______

QUESTION **2** A new score of 39 is included with the scores in question 1.

a What is the new range? ______

b What is the interquartile range? ______

c What is the standard deviation? ______

d Which measure of spread was most affected by the inclusion of the extra score? ______

e Which measure of spread was least affected? ______

f Briefly comment on the effect on the measures of spread of a score that is much larger or smaller than the existing scores.

QUESTION **3** The stem-and-leaf plot shows scores in the practical and theory parts of a competition.

a Find the range of:

i practical marks ______

ii theory marks ______

b Find the interquartile range for:

i practical marks ______

ii theory marks ______

c Find the standard deviation for:

i practical marks ______

ii theory marks ______

Practical		Theory
9	4	8
8 8 7 4 2 1	5	0 2 7
7 3 0	6	1 3 3 5 6 8
6 4 1	7	2 4 6 7 9 9
9 6 3 1 1 0	8	3 5 9
2	9	4

d Comment on similarities and differences between the two sets of marks. Refer to the measures of spread and to the shape of the distribution.

Data analysis

UNIT 6: Bivariate data

QUESTION **1** Indicate whether the datasets refer to one or two variables.

a investigating a group of plants to find the height they reach in a particular time

____

b surveying a group of babies as to their height and age ______________________________

QUESTION **2** Indicate whether each dataset has an association or a causal relationship between the two variables.

a swimming ability and athletic ability of a group of students ______________________________

b the amount of time studying and the exam results of a group of students

____

c the height of a person and their handspan ______________________________

QUESTION **3** Identify the dependent and independent variables when examining these relationships.

a determining whether the blood alcohol level of a person affects the time needed to complete certain tasks

____

b finding how the amount of meaningful sleep is affected by the amount of coffee consumed in the hours before going to bed

____

QUESTION **4** For each scatter plot describe the strength and direction of the relationship:

a

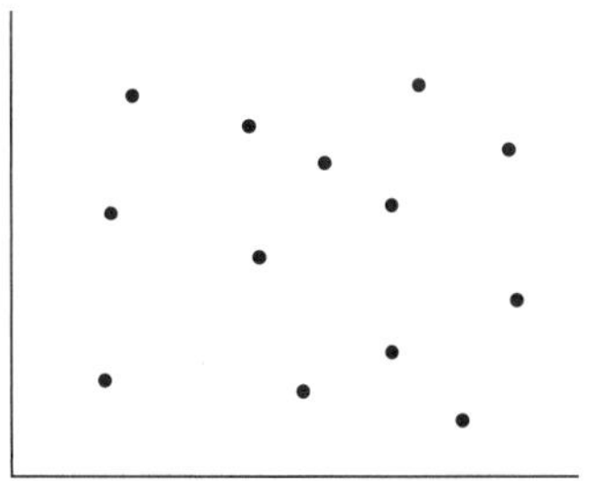

____

____

b

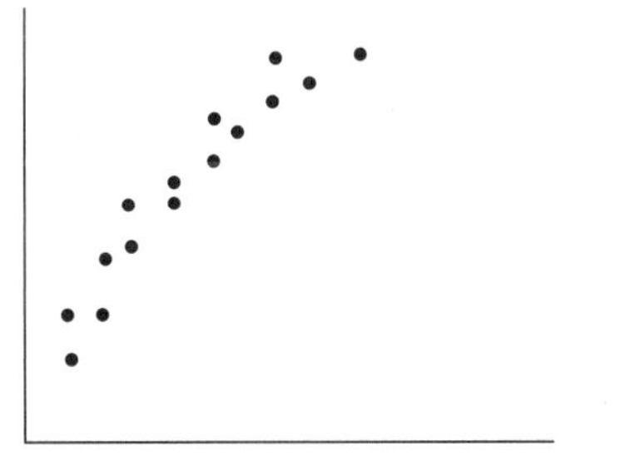

____

____

c

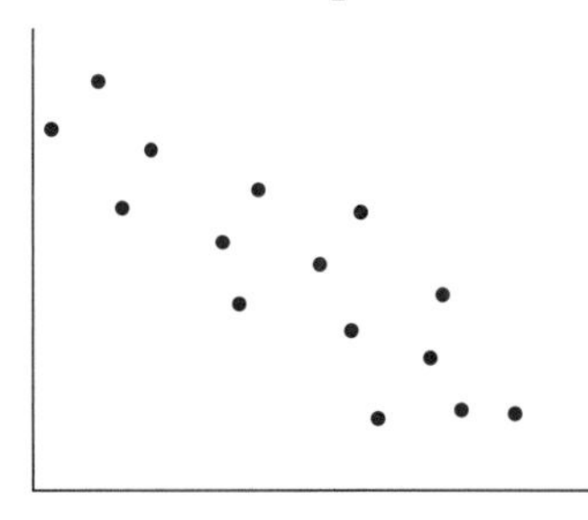

____

____

d

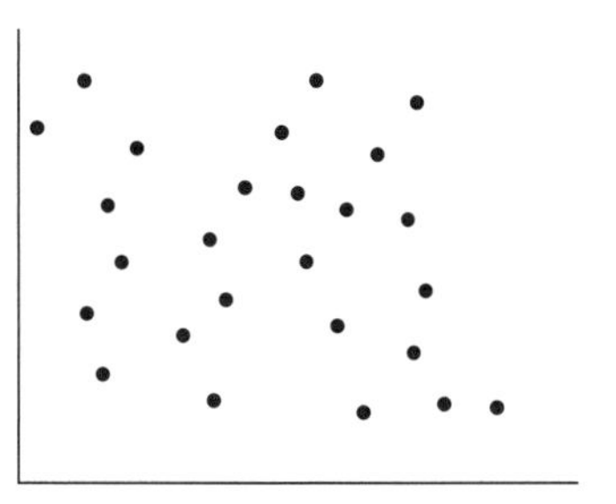

____

____

e

____

____

f

____

____

Data analysis

UNIT 7: Scatter plots (1)

QUESTION 1 This scatter plot has been drawn to show the results some students achieved in an assignment (out of ten) and in the final exam (out of 100) of their course.

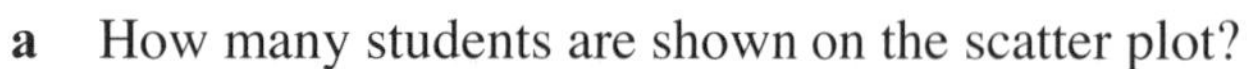

a How many students are shown on the scatter plot?

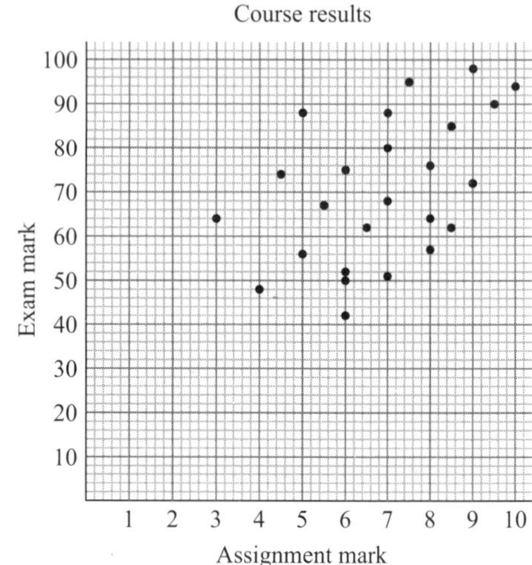

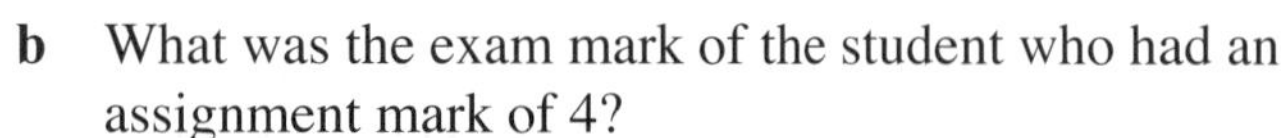

b What was the exam mark of the student who had an assignment mark of 4?

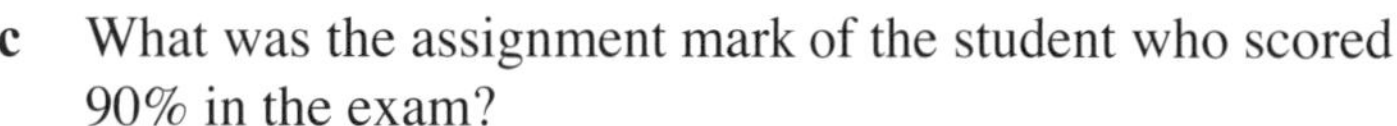

c What was the assignment mark of the student who scored 90% in the exam?

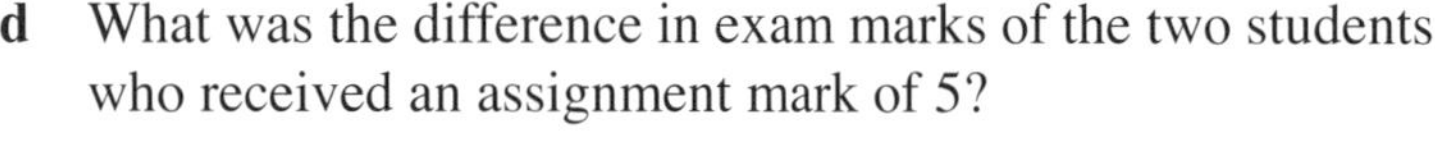

d What was the difference in exam marks of the two students who received an assignment mark of 5?

e Only one student scored the same percentage mark for both the assignment and the exam. What mark was that?

f How many of the students scored more than 80% in the exam but fewer than 6 on the assignment?

QUESTION 2 The following table shows the ages and advertised prices of a particular model of car.

Age (years)	Price ($)	Age (years)	Price ($)
3	20 500	10	5000
7	12 800	9	7000
2	26 900	6	9000
1	30 000	2	29 000
8	10 000	1	32 000
2	28 000	3	27 000
9	9000	8	11 000
5	14 000	9	7500
6	10 000	6	9500

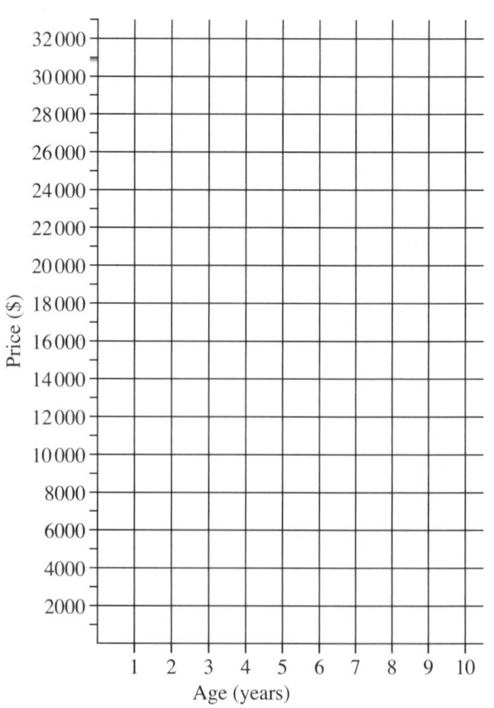

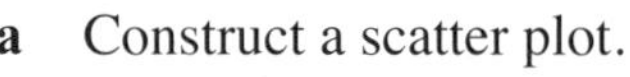

a Construct a scatter plot.

b Marcus has a 4-year-old car of this model that he wants to sell. At what price would you suggest he advertise his car? Justify your answer.

Data analysis

UNIT 8: Scatter plots (2)

QUESTION 1 In a class, the number of hours each student spent studying for an examination and the marks each one was awarded were recorded as shown in the table below.

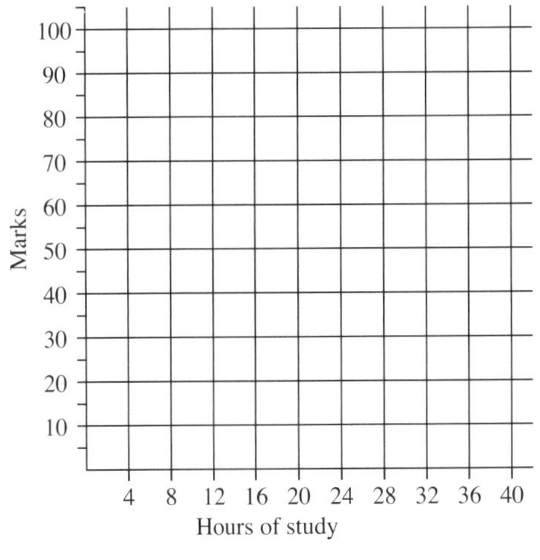

Student	Marks	Hours	Student	Marks	Hours
1	15	2	11	72	21
2	93	35	12	82	29
3	30	5	13	85	30
4	52	8	14	9	2
5	61	15	15	27	3
6	82	30	16	39	4
7	97	36	17	48	6
8	100	39	18	92	36
9	5	1	19	67	20
10	38	7	20	99	38

a Construct a scatter plot to show this data.

b Comment on any trends.

____

____

QUESTION 2 The assessment test results in Maths and Science of a class of 15 students are given in the table.

Student	Maths mark	Science mark	Student	Maths mark	Science mark
1	70	58	9	32	36
2	37	40	10	53	58
3	52	55	11	42	48
4	66	62	12	64	56
5	36	32	13	27	34
6	46	50	14	67	73
7	30	35	15	57	49
8	62	68			

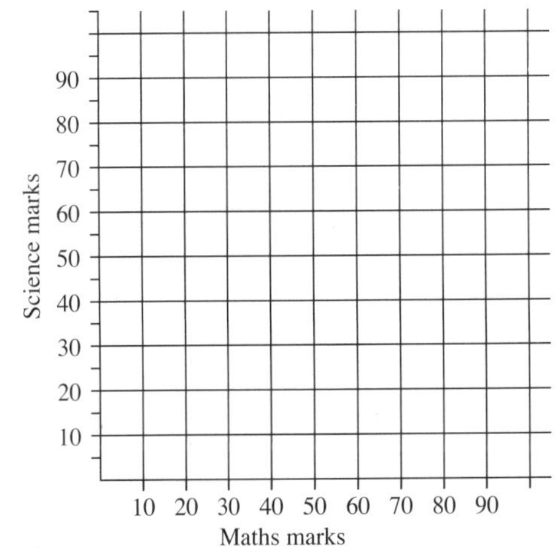

a Construct a scatter plot to show this data.

b Simone is also in the class. She scored 65 in maths but missed the science test. What would you predict her science score to be? Justify your answer.

____

____

____

Data analysis

UNIT 9: Line of best fit (1)

QUESTION **1** The data in the table to the right gives the marks out of 100 in a Maths and Biology test for a group of students.

Maths mark	Biology mark
84	74
88	82
51	47
62	55
100	90
68	59
79	73
74	64
95	90
48	42
40	34

a Plot this data on this graph.

b Draw a line of best fit.

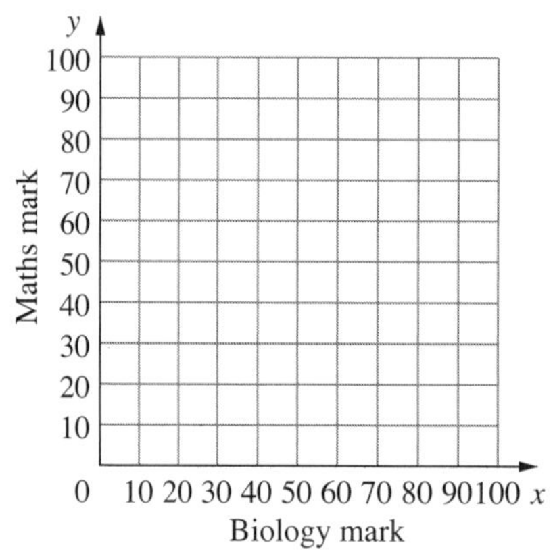

c Use your graph to find an equation for the line of best fit.

____

____

____

QUESTION **2** The table to the right shows the assessment results in Maths and Science of a class of 12 students.

Students	Maths	Science
1	75	63
2	42	47
3	57	62
4	71	69
5	41	39
6	51	57
7	35	42
8	67	75
9	37	43
10	58	65
11	47	55
12	69	63

a Draw a scatter plot of these results.

b Draw in the line of best fit.

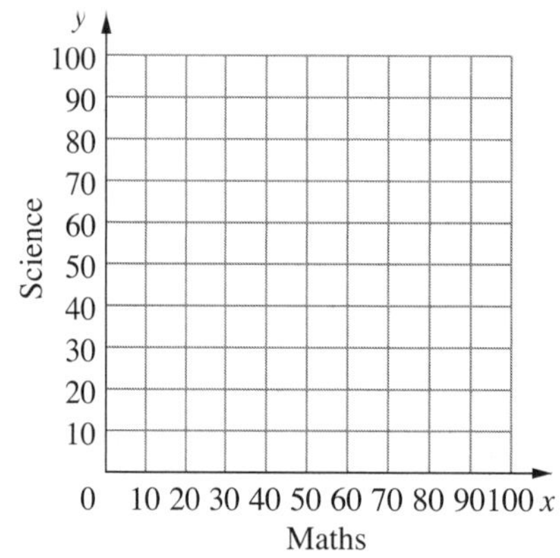

c Use this line of best fit to predict Science marks for students with the following Maths results.

i 95 **ii** 20

______________________ ______________________

d Use the graph to find the equation of the line of best fit.

____

____

____

Data analysis

UNIT 10: Line of best fit (2)

QUESTION 1 A scatter plot shows the height (h cm) of some seedlings n weeks after they were planted. A line of best fit has been drawn on the scatter plot.

a What is the equation of the line of fit?

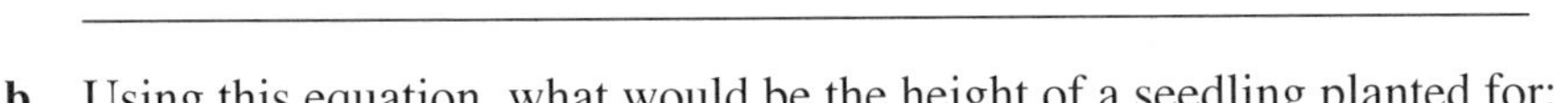

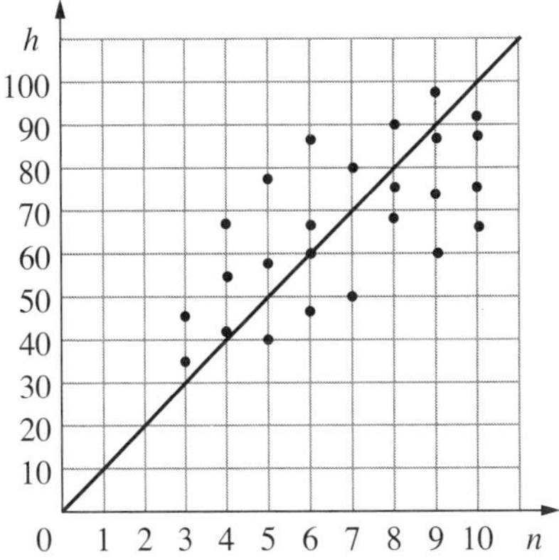

b Using this equation, what would be the height of a seedling planted for:

i $8\frac{1}{2}$ weeks? ____________________

ii 13 weeks? ____________________

c Using this equation, how many weeks would you predict that a seedling had been planted if it had a height of:

i 55 cm? ____________________

ii 160 cm? ____________________

QUESTION 2 A different line of fit was drawn on the same scatter plot from question 1.

a What is the equation of the line of fit?

____

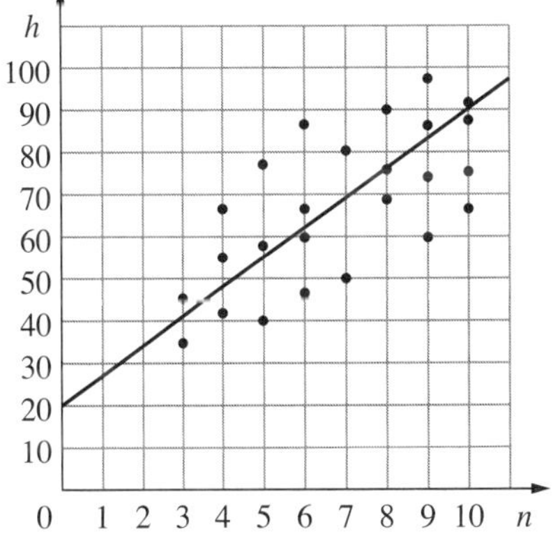

b Using this equation, what would be the height of a seedling planted for:

i $8\frac{1}{2}$ weeks? ____________________

ii 13 weeks? ____________________

c Using this equation, how many weeks ago would you predict that a seedling had been planted if it had a height of:

i 55 cm? ____________________

ii 160 cm? ____________________

d Briefly comment on the differences between predictions based on the different lines of fit.

____

____

QUESTION 3 A line of fit was drawn on this scatter plot that shows the age and value of some cars.

a What is the equation of the line of fit?

____

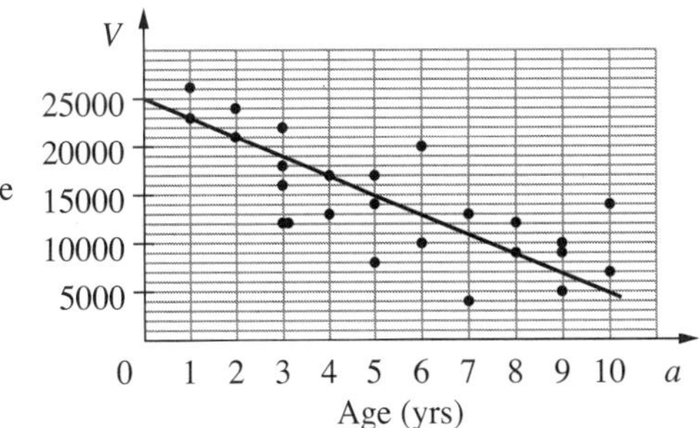

b What is the predicted value of a car that is 5 years old?

____

c What is the predicted age of a car that is valued at $5000?

____

d After what age will the equation no longer be valid? Justify your answer.

____

____

Data analysis

UNIT 11: Sampling methods

QUESTION 1 Choose from census, random, sample, stratified and systematic to fill in these blanks.

a A survey of the whole population is called a ________________.

b A ________________ of the population is often used when surveying the whole population.

c A ________________ ________________ is one where every element of the population has an equal chance of being included.

d A ________________ ________________ is one where a system or particular method has been used to choose the elements of the population that are included.

e A ________________ ________________ is one where the proportions of elements of the population with a particular characteristic match the proportions in the general population.

QUESTION 2 Kelly wants to conduct a survey of office workers from a high-rise city building. She waits outside the building and decides to question the first 60 people who leave the building after a certain time. Is this a random sample? Explain why or why not.

____

____

QUESTION 3 At a factory, every twentieth item is examined by a quality-control inspector.

a What type of sample is that?

____

b Some issues with the inspected items are found. As a result, a census is conducted. What items would be examined?

____

c Once it is believed that the problem has been fixed, what changes might be made to quality control to check?

____

QUESTION 4 A school has students from kindergarten to Year 6. The table shows the number of students in each year.

Year	K	1	2	3	4	5	6
Number of children	32	29	25	27	23	28	36

a Find the number of Year 6 students who would be chosen to participate in a stratified survey of 50 children from the school.

____

b In a different stratified survey, five Year 2 students were chosen. How many children were chosen altogether for the second survey?

____

Data analysis

UNIT 12: Evaluating statistical information

QUESTION 1 Each of these displays is misleading. Briefly explain what is wrong with each graph.

a **Sales**

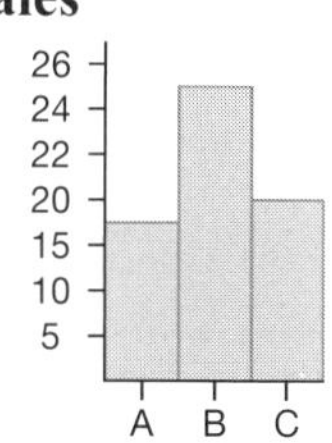

b **Sales**

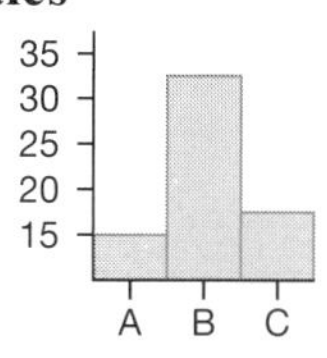

c **Sales**

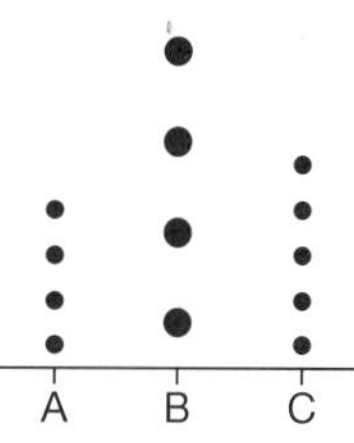

QUESTION 2 The opening sentence of an article in a newspaper said: 'Australians each drink 111 kg of wine, beer and other alcoholic beverages each year.'

a Could this statement possibly be true? Justify your answer.

b What do you think the statement should have said?

c Should you believe the rest of the article? What things should be considered?

d What was the effect of including the word 'possibly' in part **a** of this question?

QUESTION 3 A current-affairs program showed a program about the conviction of a woman for a crime. The woman protested her innocence. The program gave details about the evidence in the case and questioned that evidence. At the conclusion of the program, viewers were asked to vote 'yes' or 'no' to the question 'Should she have been convicted?'

a How important do you think the answer to this survey would be? ____________

b What should be taken into account when considering the results of the survey?

QUESTION 4 An advertisement says: 'Nine out of ten chemists recommend Carla's Cream as an effective treatment for corns.'

a What are the advertisers trying to achieve by including such a statement?

b Is it possible that the statement is correct but that Carla's Cream is not an effective treatment for corns? Comment.

Data analysis

UNIT 13: Investigating reports

QUESTION 1 A survey of some Year 10 students found that 35% of the students couldn't swim the length of a 50-m pool. A politician, appalled by this statistic, suggested that all school students should have swimming lessons. Do you think that is a good response? Justify your answer.

____

____

____

____

QUESTION 2 A particular vitamin supplement (brand X) is being advertised with a claim that 'studies show that brand X is the best'. What information about these studies would be helpful in deciding whether or not the claim is correct?

____

____

____

QUESTION 3 A phone survey just before an election asked whether voters would still vote for a particular politician if she changed her stance on an environmental issue. The survey was conducted for one of the politician's opponents. The politician claimed that she had no intention of changing her stance. Why do you think the question was being asked in the survey? Comment.

____

____

____

QUESTION 4 A current-affairs show on television showed a program about a particular crime and the leniency of the sentence given to the perpetrator. At the end of the show the viewers were asked to vote whether or not they believed sentences for crimes were long enough. What problems might there be with the results of this survey?

____

____

____

QUESTION 5 An insurance company uses probabilities when determining the size of premiums that must be paid by their customers. Two neighbours, with exactly the same type of car, have to pay completely different amounts for insurance with the same company. Why might this be?

____

____

QUESTION 6 The results of a census in a local area showed that the area had a high percentage of young families. What implications might this have for government decision-makers?

____

____

Data analysis

TOPIC TEST — PART A

Instructions
- This part consists of 10 multiple-choice questions.
- Fill in only ONE CIRCLE for each question.
- Each question is worth 1 mark.

Time allowed: 15 minutes **Total marks: 10**

Marks

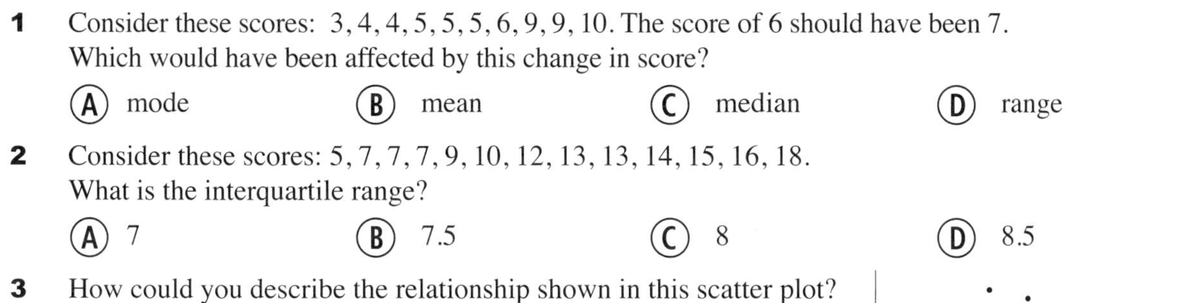

1 Consider these scores: 3, 4, 4, 5, 5, 5, 6, 9, 9, 10. The score of 6 should have been 7. Which would have been affected by this change in score?

(A) mode (B) mean (C) median (D) range 1

2 Consider these scores: 5, 7, 7, 7, 9, 10, 12, 13, 13, 14, 15, 16, 18. What is the interquartile range?

(A) 7 (B) 7.5 (C) 8 (D) 8.5 1

3 How could you describe the relationship shown in this scatter plot?

(A) weak positive (B) weak negative
(C) strong positive (D) strong negative 1

4 Consider these scores 8, 4, 7, 10, 6, 3, 6. The score 6 is the:

(A) lower quartile (B) median (C) upper quartile (D) upper extreme 1

5 Consider this box plot. The display is:

(A) symmetric (B) negatively skewed
(C) positively skewed (D) bimodal 1

6 Referring to the dot plot, the mode is:

(A) 4 (B) 5
(C) 6 (D) 8 1

7 Referring to the dot plot, the lower quartile is:

(A) 2 (B) 3 (C) 3.5 (D) 4 1

8 For a set of scores the five-number summary is [8, 12, 15, 19, 22]. The interquartile range is:

(A) 3 (B) 4 (C) 7 (D) 14 1

9 Referring to the five-number summary in question 8, what percentage of scores will lie between 8 and 19?

(A) 25% (B) 50% (C) 75% (D) 80% 1

10 Which graph is misleading?

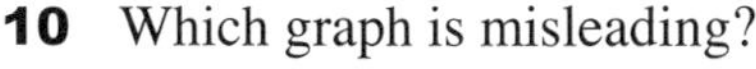

(A)
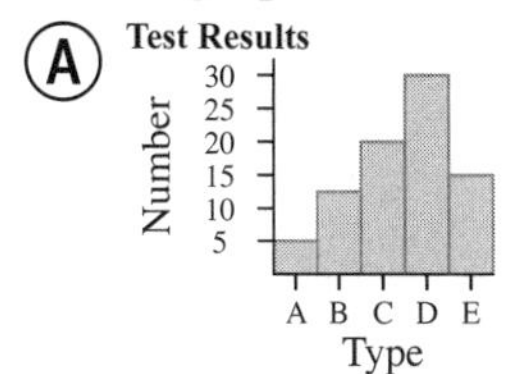

(B)
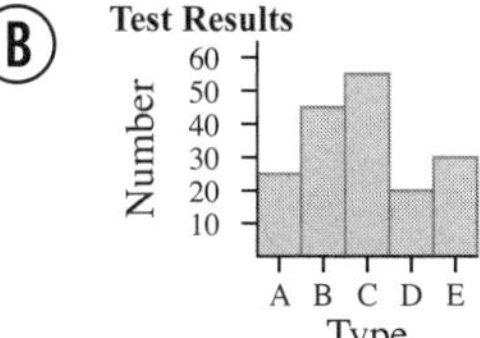

(C)
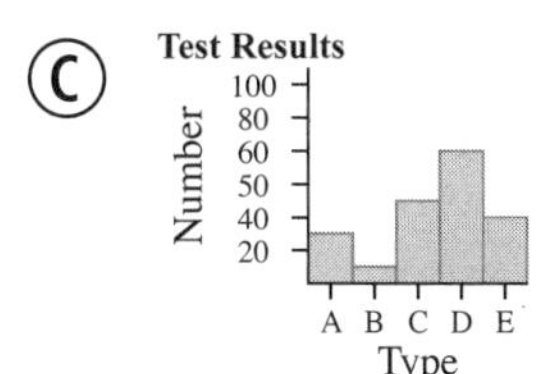

(D)
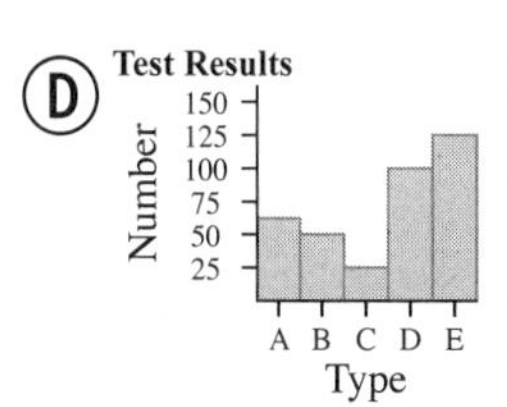

1

Total marks achieved for PART A

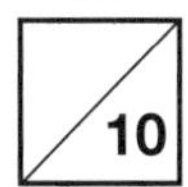

Data analysis

TOPIC TEST PART B

Instructions
- This part consists of 4 questions.
- Write only the answer in the answer column.
- For any working use the question column.

Time allowed: 20 minutes **Total marks: 15**

Questions **Marks**

1 A stem-and-leaf plot has been drawn to illustrate the results achieved by a class in an exam.

Exam results

Stem	Leaf
4	4 8
5	0 4 4 7
6	1 3 5 8 9 9 9
7	0 2 2 3 5 8
8	1 1 3 6 6 7 8 9
9	0 3 4 5

a What is the range? ______ 1

b What is the median? ______ 1

c What is the interquartile range? ______ 1

d What information could be gained from the stem-and-leaf plot that could not be gained from a box plot? ______ 1

e What information could be easily gained from a box plot? ______ 1

2 The results for a competition are

23	18	11	36	22	14	20	19	25	29	24	25	30	22
21	19	22	20	23	21	22	21	20	20	24	25	29	30

a Construct a dot plot of the data. 1

Find the: **b** range ______

c interquartile range ______ 1

d standard deviation ______

e Briefly comment on the relative merits of each measure of spread.

______ 1

3 This scatter plot was drawn to show the profit ($\$p$) made when n toys have been sold at a market.

a Draw a line of best fit on the plot. 1

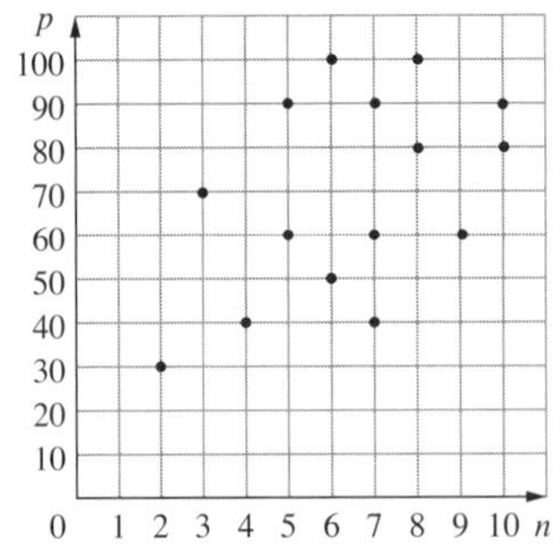

b What is the equation of your line of best fit?

______ 1

c How much profit does the equation predict will be made if 17 toys are sold? ______ 1

d About how many toys sell for a \$320 profit?

______ 1

4 A survey of 15 Year 10 students found that only 20% eat the recommended amount of fruit each day. Is this something to be concerned about? Justify your answer.

______ 1

Total marks achieved for PART B /15

CHAPTER 12

Probability

UNIT 1: Review of basic probability

QUESTION 1 This spinner is spun.

a What number is most likely to be spun? ____________________

b What number is least likely to be spun? ____________________

How could you describe the probability of spinning:

c 4? ____________________ **d** 5? ____________________

e a number greater than 1? ____________________ **f** a number less than 6? ____________________

QUESTION 2 A fair dice is thrown. What is the probability of it showing:

a 5? ____________________ **b** an even number? ____________________ **c** a number greater than 2? ____________________

QUESTION 3 A bag holds 3 red, 2 green and 5 blue pegs. One peg is selected at random. What is the probability that the peg is:

a red? ____________ **b** green? ____________ **c** blue? ____________

d yellow? ____________ e not red? ____________ f red or blue? ____________

QUESTION 4 A card is chosen at random from a regular pack of playing cards. What is the probability that the card is:

a the ace of spades? ____________ **b** a queen? ____________

c red? ____________ **d** a club? ____________

e a black king? ____________ **f** not a diamond? ____________

QUESTION 5 There are 100 tickets in a hat: 35 are blue, 40 are yellow and the rest are white. One ticket is drawn from the hat at random. What is the probability that it is:

a yellow? ____________ **b** white? ____________

c not white? ____________ **d** not blue? ____________

e yellow or white? ____________ **f** neither blue nor yellow? ____________

QUESTION 6 A letter is chosen at random from the alphabet. What is the probability that the letter is:

a J? ____________ **b** F or G? ____________

c not K? ____________ **d** X, Y or Z? ____________

e a vowel (A, E, I, O or U)? ____________ **f** not a vowel? ____________

Probability

UNIT 2: Tree diagrams

QUESTION 1 A coin is tossed three times and the results are noted. Use the tree diagram to find the probability of:

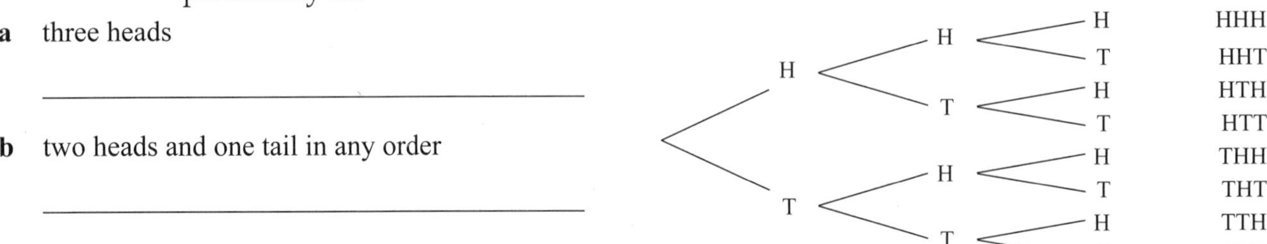

a three heads ____________________

b two heads and one tail in any order ____________________

c at least one head ____________

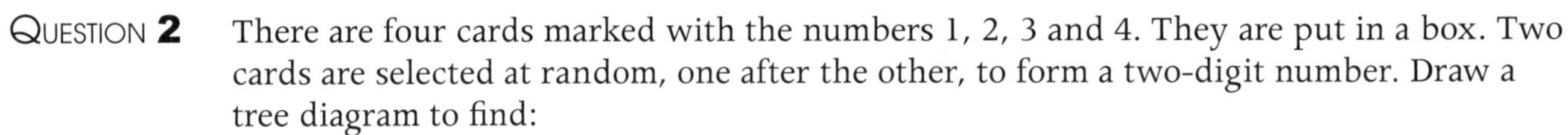

QUESTION 2 There are four cards marked with the numbers 1, 2, 3 and 4. They are put in a box. Two cards are selected at random, one after the other, to form a two-digit number. Draw a tree diagram to find:

a how many different two-digit numbers can be formed ____________________

b the probability that the number formed is less than 34 ____________________

c the probability that the number formed is divisible by 3 ____________________

d the probability that the number formed is even ____________________

QUESTION 3 Three red balls and two blue balls are placed in a bag. Two balls are selected at random, without replacement. What is the probability of having:

a two red balls? ____________

b two blue balls? ____________

c one red ball and one blue ball? ____________

QUESTION 4 In a family of three children, use a tree diagram to find the probability of the following.

a three boys ____________

b two boys and one girl ____________

c one boy and two girls ____________

d the eldest child being a boy ____________

e the youngest child being a girl ____________

f three girls ____________

Probability

UNIT 3: Tables, diagrams and lists

QUESTION 1 Two dice are rolled. The smaller number is subtracted from the larger number to form the score. (If the numbers are the same, the score is zero.)

a Complete the table to show the possible scores.

2nd dice / 1st dice –	1	2	3	4	5	6
1						
2						
3						
4						
5						
6						

What is the probability that the score is:

b 3? ______________

c 6? ______________

d less than 4? ______________

QUESTION 2 Suppose we wish to throw a total of 6. Which is the better chance: rolling one dice or two dice? Use a table to help find the answer.

____

____

____

QUESTION 3 Gemma has three cards: one card shows the number 1, a second card shows 2 and a third card shows 3. Gemma places the three cards in a row to form a three-digit number.

a List the possible numbers.

____

____

What is the probability that the number formed:

b is 123? ______________ **c** is even? ______________ **d** starts with 3? ______________

e is greater than 200? ______________ **f** is less than 220? ______________

QUESTION 4 A family consists of four children: Holly, Alex, James and George. A pair of the children are chosen at random to go shopping.

a Draw a tree diagram to show all possible shopping pairs.

b Find the probability of choosing a pair that includes:

i Holly **ii** George but not Holly

______________ ______________

______________ ______________

Probability

UNIT 4: Venn diagrams

QUESTION 1 The Venn diagram shows the number of students at a school who played softball or netball.

S N 17
15 21 37

a How many students were at the school? ____________

b How many students played netball? ____________

c How many students played softball? ____________

What is the probability that a randomly selected student from the school played:

d both softball and netball? ____________

e softball or netball? ____________

QUESTION 2 In a class of 30 students who study either French or German or both, 18 study French and 22 study German.

a Draw a Venn diagram to represent this.

b How many study both French and German?

____

c If a student is chosen at random from this class, what is the probability that the student studies only French?

____

QUESTION 3 A group of 50 students on a music course were asked whether they played piano, guitar or violin. The results are shown in the Venn diagram. How many of the students play:

P G
6 7 13
1
4 2
3
V 14

a none of those instruments? ____________

b all three of the instruments? ____________

c one of the instruments only? ____________

d piano? ____________

e both guitar and violin? ____________

f piano and guitar but not violin? ____________

g either piano or violin or both?

____

h either guitar or piano but not both?

____

QUESTION 4 A group of 54 students were asked about their pets. It was found that 8 had no pets while 27 had dogs. Eleven had just a cat and 5 had another type of pet but no dog or cat. Altogether, 10 had both a dog and cat and 6 of those also had another pet. Eight had both a dog and another pet (not a cat).

a Fill in the Venn diagram with this information.

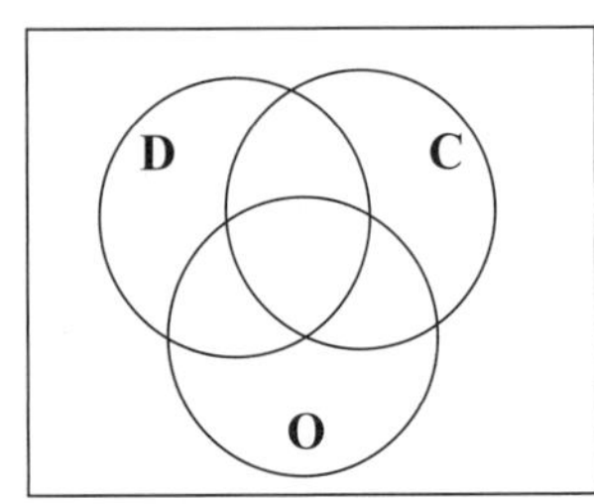

b What is the probability that the owner of a cat also had a dog? ____________

c What is the probability that the owner of a dog also had a cat? ____________

d What is the probability that a student had both a dog and a cat but no other pets?

____

Probability

UNIT 5: Two-way tables

QUESTION 1 A survey was taken of the numbers of people in cars. The results are shown in the table.

	Drivers	Passengers	Total
Men	72	38	
Women	56	44	
Total			

a Fill in all the totals on the table.

b How many women were passengers? ____________

c How many men were there altogether? ____________

What is the probability that:

d a man was a driver? ____________

e a driver was a man? ____________

f a woman was a passenger? ____________

g a passenger was a woman? ____________

h a person was a male driver? ____________

i a person was a female passenger? ____________

QUESTION 2 A survey was taken of 100 people to determine the number of people who play cricket and football. It was found that 19 play both cricket and football, 28 play cricket but not football and 31 play football but not cricket.

a Fill in the table with the given information.

	Plays football	Does not play football	Total
Plays cricket			
Does not play cricket			
Total			

Using the results of the survey, what is the probability that:

b a person plays football?

____

c a person who plays football also plays cricket?

____

d a person who plays cricket also plays football?

____

e a person plays neither cricket nor football?

____

f a person plays both cricket and football?

____

g a person who does not play cricket also does not play football? ____________

h a person who does not play football plays cricket? ____________

QUESTION 3 Complete the totals and show the information in this table in the Venn diagram.

	A	Not A	Total
B	37	41	
Not B	24	18	

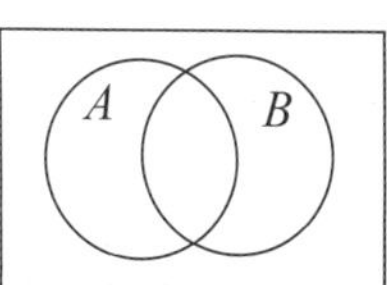

QUESTION 4 Complete the table from the information in the Venn diagram.

	X	Not X	Total
Y			
Not Y			

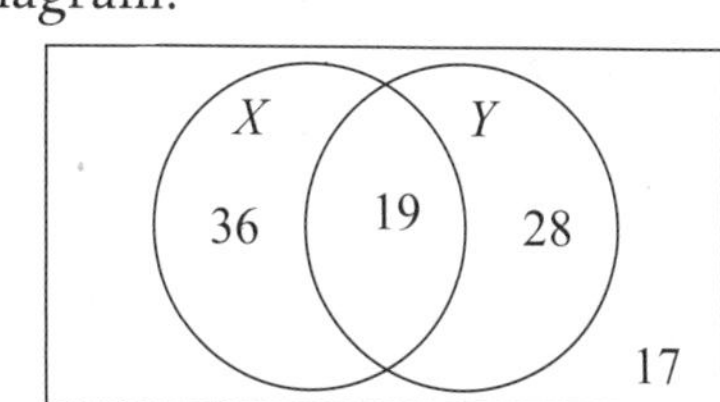

Probability

UNIT 6: Set notation

Question 1 Complete the following statements.

a The symbol $\cup$ is the ____________ and $A \cup B$ means ____________.

b The symbol $\cap$ is the ____________ and $A \cap B$ means ____________.

c $\bar{A}$ (or A' or A^c) is the ____________ of A and means ____________.

Question 2 Numbers are selected from the whole numbers between 1 and 10 inclusively. A is the set of even numbers, B is the set of multiples of 3 and C is the set of prime numbers. Show, using set notation:

a A ____________ b B ____________ c C ____________

d $\bar{A}$ ____________ e $\bar{B}$ ____________ f $\bar{C}$ ____________

g $A \cup B$ ____________ h $B \cup C$ ____________

i $A \cup C$ ____________ j $A \cap B$ ____________

k $B \cap C$ ____________ l $A \cap C$ ____________

m $\bar{A} \cup B$ ____________ n $B \cap \bar{C}$ ____________

o $A \cup (B \cup C)$ ____________ p $A \cap (B \cup C)$ ____________

q $(A \cap B) \cup C$ ____________ r $B \cap (\bar{A} \cap C)$ ____________

Question 3 On the given Venn diagrams shade the region represented by:

a $A \cup C$

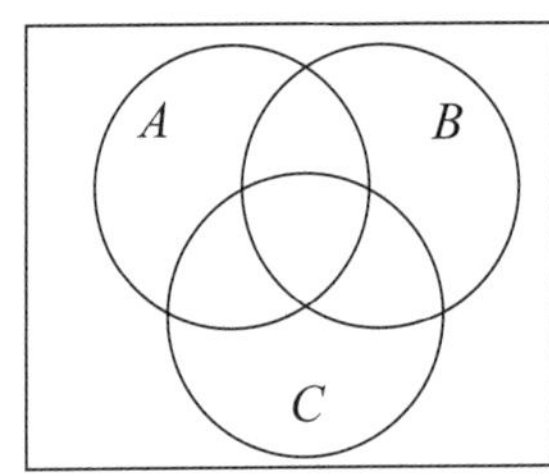

b $B \cap C$

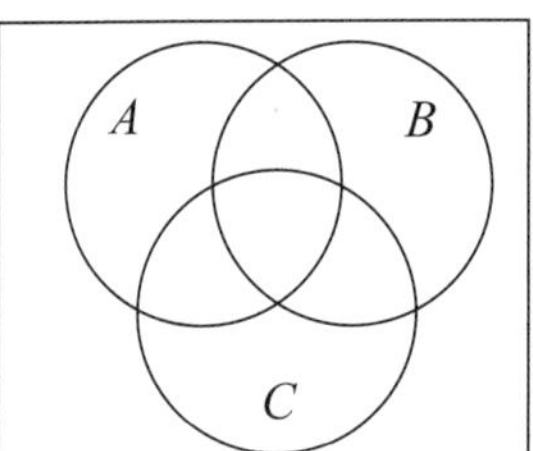

c $A \cup \bar{B}$

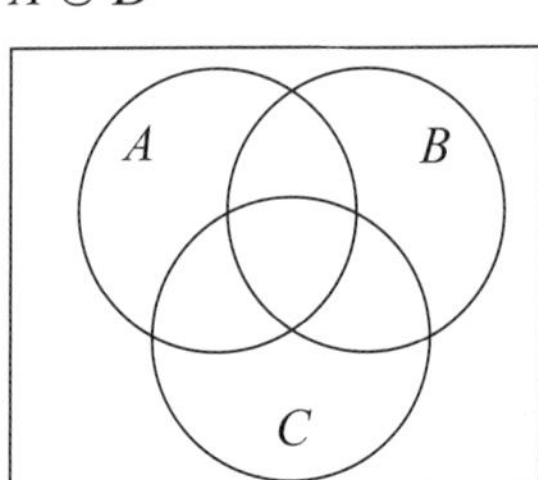

d $A \cap (B \cap C)$

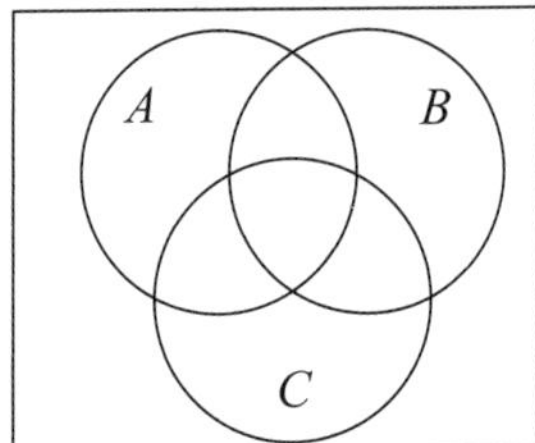

e $B \cup (A \cap C)$

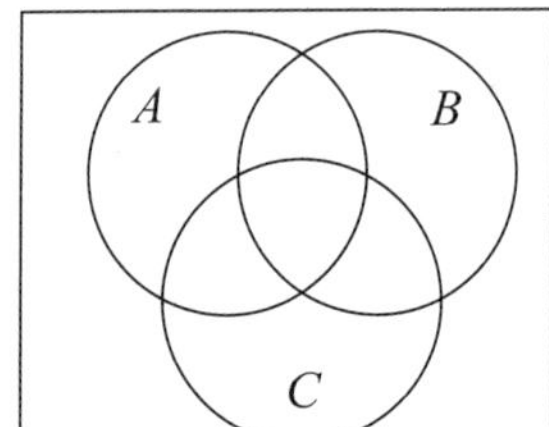

f $C \cap (A \cup B)$

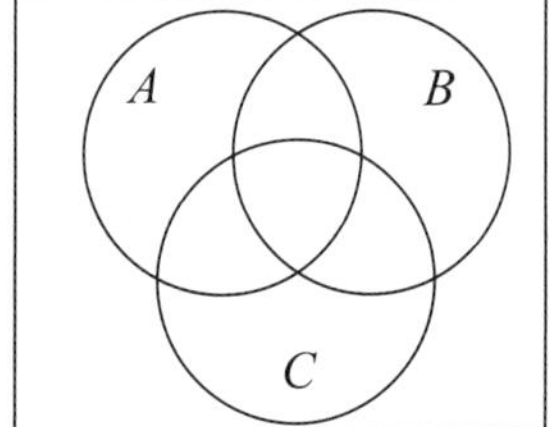

Probability

UNIT 7: Independent events

QUESTION 1 Complete.

Two events are independent if the outcome of the first event does ______________________ the outcome of the second event.

QUESTION 2 Determine whether the pair of events are dependent or independent.

a tossing two coins ______________

b tossing a coin and throwing a dice ______________

c taking and eating two jellybeans from a bowl, one after the other ______________

d selecting two cards from a pack, one after the other, with replacement ______________

e selecting two cards from a pack, one after the other, without replacement ______________

QUESTION 3 A dice is rolled and a coin is tossed. What is the probability of getting:

a a head on the coin? __________ **b** a 4 on the dice? __________ **c** a 4 and a head? __________

QUESTION 4 A coin is tossed three times. What is the probability that:

a the first toss is a head? ______________ **b** the second toss is a head? ______________

c the third toss is a head? ______________ **d** all three tosses are heads? ______________

QUESTION 5 A bag holds 5 red, 3 blue and 2 green counters. A counter is selected at random, its colour noted, and it is then replaced. A second counter is then selected at random. What is the probability that:

a the first counter is red? ______________ **b** the second counter is red? ______________

c both counters are red? ______________ **d** both counters are blue? ______________

______________ ______________

e both counters are green? ______________ **f** one is red and one is blue? ______________

g neither is green? ______________ **h** at least one is green? ______________

QUESTION 6 A basket holds 6 white, 4 black and 2 grey pegs. Three pegs are taken, one after the other, with replacement. What is the probability that:

a all the pegs are white? **b** all the pegs are black? **c** all the pegs are grey?

______________ ______________ ______________

______________ ______________ ______________

d none are black? **e** at least one is black? **f** at least one is grey?

______________ ______________ ______________

______________ ______________ ______________

Probability

UNIT 8: Dependent events

QUESTION 1 Complete.

Two events are dependent if the outcome of the first event ________________ the outcome of the second event.

QUESTION 2 Determine whether the pair of events are dependent or independent:

a throwing two dice __________

b spinning a spinner twice __________

c winning two prizes in a raffle __________

d taking two marbles from a bag, one after the other, without replacement __________

e taking two marbles from a bag, one after the other, with replacement __________

QUESTION 3 Jesse buys five tickets in a raffle. 1000 tickets are sold altogether. There are two prizes. A ticket is drawn for first prize and this ticket is discarded before a second ticket is drawn. What is the probability that Jesse wins:

a first prize? __________

b second prize if he didn't win first prize? __________

c second prize if he won first prize? __________

d both prizes? __________

QUESTION 4 A bag holds 5 red, 3 blue and 2 green counters. A counter is selected at random and is not replaced. A second counter is then selected at random. What is the probability that:

a the first counter is red? __________

b the second counter is also red? __________

c both counters are red? __________

d both counters are blue? __________

e both counters are green? __________

f one is red and one is blue? __________

g neither is green? __________

h at least one is green? __________

QUESTION 5 A basket holds 6 white, 4 black and 2 grey pegs. Three pegs are taken, one after the other, without replacement. What is the probability that:

a all the pegs are white?

____

b all the pegs are black?

____

c all the pegs are grey?

____

d none are black?

____

e at least one is black?

____

f at least one is grey?

____

Probability

UNIT 9: Multi-stage events (1)

QUESTION 1 A coin is tossed and a dice is thrown. What is the probability of getting:

a a 6 and a tail? ________________

b a 6 or a tail or both? ________________

c a number less than 5 and a head? ________________

d a head and an even number? ________________

QUESTION 2 Two fair dice are thrown. Use a table to find the probability that the sum of the 2 numbers thrown is:

a 10 ___________ **b** odd ___________

c even ___________ **d** a prime number ___________

e a multiple of 5 ___________ **f** greater than 9 ___________

QUESTION 3 A coin is tossed twice. What is the probability of getting a head and a tail in any order?

____

QUESTION 4 Three cards marked with the numbers 5, 6 and 7 are put in a box. Two cards are selected at random, one after the other, to form a two-digit number.

a How many different two-digit numbers can be formed? ___________

b What is the probability that the number formed is less than 67? ___________

c What is the probability that the number formed is divisible by 5? ___________

QUESTION 5 There are 3 children in a family. What is the probability of:

a there being 3 boys? ___________

b there being 1 boy and 2 girls? ___________

c the youngest child being a girl? ___________

d the eldest child being a girl? ___________

Probability

UNIT 10: Multi-stage events (2)

QUESTION **1** A team of four players (A, B, C and D) is to select a captain and a vice-captain.

a Write all the possible outcomes.

b Find the probability that player A will be either captain or vice-captain.

QUESTION **2** A poker machine has three wheels. The first wheel has the numbers 1, 2 and 3 on it. The other wheels each have the letters A, A and B on them. When the machine is played the wheels spin and line up randomly. The machine is played once. What is the probability of getting:

a 3 on the first wheel?

b 2BB?

c 1AA?

d AB or BA on the second and third wheels?

QUESTION **3** Three coins are tossed simultaneously. Find the probability of throwing:

a 3 heads

b 3 tails

c 3 heads or 3 tails

d 2 heads and 1 tail in any order

QUESTIONS **4** The probability of a cure with drug A is 0.6 and the probability of a cure with drug B is 0.8. If drug A is administered to one patient and drug B to another patient, what is the probability that neither patient will be cured?

QUESTON **5** The probability that a shooter will not hit a target in a single shot is 1 in 16. In a competition he fired 2 shots. Find the probability that both missed the target.

QUESTION **6** Clare Rainbow decides to have a holiday for 3 days at a resort. The probability of a day being sunny is 0.7 and the probability of a day being rainy is 0.3. Find the probability that Clare will have 3 sunny days for the holiday.

QUESTIN **7** Sharif buys 3 tickets in a raffle in which there is a total of 20 tickets. There are 2 prizes. Find the probability that he wins:

a the first prize

b the first prize only

c both prizes

d no prizes

e at least 1 prize

f 1 prize only

Probability

UNIT 11: Conditional statements

QUESTION **1** A fair dice is rolled.

a What is the probability that it shows 2?

b It is known that the number rolled is less than 4. What is the probability that it is 2?

c If the number was also greater than 1, what is the probability that it is 2?

d If the result was an even number less than 4, what is the probability that it is 2?

QUESTION **2** A card is chosen at random from a standard pack of playing cards. It is a picture card. What is the probability that the card is:

a a queen?

b the king of spades?

c a black jack?

QUESTION **3** There are three pens in a box. Two are black and the other is blue. Two pens are chosen, one after the other, without replacement. What is the probability that:

a the first pen is black?

b both pens are black?

c the pens are different colours?

It is known that the first pen was black. What is the probability that:

d the second pen is black?

e both pens are black?

f the pens are different colours?

QUESTION **4** There are 5 red, 4 blue and 3 green balls in a bag. Without looking, two balls are taken from the bag one after the other.

If the first ball is replaced before the second one is taken, what is the probability that:

a both balls are red?

b both balls are green?

c at least one ball is green?

If the first ball is not replaced before the second one is taken, what is the probability that:

d both balls are red?

e both balls are green?

f at least one ball is green?

If the first ball is not replaced and it was not green, what is the probability that:

g both balls are red?

h both balls are green?

i at least one ball is green?

QUESTION **5** There are 500 tickets sold in a raffle. 200 tickets are green, 120 are blue and the rest are white. Ken has 10 tickets and they are all white. The first prize is drawn and it is white. What is the probability that Ken wins first prize?

Probability

UNIT 12: Mistakes and misconceptions

QUESTION 1 'On any day there might be rain or there might not be any rain. Therefore there is a 50-50 chance of rain on any day.'

a Is this statement correct? ____________

b Explain why or why not. ____

____

QUESTION 2 There are four children in a family and they are all boys. A fifth baby is expected. 'Because having five boys in a family is very unusual, the next baby is more likely to be a girl than a boy.'

a Is this statement correct? ____________

b Explain why or why not. ____

____

QUESTION 3 'If I randomly choose a letter from the alphabet, there is a 1 in 26 chance that it will be x.'

a Is this statement correct? ____________

b Explain why or why not. ____

____

QUESTION 4 'If I open a book and randomly choose a letter from that page, there is a 1 in 26 chance that it will be x.'

a Is this statement correct? ____________

b Explain why or why not. ____

____

QUESTION 5 A fair coin is tossed five times and shows tails each time. It is tossed a sixth time. 'It has a greater chance of being a tail than a head.'

a Is this statement correct? ____________

b Explain why or why not. ____

____

QUESTION 6 Bill wanted to know the probability of getting rain on at least one of the next three days. He looked on the internet and found that there was a 10% chance of rain on each of the days. He multiplied $0.1 \times 0.1 \times 0.1$ and concluded that the chance of 0.1% meant that there was a very, very small chance of rain on at least one of the days.

a What is the correct chance of getting rain on at least one of the three days?

____

____

b Briefly comment on what Bill was doing wrong. ____

____

____

Probability

TOPIC TEST **PART A**

Instructions
- This part consists of 10 multiple-choice questions.
- Fill in only ONE CIRCLE for each question.
- Each question is worth 1 mark.

Time allowed: 15 minutes **Total marks: 10**

Marks

1 If two coins are tossed together, what is the probability of two tails? [1]

Ⓐ $\frac{1}{2}$ Ⓑ $\frac{1}{3}$ Ⓒ $\frac{1}{4}$ Ⓓ $\frac{1}{6}$

2 A fair dice is rolled. What is the probability of getting a 6 or a 1? [1]

Ⓐ $\frac{1}{2}$ Ⓑ $\frac{1}{3}$ Ⓒ $\frac{1}{4}$ Ⓓ $\frac{1}{6}$

3 There are 5 red, 3 blue and 2 green balls in a box. Two balls are taken, one after the other, without replacement. What is the probability that the balls are both red? [1]

Ⓐ $\frac{1}{2}$ Ⓑ $\frac{1}{3}$ Ⓒ $\frac{1}{4}$ Ⓓ $\frac{2}{9}$

4 There are 5 red, 3 blue and 2 green balls in a box. Two balls are taken, one after the other, with replacement. What is the probability that the balls are both red? [1]

Ⓐ $\frac{1}{2}$ Ⓑ $\frac{1}{3}$ Ⓒ $\frac{1}{4}$ Ⓓ $\frac{2}{9}$

5 The probability that a basketball player scores a goal from the free-throw line is 0.3. What is the probability that the player gets 2 goals from 2 free throws? [1]

Ⓐ 0.3 Ⓑ 0.6 Ⓒ 0.06 Ⓓ 0.09

6 A card is taken from a standard pack of playing cards. It is a club. What is the probability that it is a ten? [1]

Ⓐ $\frac{1}{13}$ Ⓑ $\frac{1}{52}$ Ⓒ $\frac{4}{13}$ Ⓓ $\frac{1}{4}$

7 A fair coin is tossed five times. It shows heads each time. How could you describe the probability that it shows heads on a sixth toss? [1]

Ⓐ unlikely Ⓑ likely Ⓒ fifty-fifty Ⓓ certain

8 A dice is thrown twice. What is the probability of getting at least one six? [1]

Ⓐ $\frac{1}{36}$ Ⓑ $\frac{1}{6}$ Ⓒ $\frac{11}{36}$ Ⓓ $\frac{1}{4}$

9 A box holds 2 blue and 1 red pen. One pen is taken at random, used and replaced. A second pen is then taken at random. What is the probability that both pens were red? [1]

Ⓐ 0 Ⓑ $\frac{1}{3}$ Ⓒ $\frac{1}{9}$ Ⓓ $\frac{2}{9}$

10 Bella buys five tickets in a raffle. One hundred tickets are sold and there are two prizes. What is the probability that Bella wins both prizes? [1]

Ⓐ $\frac{1}{250}$ Ⓑ $\frac{1}{495}$ Ⓒ $\frac{1}{500}$ Ⓓ $\frac{2}{495}$

Total marks achieved for PART A

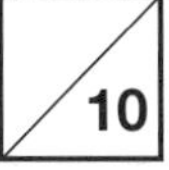

Probability

TOPIC TEST **PART B**

Instructions
- This part consists of 4 questions.
- Write only the answer in the answer column.
- For any working use the question column.

Time allowed: 20 minutes **Total marks: 15**

Questions	Answers	Marks
1 In an experiment, a card is drawn from a pack of playing cards and a coin is tossed. What is the probability of getting:		
a an ace and a head? ______	______	1
b the queen of hearts and a tail? ______	______	1
2 A box contains three white and seven red balls. A ball is drawn from the box and is not replaced. Then a second ball is drawn. Find the probability of drawing:		
a red then white ______	______	1
b white then red ______	______	1
c 2 white balls ______	______	1
d 2 red balls ______	______	1
e a white and red in any order ______	______	1
f at least one red ______	______	1
3 A coin is tossed 3 times. What is the probability of:		
a 3 tails? ______	______	1
b at least one head? ______	______	1
If the first toss was a tail, what is the probability of:		
c 3 tails? ______	______	1
d at least one head? ______	______	1
4 Three dice are thrown together.		
a What is the probability of three 6s? ______	______	1
It is known that all the tosses produced numbers greater than 3. Kathy said, incorrectly, that the probability of three 6s will be twice what it was before.		
b What is the correct probability? ______	______	1
c Briefly explain why Kathy is wrong. ______	______	1

Total marks achieved for PART B 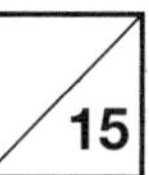

Exam Paper 1

Instructions for all parts • Attempt all questions. **Total marks: 100**
Time allowed: $1\frac{1}{2}$ hours • Allow about 45 minutes for each part.

EXAM PAPER 1 PART A

Fill in only one circle for each question.

Marks

1 Which one of the following is **not** equal to $3x$?

(A) $x + x + x$ (B) $3 \times x$ (C) $4x - x$ (D) $3x^2 - x$

1

2 Simplify $4^2 \times 4^3$.
(A) 4^5 (B) 4^6 (C) 16^5 (D) 16^6

1

3 Claudette wrote the following lines of working to solve this equation: $4x - 8 = 15$.
Line 1: $4x = 15 + 8$; Line 2: $4x = 23$; Line 3: $x = \frac{23}{4}$; Line 4: $x = 5\frac{3}{23}$
In which line did she make an error?
(A) Line 1 (B) Line 2 (C) Line 3 (D) Line 4

1

4 Simplify $-6a + 8b - 3a - 6b$.
(A) $7ab$ (B) $-9a + 2b$ (C) $-3a - 2b$ (D) $-6a + 2b$
1

5 A 24-cm length of wire is bent to form a rectangle. If the width of the rectangle is 5 cm, find its area.
(A) 35 cm² (B) 25 cm² (C) 27 cm² (D) 49 cm²
1

6 What is the median of this set of scores?
(A) 3 (B) 4
(C) 5 (D) 6

Score	Frequency
3	1
4	2
5	8
6	4

1

7 What is the area of triangle BEC?
(A) 289 cm² (B) 225 cm²
(C) 127.5 cm² (D) 60 cm²

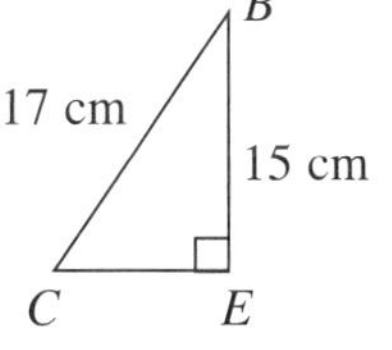

1

8 Given that $v = u + at$, $v = 10.8$, $u = 8.3$ and $a = 4.2$, find the value of t correct to two significant figures.
(A) 0.50 (B) 0.59 (C) 0.595 (D) 0.60
1

9 If $2x - 3 = 31$, we know that x equals:
(A) 7 (B) 14 (C) 17 (D) 34
1

10 Which triangles are congruent?
(A) I and II only (B) I and III only
(C) II and III only (D) I, II and III

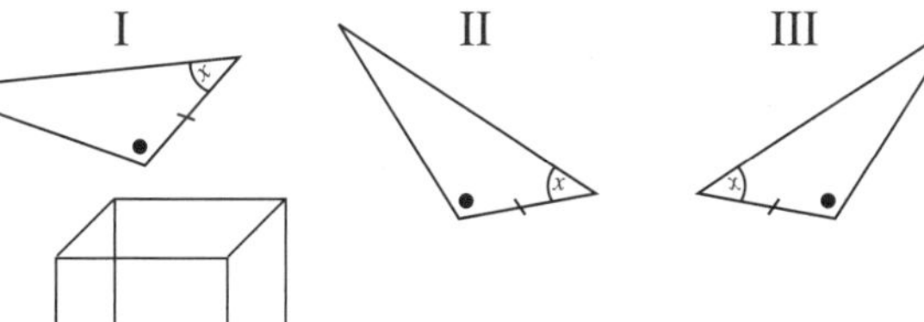

1

11 What is the surface area of this cube?
(A) 36 cm² (B) 72 cm²
(C) 108 cm² (D) 216 cm²

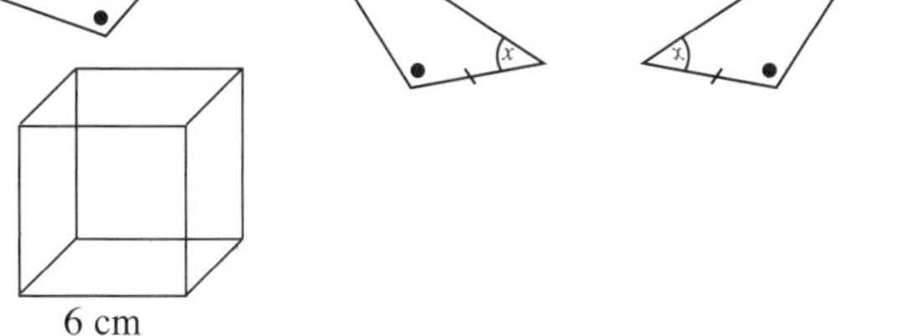

1

12 Which expression is **not** equal to $8x$?
(A) $x \times x \times x \times x \times x \times x \times x \times x$ (B) $12x - 4x$
(C) $16x \div 2$ (D) $x + x + x + x + x + x + x + x$
1

Continued on the next page

EXAM PAPER 1 PART A

Fill in only one circle for each question.

Marks

13 For the set of scores 60, 70, 80, 60, 90, find the difference between the mean and the mode.
Ⓐ 10 Ⓑ 12 Ⓒ 20 Ⓓ 30 

14 In the diagram $AB = BC = CA$ and $\angle CDE$ is a right angle.
What is the value of x?

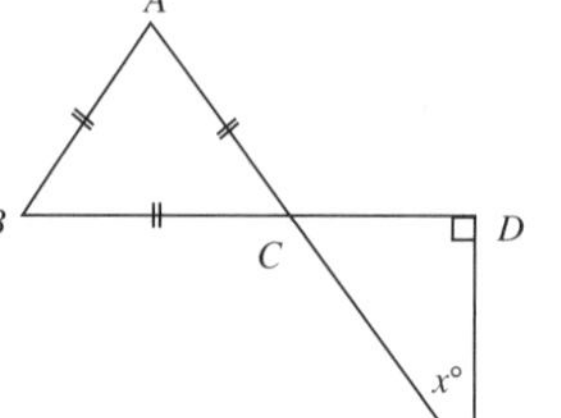

Ⓐ 30 Ⓑ 40
Ⓒ 45 Ⓓ 60 

15 Simplify $\frac{m^8 n}{m^4 n^4}$.
Ⓐ $m^4 n$ Ⓑ $\frac{m^4}{n^3}$ Ⓒ $\frac{m^4}{n^4}$ Ⓓ $m^4 n^3$ 

16 The solution of $x - 3 > 2$ is:
Ⓐ $x < 1$ Ⓑ $x > 1$ Ⓒ $x < 5$ Ⓓ $x > 5$ 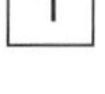

17 Which of the following is a linear equation?
Ⓐ $y = x^2 - 4$ Ⓑ $y^2 = \frac{x}{2}$ Ⓒ $y = 8 - 3x$ Ⓓ $y = \sqrt{x + 7}$ 1

18 If $\sin \theta = \frac{1}{2}$, find the size of angle θ.
Ⓐ 60° Ⓑ 50° Ⓒ 45° Ⓓ 30° 1

19 Solve the equation $2x + 5 = 85$.
Ⓐ $x = 8$ Ⓑ $x = 30$ Ⓒ $x = 40$ Ⓓ $x = 45$ 

20 A pipe has an inner radius of 10 cm and an outer radius of 20 cm.
The shaded area in square centimetres is given by:

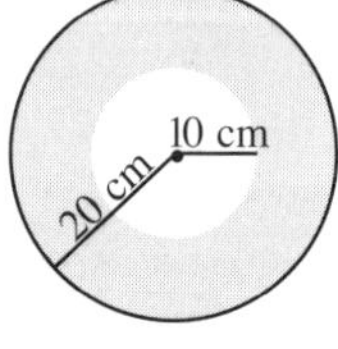

Ⓐ 100π Ⓑ 200π
Ⓒ 300π Ⓓ 400π 1

21 The nine letters of the word AUSTRALIA are written on separate cards and placed in a bag. One card is chosen at random. What is the probability of choosing A or R or T?
Ⓐ $\frac{1}{3}$ Ⓑ $\frac{4}{9}$ Ⓒ $\frac{5}{9}$ Ⓓ $\frac{2}{3}$ 1

22 Find which of the following expressions has a value of 7.
Ⓐ $7 + 7 \div 7$ Ⓑ $-(-7)^0$ Ⓒ $\frac{3}{21}$ Ⓓ $7^7 \div 7^6$ 1

23 In a single throw of a dice, the probability of rolling an odd number is:
Ⓐ $\frac{1}{5}$ Ⓑ $\frac{1}{4}$ Ⓒ $\frac{1}{3}$ Ⓓ $\frac{1}{2}$ 1

24 $(x + 5)(x - 2) =$
Ⓐ $x^2 + 7x - 10$ Ⓑ $x^2 - 7x + 10$ Ⓒ $x^2 - 3x - 10$ Ⓓ $x^2 + 3x - 10$ 1

25 Consider the scores 1, 3, 7, 8, 10, 13, 15, 16, 17, 20, 22, 23, 28. Find the interquartile range.
Ⓐ 13.5 Ⓑ 14 Ⓒ 14.5 Ⓓ 15 1

Continued on the next page

Fill in only one circle for each question.

Marks

26 Which expression will give the value of x?

Ⓐ $9 \sin 24°$ Ⓑ $\dfrac{9}{\sin 24°}$

Ⓒ $9 \tan 24°$ Ⓓ $\dfrac{9}{\tan 24°}$

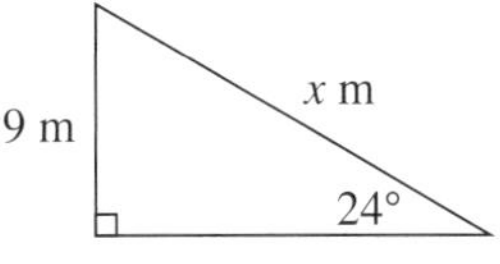

27 $\dfrac{5x}{12} \times \dfrac{3x}{10} =$

Ⓐ $\dfrac{x^2}{8}$ Ⓑ $\dfrac{25x^2}{4}$ Ⓒ $\dfrac{x^2}{4}$ Ⓓ $\dfrac{8x^2}{15}$

28 Which is correct?

Ⓐ $x = 73$ and $y = 79$ Ⓑ $x = 107$ and $y = 79$

Ⓒ $x = 73$ and $y = 101$ Ⓓ $x = 107$ and $y = 101$

73° 79° x° y°

29 $3a^2b^4 \times 2a^3b^2 =$

Ⓐ $5a^5b^6$ Ⓑ $6a^5b^6$ Ⓒ $5a^6b^8$ Ⓓ $6a^6b^8$

30 Which is closest to the curved surface area of a cylinder of height 35 cm and diameter 18 cm?

Ⓐ 1980 cm^2 Ⓑ 3960 cm^2 Ⓒ 990 cm^2 Ⓓ 8900 cm^2

31 Three coins are tossed together. Find the probability of at least one tail.

Ⓐ $\frac{1}{8}$ Ⓑ $\frac{1}{2}$ Ⓒ $\frac{3}{4}$ Ⓓ $\frac{7}{8}$

32 A solution to the equation $2x^2 - 32 = 0$ is:

Ⓐ $x = 2$ Ⓑ $x = -4$ Ⓒ $x = 8$ Ⓓ $x = 16$

33 These two triangles are:

Ⓐ similar but not congruent   Ⓑ congruent but not similar

Ⓒ neither congruent nor similar   Ⓓ both congruent and similar

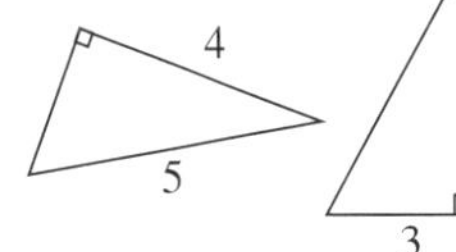

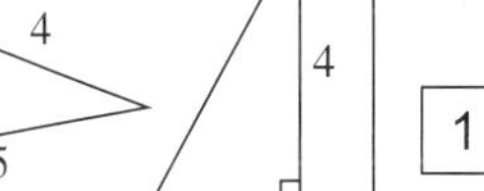

34 Find the equation of a circle, centre the origin, radius 9 units.

Ⓐ $x^2 + y^2 = 9$ Ⓑ $x^2 + y^2 = 3$ Ⓒ $x^2 + y^2 = 81$ Ⓓ $(x + y)^2 = 9$

35 Find the solution to $-x \leq -2$ when graphed on the number line.

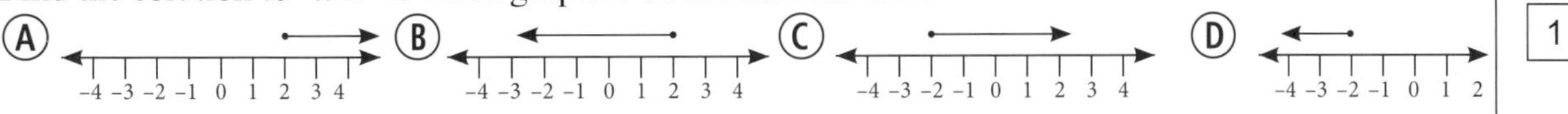

36 $a^2 - 6a + 5 =$

Ⓐ $(a - 2)(a - 3)$ Ⓑ $(a - 3)(a + 2)$ Ⓒ $(a - 1)(a + 5)$ Ⓓ $(a - 5)(a - 1)$

37 How could you describe the relationship shown by this scatter plot?

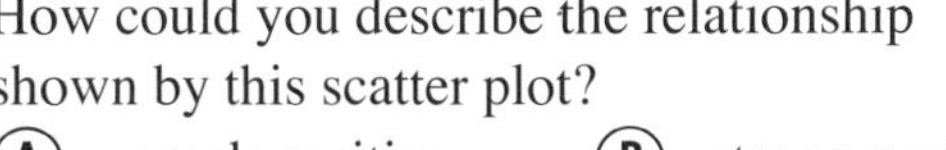

Ⓐ weak positive   Ⓑ strong positive   Ⓒ weak negative   Ⓓ strong negative

38 Which of these lines is parallel to $y = 5 - x$?

Ⓐ $y = 5x + 3$ Ⓑ $y = x + 3$ Ⓒ $y = -x + 3$ Ⓓ $y = 5 - 3x$

Continued on the next page

EXAM PAPER 1 PART A

Fill in only one circle for each question.

Marks

39 $x =$ 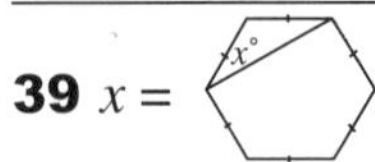

Ⓐ 30 Ⓑ 36 Ⓒ 40 Ⓓ 45 | 1

40 A rectangular prism is 4 m long, 2 m wide and 90 cm high. Find its volume.

Ⓐ 7.2 m^3 Ⓑ 720 m^3 Ⓒ 720 cm^3 Ⓓ 7200 cm^3 | 1

41 Which could be the graph of $y = -3^x$?

Ⓐ 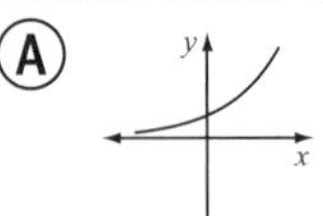 Ⓑ Ⓒ 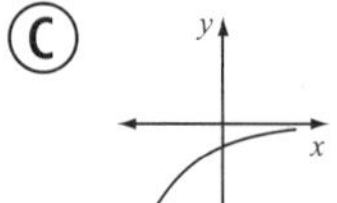 Ⓓ 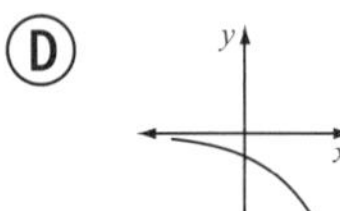 

42 If $s = 4\pi r^2$, which is closest to the value of s when $r = 14$?

Ⓐ 3095 Ⓑ 7738 Ⓒ 47 Ⓓ 2463 | 1

43 This box plot was drawn for a set of scores. The middle 50% of scores are between:

Ⓐ 11 and 14 Ⓑ 11 and 18 Ⓒ 14 and 18 Ⓓ 14 and 22 | 1

44 Which is **not** the equation of a parabola?

Ⓐ $y = 9 - x^2$ Ⓑ $y = x^2 + 9$ Ⓒ $y^2 = x^2$ Ⓓ $y = 2x^2 - 1$ | 1

45 Find the simultaneous solution of the equations $y = 2x + 1$ and $y = 3x - 1$.

Ⓐ $x = 1$ and $y = 3$ Ⓑ $x = 1$ and $y = 2$ Ⓒ $x = 2$ and $y = 5$ Ⓓ $x = 2$ and $y = 3$ | 1

46 $\left(\frac{6x^{-4}}{3x^2}\right)^{-1} =$

Ⓐ $\frac{x^6}{2}$ Ⓑ $\frac{1}{2x^6}$ Ⓒ $\frac{x^2}{2}$ Ⓓ $\frac{1}{2x^2}$

47 $x =$

Ⓐ $7\frac{1}{2}$ Ⓑ $8\frac{1}{3}$ Ⓒ 9 Ⓓ $9\frac{3}{4}$ | 1

48 Find the solution to $\frac{x}{3} + \frac{x}{4} = 12$.

Ⓐ $x = 7$ Ⓑ $x = 12$ Ⓒ $x = 20\frac{4}{7}$ Ⓓ $x = 42$ | 1

49 Which of the following shapes does **not** have an area of 80 cm^2?

Ⓐ 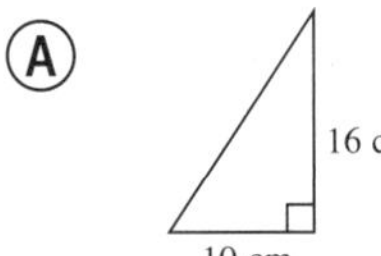

Ⓑ 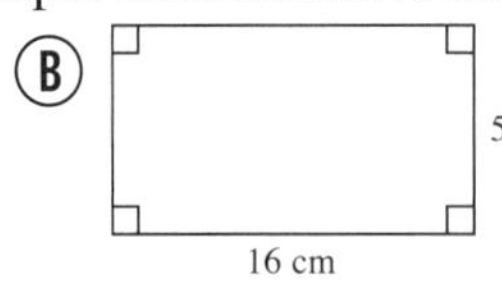

Ⓒ 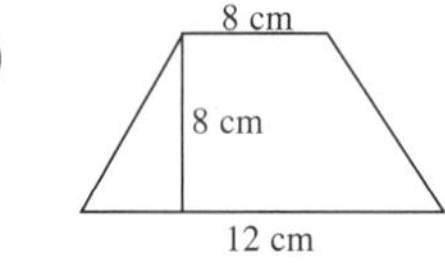

Ⓓ 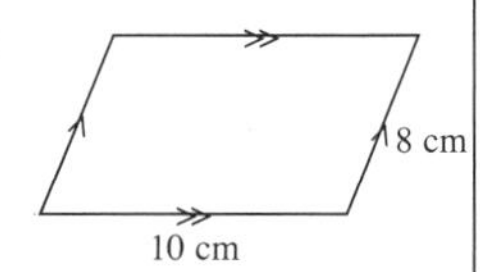

50 In which of the following is x **not** necessarily equal to 45?

Ⓐ 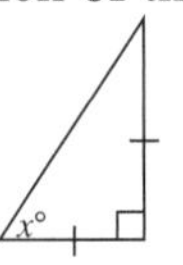

Ⓑ 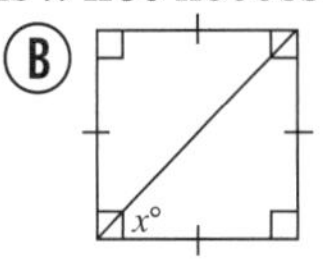

Ⓒ 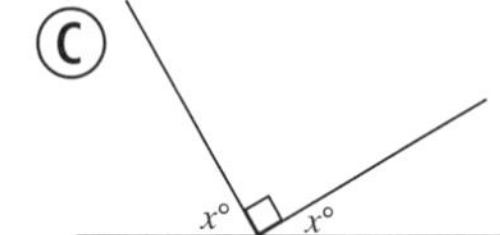

Ⓓ 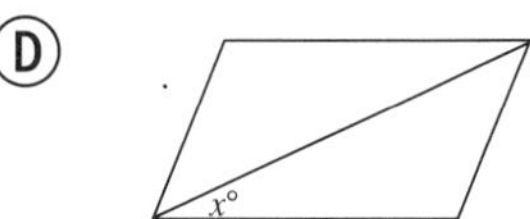

Total marks achieved for PART A /50

EXAM PAPER 1 PART B

Show all working for each question.

Marks

1 **a** Find the simple interest earned if \$7000 is invested at 6% p.a. for 3 years. 1

b Find the amount, to the nearest dollar, to which \$7000 would accumulate if invested at 6% interest compounded annually for 3 years. 1

c How much more interest was earned with compound interest than with simple interest? 1

2 There are 5 red, 4 blue and 3 white balls in a bag. Two balls are taken from the bag, one after the other, without looking. Find the probability of getting 2 white balls if the balls are:

a not replaced 1

b replaced 1

3 Find:

a $\frac{x}{2} + \frac{x}{3}$ 1

b $\frac{3a}{5} - \frac{2a}{3}$ 1

4 Given the formula $S = V(1 - r)^n$, find:

a S when $V = 25\,000$, $r = 0.2$ and $n = 4$ 1

b V if $S = 28\,900$, $r = 0.15$ and $n = 2$ 1

Continued on the next page

EXAM PAPER 1 PART B

Show all working for each question.

Marks

5 The area of the triangular face of this prism is 2150.5 cm^2.

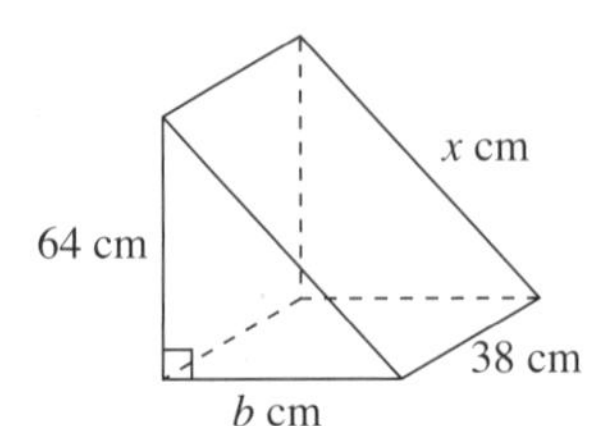

a Find the length, b cm, of the base of the triangle. 1

b Find the length, x cm, of the hypotenuse. 1

c Find the surface area of the prism. 1

6 Factorise fully.

a $2x^3y^4 - 6x^4y^2$ 1

b $x^2 - 4x + 3$ 1

c $a^2 + 7a - 18$ 1

d $x^2 - 36$ 1

e $m^2 + 5m - mn - 5n$ 1

7 Q is 56 km from P on a bearing of 064°. R is 31 km from P on a bearing of 154°.

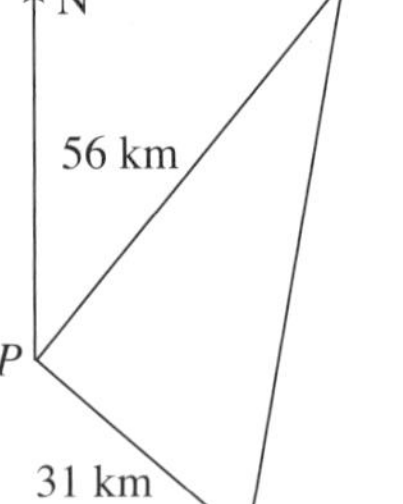

a Find the compass bearing of P from Q. 1

b Find the size of $\angle QPR$. 1

c Find the size of $\angle PQR$ to the nearest degree. 1

d Find the true bearing of R from Q. 1

e Find the distance to the nearest kilometre from Q to R. 1

Continued on the next page

EXAM PAPER 1 PART B

Show all working for each question.

Marks

8 Solve.

a $\dfrac{5x+4}{3} + \dfrac{3x+5}{4} = 5$

b $3x^2 = 75$

1

1

9 Solve $3x + 2y = 11$ and $2x - y = 12$ simultaneously.

10 a Solve $12 - 5x \geq 2$

b Graph the solution on the number line provided.

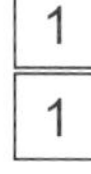

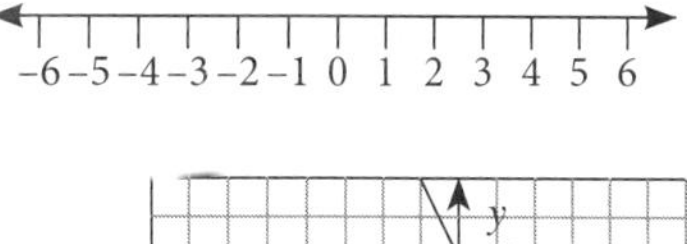

11 a Find the gradient of line l.

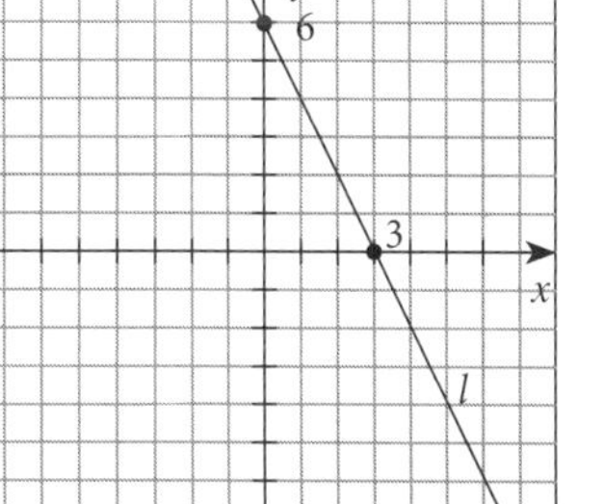

b Find the gradient of any line parallel to l.

1

c Find the gradient of any line perpendicular to l.

1

d Line m is perpendicular to line l and intersects it on the x-axis. Graph line m on the diagram.

1

e Find the equation of line m. ______________

1

f Find the area of the triangle formed by lines l and m and the y-axis.

1

Continued on the next page

EXAM PAPER 1 PART B

Show all working for each question.

Marks

12 Expand.

a $(a + 5)(a + 4)$

b $(2x - 1)(3x + 5)$

13 This water trough is in the shape of a half cylinder. The width of the trough is 40 cm and the length is 2.6 m.

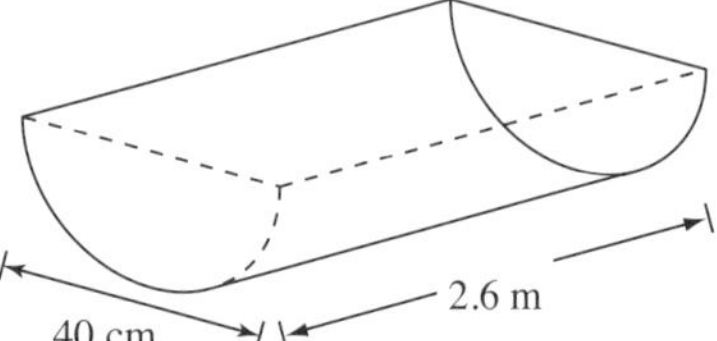

a Find the area of the semicircular cross-section. Give the answer in square metres correct to two decimal places.

1

b Find the volume of water the trough will hold to the nearest 10 litres. ($1\ m^3 = 1000$ L)

14 $ABCD$ is a square. $AP = BQ = DR$.

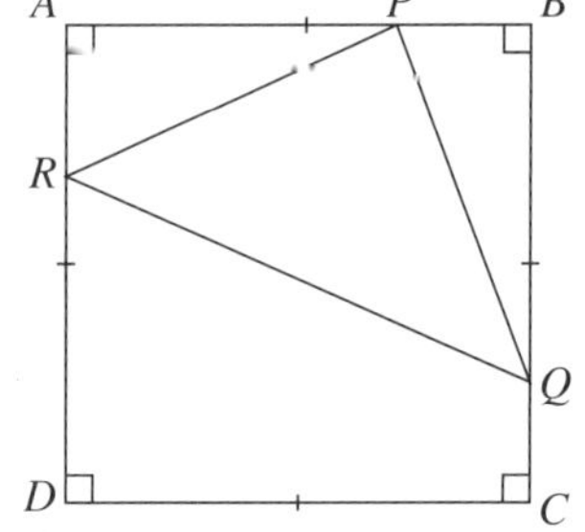

a Explain why $AR = PB$.

1

b Which test shows that $\Delta RAP \equiv \Delta PBQ$?

c Find the size of $\angle RPQ$.

d Find the size of $\angle PQR$.

Continued on the next page

Show all working for each question.

Marks

15 This graph shows the estimated population, P, of birds on an island at time t, where t is the time in years since 1960.

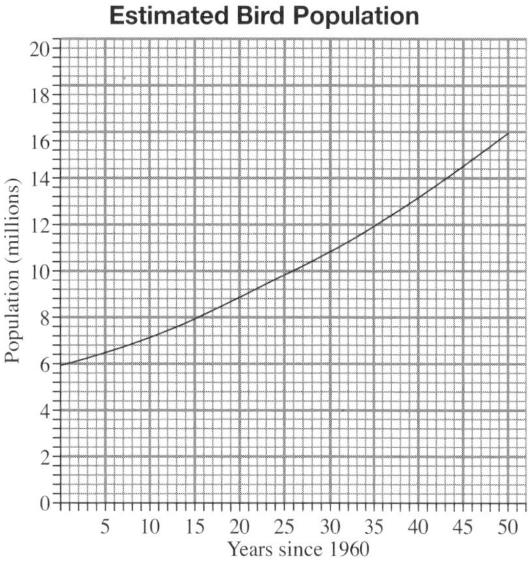

Estimated Bird Population

a What was the estimated population in 1995? ____________ 1

b When was the population 8 million? ____________ 1

c How much did the population increase between 1960 and 2000? ____________ 1

d What would you predict the population would be in 2020? ____________ 1

e What type of graph is this? ____________ 1

16 This dot plot was drawn to show some scores in a game.

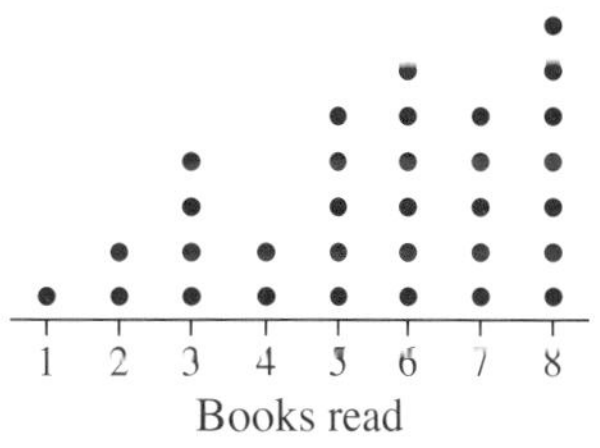

a Find the median

____________ 1

b Find the interquartile range.

____

____________ 1

c Draw a box plot (using the scale below) to illustrate the game scores.

1 2 3 4 5 6 7 8 1

d Briefly comment on the shape of both dot plot and box plot.

____

____

____________ 1

Total marks achieved for PART A /50

Exam Paper 2

Instructions for all parts • Attempt all questions. **Total marks: 100**
Time allowed: $1\frac{1}{2}$ hours • Allow about 45 minutes for each part.

EXAM PAPER 2 PART A

Fill in only one circle for each question.

Marks

1 $a^2 + a^2 =$
(A) a^4 (B) $2a^2$ (C) $2a^4$ (D) a — 1

2 If $5x - 7 = 88$, find the value of x.
(A) $\frac{81}{5}$ (B) $12\frac{4}{7}$ (C) 19 (D) 17 — 1

3 Which pair of values satisfies the equation $x + y = 7$ and $x - y = 3$?
(A) $x = 5, y = -2$ (B) $x = 5, y = 2$ (C) $x = -5, y = 2$ (D) $x = -5, y = -2$ — 1

4 What is the value of x?
(A) 65 (B) 125
(C) 55 (D) 110 — 1

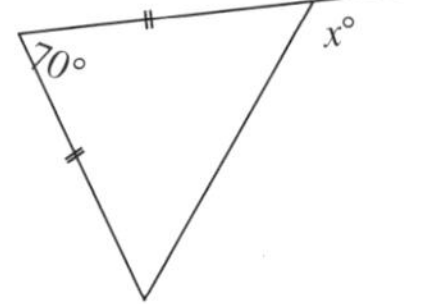

5 Which expression does **not** equal $4m$?
(A) $m \times m \times m \times m$ (B) $4 \times m$ (C) $5m - m$ (D) $m + m + m + m$ — 1

6 Calculate the area of the rhombus.
(A) 48 cm² (B) 24 cm²
(C) 40 cm² (D) 30 cm² — 1

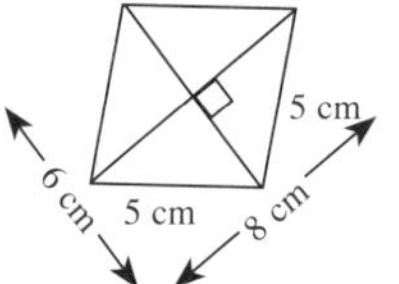

7 Which of the following is equal to m^4?
(A) $4m$ (B) $m + m + m + m$ (C) $4m^2$ (D) $m \times m \times m \times m$ — 1

8 The nine letters of the word FANTASTIC are written on separate cards and placed in a box. One card is chosen at random. Find the probability of selecting the letter A or the letter T.
(A) $\frac{1}{9}$ (B) $\frac{2}{9}$ (C) $\frac{3}{9}$ (D) $\frac{4}{9}$ — 1

9 AC and BD are straight lines. Find the value of y.
(A) 15° (B) 18°
(C) 25° (D) 28° — 1

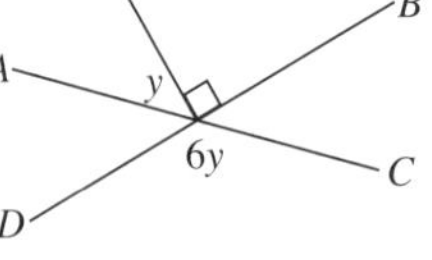

10 $-12x + x - 5x =$
(A) $-16x$ (B) $16x$ (C) $-8x$ (D) $8x$ — 1

11 The area of this shape is closest to:
(A) 249.1 cm² (B) 274.3 cm²
(C) 324.5 cm² (D) 236.6 cm² — 1

12 Which statement is correct for the diagram?
(A) $\sin\theta = \frac{12}{13}$ (B) $\sin\theta = \frac{5}{13}$
(C) $\cos\theta = \frac{5}{13}$ (D) $\tan\theta = \frac{12}{5}$ — 1

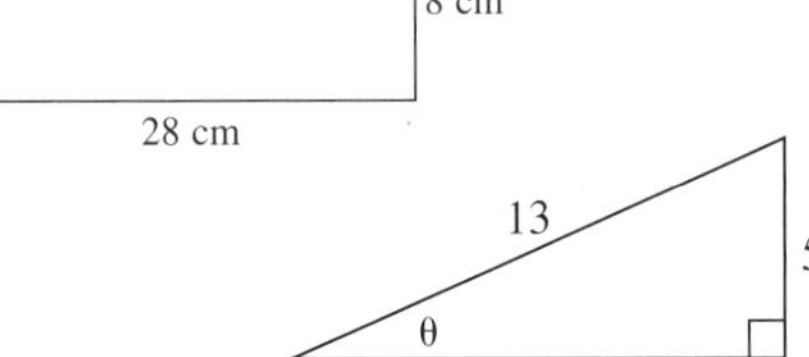

Continued on the next page

EXAM PAPER 2 PART A

Fill in only one circle for each question.

Marks

13 Solve for x: $8(x - 1) = 3x + 32$

(A) $x = 2.5$ (B) $x = 8$ (C) $x = 6.2$ (D) $x = 4.8$

14 $x =$

(A) 60 (B) 72

(C) 80 (D) 82

15 Which congruence test could be used to show the following pair of triangles are congruent?

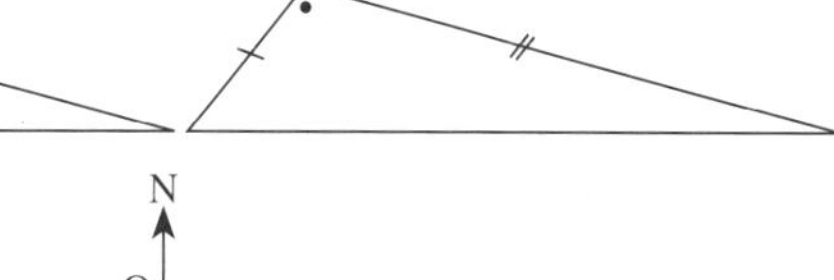

(A) SSS (B) AAS

(C) SAS (D) RHS

16 In this diagram, the true bearing of P from O is:

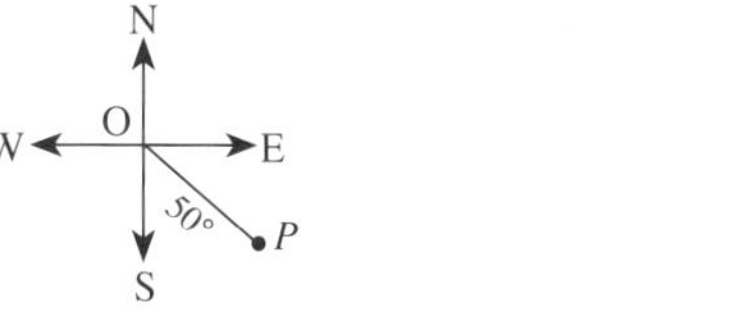

(A) 040°T (B) 050°T

(C) 130°T (D) 320°T

17 A bag contains 5 blue, 6 white and 9 black balls. If a ball is drawn at random, find the probability that it is either black or white.

(A) $\frac{7}{10}$ (B) $\frac{11}{20}$ (C) $\frac{3}{4}$ (D) $\frac{4}{3}$

18 Which is correct?

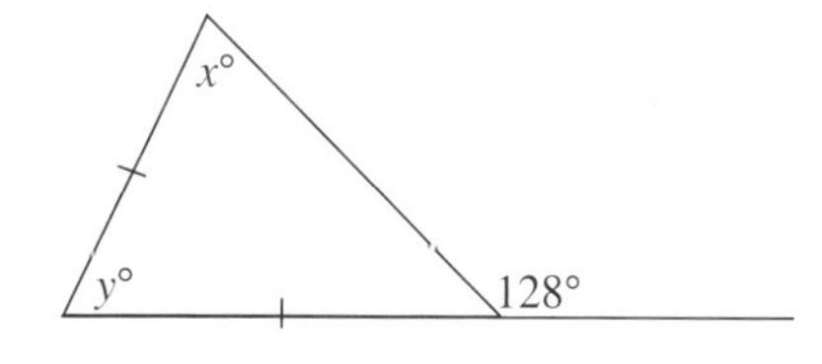

(A) $x = 52°$ $y = 52°$ (B) $x = 76°$ $y = 52°$

(C) $x = 52°$ $y = 76°$ (D) $x = 76°$ $y = 76°$

19 In ΔABC, sides AC and BC are equal and side AB is shorter than side AC. Which statement is true?

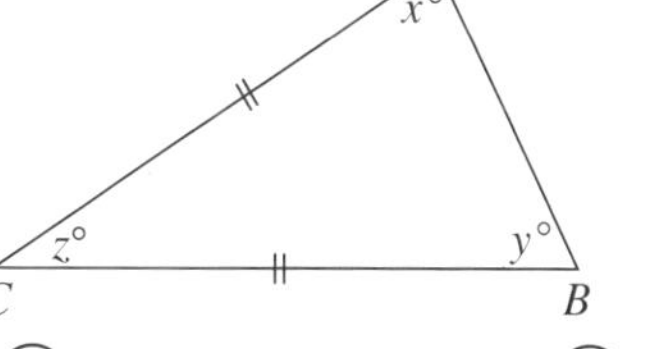

(A) $x = y$ (B) $x = z$

(C) $y = z$ (D) $x = y = z$

20 If $\frac{2x + 1}{5} = 7$, find the value of x.

(A) 18 (B) 17 (C) 16 (D) 15

21 Solve $4 - t \geq 12$.

(A) $t \geq 8$ (B) $t \geq -8$ (C) $t \leq -8$ (D) $t \leq 8$

22 $2x^{-2} = ?$

(A) $\frac{4}{x}$ (B) $\frac{1}{2x^2}$ (C) $\frac{1}{4x^2}$ (D) $\frac{2}{x^2}$

23 The area of a parallelogram is given by Area = base × perpendicular height. The height of the parallelogram is increased by 20% and its base is decreased by 20%. What fraction is the new area of the original area?

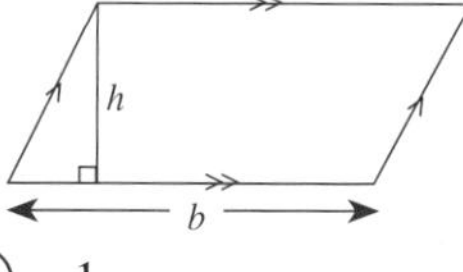

(A) $\frac{4}{5}$ (B) $\frac{24}{25}$ (C) $\frac{6}{5}$ (D) 1

24 A rhombus is drawn with angles as shown. Find the value of x.

(A) 40 (B) 37

(C) 74 (D) 80

Continued on the next page

EXAM PAPER 2 PART A

Fill in only one circle for each question.

Marks

25 This is a parallelogram with one of its diagonals drawn.
Which of the following statements is true?

Ⓐ Each diagonal bisects the angle through which it passes.
Ⓑ The diagonals are equal.
Ⓒ The diagonals bisect each other.
Ⓓ The diagonals meet at right angles.

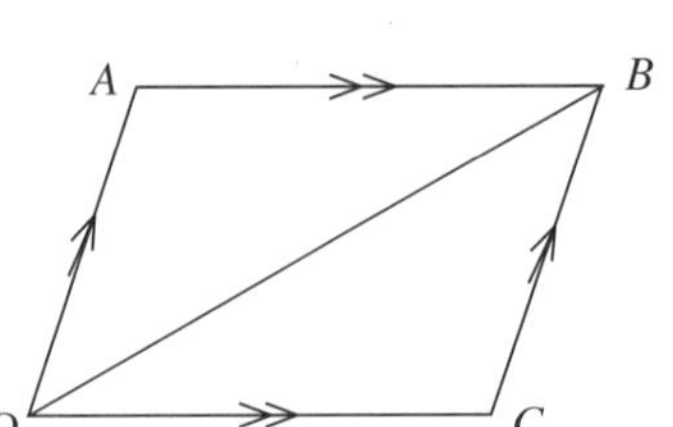

1

26 $(x-3)(x-8) =$

Ⓐ $x^2 - 5x - 24$ Ⓑ $x^2 - 5x + 24$ Ⓒ $x^2 - 11x - 24$ Ⓓ $x^2 - 11x + 24$ 1

27 The compound interest earned when \$2000 is invested at 9% p.a. for 3 years is closest to:

Ⓐ \$540 Ⓑ \$590 Ⓒ \$2540 Ⓓ \$2590 1

28 $\frac{2x}{3} + \frac{x}{5} =$

Ⓐ $\frac{2x}{15}$ Ⓑ $\frac{3x}{8}$ Ⓒ $\frac{2x}{3}$ Ⓓ $\frac{13x}{15}$ 1

29 Which could be the graph of $y = 1 - x^2$?

Ⓐ 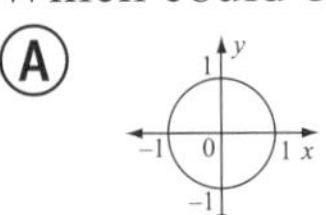 Ⓑ 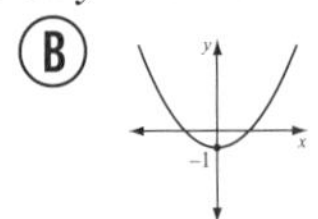 Ⓒ 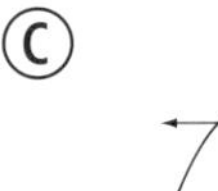 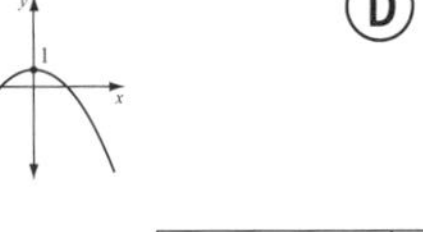 Ⓓ 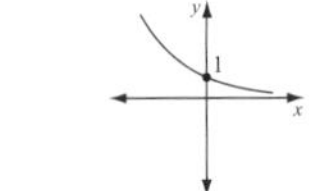 1

30 What is the interquartile range?

Ⓐ 4 Ⓑ 6 Ⓒ 7 Ⓓ 9

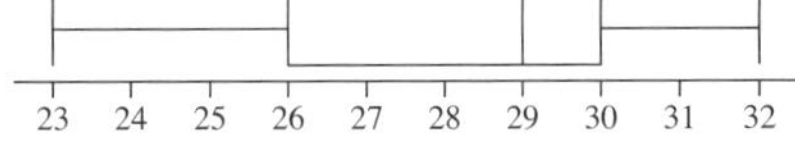

1

31 What is the gradient of any line perpendicular to $y = -\frac{2}{3}x + 4$?

Ⓐ $-\frac{2}{3}$ Ⓑ $\frac{2}{3}$ Ⓒ $-\frac{3}{2}$ Ⓓ $\frac{3}{2}$ 1

32 $\frac{6x^6}{12x^8} =$

Ⓐ $2x^2$ Ⓑ $\frac{1}{2x^2}$ Ⓒ $\frac{2}{x^2}$ Ⓓ $\frac{x^2}{2}$ 1

33 To the nearest degree, $Q =$

Ⓐ 31° Ⓑ 37° Ⓒ 53° Ⓓ 59°

15 m
Q
9 m

1

34 Find the probability of winning all 3 prizes in a raffle if you buy 5 tickets and 100 tickets are sold.

Ⓐ $\frac{1}{16\,170}$ Ⓑ $\frac{1}{8000}$ Ⓒ $\frac{5}{38\,808}$ Ⓓ $\frac{1}{20}$ 1

35 What is the size of each angle of a regular decagon?

Ⓐ 120° Ⓑ 135° Ⓒ 144° Ⓓ 150° 1

36 $a^2 + 10a - 24 =$

Ⓐ $(a-4)(a-6)$ Ⓑ $(a+4)(a+6)$ Ⓒ $(a+12)(a-2)$ Ⓓ $(a-12)(a+2)$ 1

37 C is due east of B. Find the bearing of C from A.

Ⓐ 110° Ⓑ 140°
Ⓒ 220° Ⓓ 250°

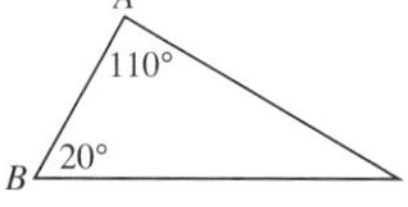

1

Continued on the next page

EXAM PAPER 2 PART A

Fill in only one circle for each question.

Marks

38 $(2x^3y^4)^3 =$
(A) $2x^6y^7$ (B) $2x^9y^{12}$ (C) $8x^6y^7$ (D) $8x^9y^{12}$ 1

39 Given $B = \frac{m}{h^2}$, the value of B when $m = 81$ and $h = 1.8$ is:
(A) 6.7 (B) 25 (C) 27 (D) 2025 1

40 $\Delta ZXY \equiv \Delta ABC$. Find the size of $\angle YXZ$.
(A) 40° (B) 60° (C) 80° (D) 100° 1

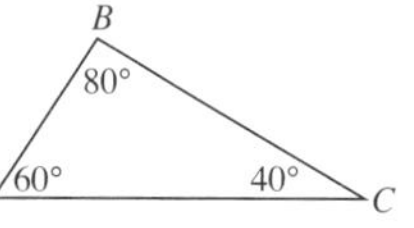

41 $(2x + 3)(5x + 7) =$
(A) $10x^2 + 29x + 21$ (B) $10x^2 + 12x + 21$ (C) $7x^2 + 8x + 10$ (D) $7x^2 + 15x + 10$ 1

42 The shaded face of this hexagonal prism has area 24 cm².
The perpendicular height of the prism is 5 cm. What is the volume?
(A) 120 cm³ (B) 144 cm³
(C) 100 cm³ (D) 720 cm³ 1

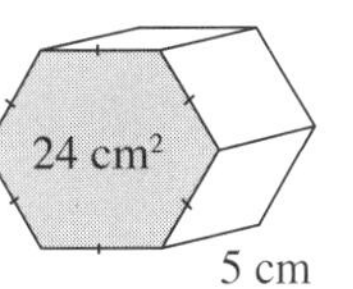

43 $\frac{a}{3} \div \frac{1}{6} =$
(A) $\frac{a}{2}$ (B) $\frac{a}{18}$ (C) $\frac{2a}{3}$ (D) $2a$ 1

44 Which of these points does NOT lie on the circle $x^2 + y^2 = 100$?
(A) $(-6, 8)$ (B) $(10, 0)$ (C) $(8, 6)$ (D) $(7, 7)$ 1

45 $x^2 - 1 -$
(A) $(x - 1)^2$ (B) $x(x - 1)$ (C) $(x - 1)(x + 1)$ (D) $(x + 1)^2$ 1

46 The probability of winning a game is $\frac{3}{8}$. Two games are played. Find the probability of winning both games.
(A) $\frac{3}{8}$ (B) $\frac{3}{28}$ (C) $\frac{9}{64}$ (D) $\frac{3}{4}$ 1

47 Find the value of x.
(A) 30 (B) 50
(C) 60 (D) 70 1

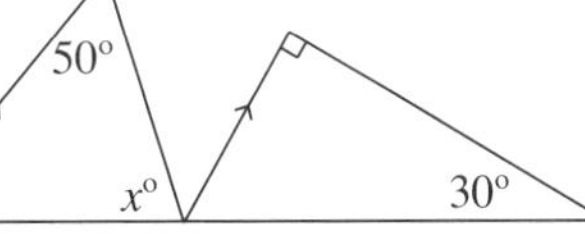

48 Which one of these lines is NOT parallel to the other three?
(A) $y = \frac{x}{2} + 5$ (B) $y = \frac{x + 1}{2}$ (C) $y = 9 - \frac{x}{2}$ (D) $y = \frac{1}{2}x - 4$ 1

49 Which equation could have been used to get the values in this table?

x	−2	−1	0	1	2
y	4	2	1	$\frac{1}{2}$	$\frac{1}{4}$

(A) $y = 2^x$ (B) $y = 2^{-x}$ (C) $y = -2^x$ (D) $y = -2^{-x}$ 1

50 3, 5, 5, 5, 6, 8, 8, 9, 10, 11, 13, 13, 14, 15, 17, 18, 20.
Consider the lower quartile (Q_1) and upper quartile (Q_3) for these scores. Which is correct?
(A) $Q_1 = 5$ and $Q_3 = 15$ (B) $Q_1 = 5$ and $Q_3 = 14.5$
(C) $Q_1 = 5.5$ and $Q_3 = 15$ (D) $Q_1 = 5.5$ and $Q_3 = 14.5$ 1

Total marks achieved for PART A /50

EXAM PAPER 2 PART B

Show all working for each question.

Marks

1 In this diagram $ST \parallel QR$.

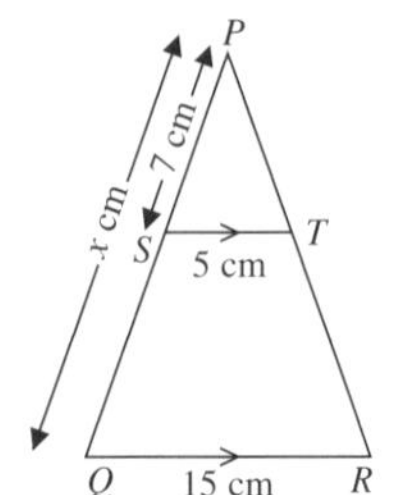

a Name two similar triangles. ______ 1

b Which test can be used to prove these triangles are similar? ______ 1

c Find the value of x.

______

______

______ 1

2 $ABCD$ is a rectangle. E is the midpoint of AB. $AE = EB = 12$ cm and $BC = 9$ cm

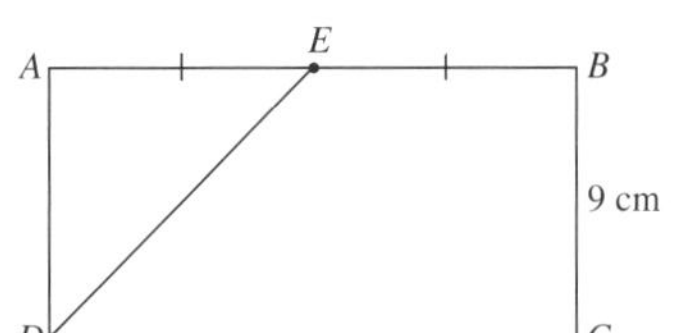

a Find the length of DE.

______

______

______ 1

b What type of quadrilateral is $EBCD$? ______ 1

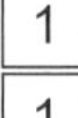

ΔAED is removed. $EBCD$ is the cross-section of a prism. The perpendicular height of the prism is 8 cm.

c Find the area of $EBCD$.

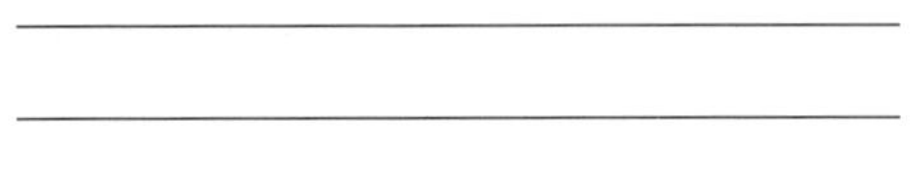

______ ______

______ ______

______ ______ 1

d Find the volume of the prism.

______

______

______ 1

e Find the surface area of the prism.

______

______

______ 1

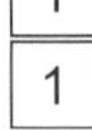

3 **a** Complete the table of values for $y = \frac{1}{2}x^2 - 4$ 1

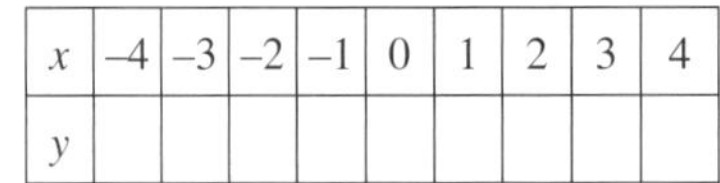

x	−4	−3	−2	−1	0	1	2	3	4
y									

b Graph the curve on the number plane. 1

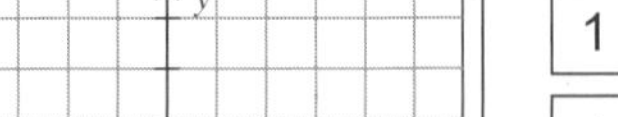

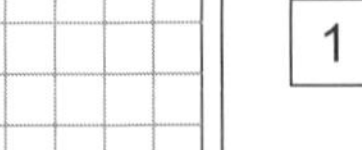

c On the same diagram graph the line $y = x$. 1

d Using the graph, how many solutions are there to the equations $\frac{1}{2}x^2 - 4 = x$? ______ 1

e Find for what value(s) of x does $\frac{1}{2}x^2 - 4 = x$.

______ 1

Continued on the next page

EXAM PAPER 2 PART B

Show all working for each question.

Marks

4 Find, in simplest form.

a $\frac{3x}{8} + \frac{5x}{12}$ 1

b $\frac{3x}{5} \times \frac{x}{6}$ 1

5 Expand and simplify.

a $(x + 2)(x + 5) + (x + 4)(x - 3)$ 1

b $3x^2y^3(4xy^2 + 5x^3y)$ 1

6 Factorise fully.

a $6a^3b^4 - 2a^2b^3$ 1

b $x^2 + 8x - 20$ 1

c $p^2 + pq + pr + qr$ 1

d $6x^2 - 6$ 1

7 a Find the compound interest earned if \$8000 is invested for a year at 6% p.a. interest compounded quarterly. 1

b Find how much more (if any) interest is earned than if simple interest of 6% p.a. was paid on the \$8000 for a year. 1

8 Solve.

a $\frac{5a + 2}{3} = \frac{3a - 5}{2}$ 1

b $\frac{x}{4} - \frac{x}{5} = 9$ 1

c $7x + 2 \geq 2x - 8$ 1

d $8x + 5 - 9x < 4$ 1

e $x^2 = 81$ 1

f $x^2 - 6x + 8 = 0$ 1

Continued on the next page

EXAM PAPER 2 PART B

Show all working for each question.

Marks

9 Solve simultaneously.

a $7a + 2b = 34$
$5a - 2b = 14$

b $3x + y = 10$
$y = 2x - 5$

1

1

10 $ABCD$ is a quadrilateral. $AB = AD$ and $BC = DC$.

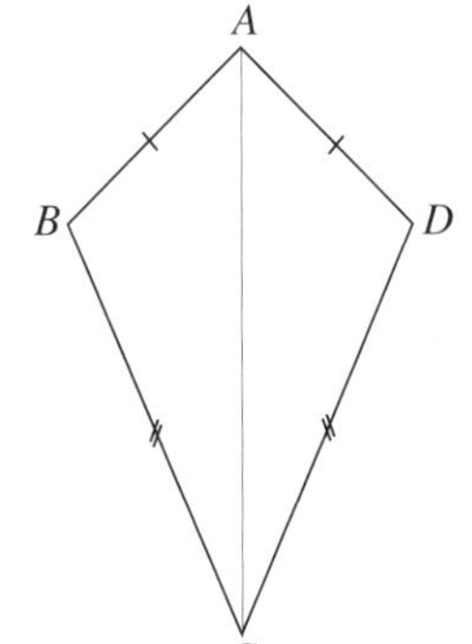

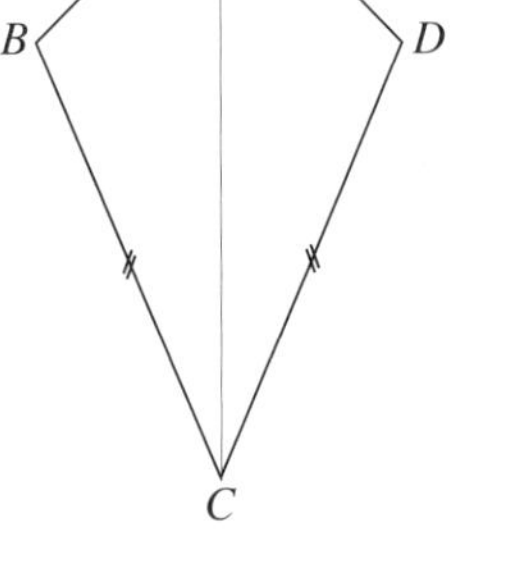

a Which test can be used to show $\Delta ABC \equiv \Delta ADC$? ______ 1

b Explain why $\angle BAC = \angle DAC$. 1

The diagonals meet at E.

c Which test can be used to show $\Delta ABE \equiv \Delta ADE$? 1

d Find the size of $\angle AEB$. ______ 1

e What property does this prove? ______ 1

11 This scatter plot shows the marks of students in Class 10Y in quizzes in both arithmetic and spelling, marked out of ten.

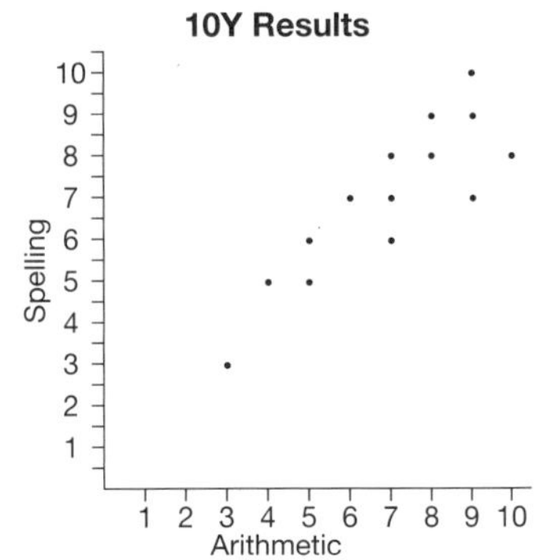

a Would the relationship between the two marks be described as:

i positive or negative? ______ 1

ii strong or weak? ______ 1

b Toby is in 10Y and scored 6 in arithmetic. What did he score in spelling? ______ 1

c Rosie is in 10Y and scored 10 in spelling. What did she score in arithmetic? ______ 1

d Olivia is also in 10Y. Her result is not shown on the scatter plot because she was sick and missed the spelling test. She scored 7 in arithmetic. What mark would you give Olivia as an esimate for spelling? Justify your answer.

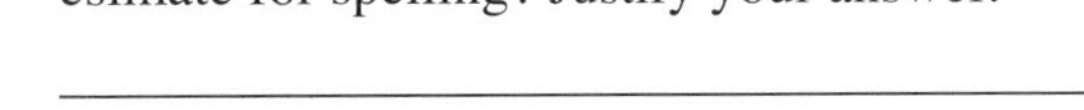

1

Continued on the next page

EXAM PAPER 2 PART B

Show all working for each question.

Marks

12 These two box plots have been drawn to show the results of quizzes in arithmetic and spelling for Class 10P.

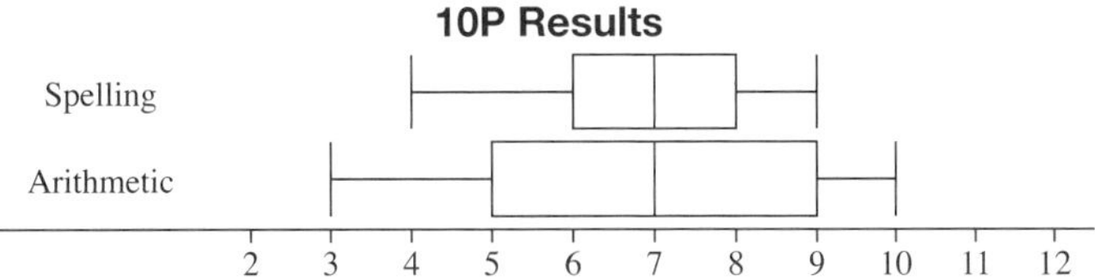

a What was the lowest mark and in which test was it scored? ______ 1

b Which measure was the same for both tests? ______ 1

c Compare the two box plots, referring to the shape and measures of spread.

______

______ 1

13 The surface area of a cube is 486 cm^2.

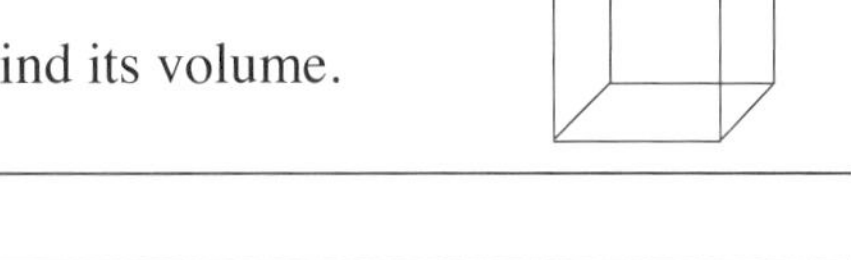

a Find the length of each side.

b Find its volume.

1

1

14 The probability of winning a game is 0.2. Two games are played. Find the probability of winning:

a neither game

b at least one game

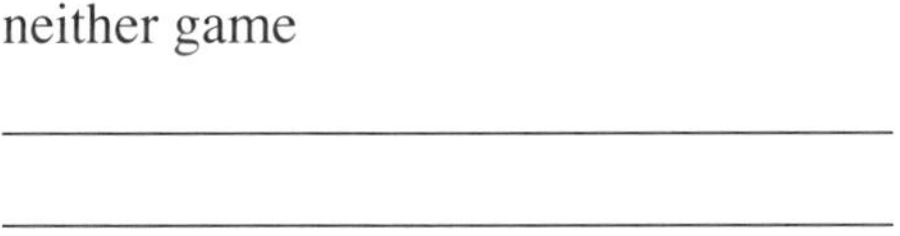

1

1

15 From the top of a building the angle of depression of the base of a second building is 65°. The angle of elevation of the top of the second building is 45°. The top of the first building is 35 m above the ground. Find:

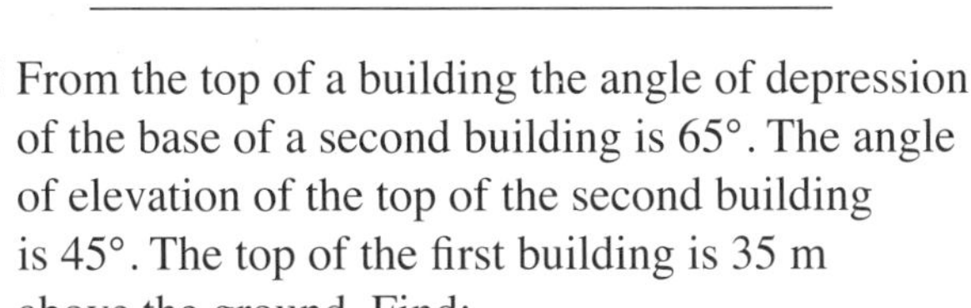

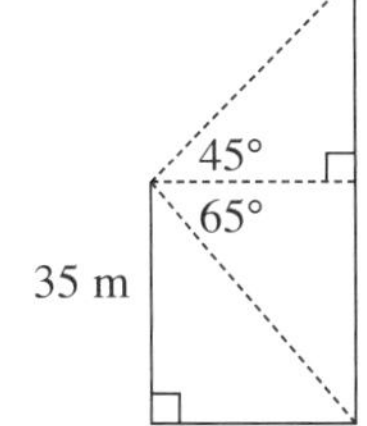

a the distance between the buildings

b the height of the second building

1

1

Total marks achieved for PART A /50

Answers

Note: Answers given in a different form, or rounded differently, might still be correct.

Chapter 1 – Financial mathematics

Page 1 **1 a** \$480 **b** \$2100 **c** \$5040 **d** \$24 375 **2 a** \$900 **b** \$3840 **c** \$3150 **d** \$21 000 **e** \$845 **f** \$1237.50 **g** \$1125 **h** \$3300 **3 a** \$2250 **b** \$9750

Page 2 **1 a** \$3600 **b** \$3600 **2 a** \$25 000 **b** \$12 500 **3 a** 1 **b** 3 **4 a** 5% p.a. **b** 8% p.a. **5 a** 6 **b** 4.8% p.a. **6 a** \$14 000 **b** \$19 390

Page 3 **1 a** \$900 **b** \$3600 **c** \$1008 **d** \$4608 **e** \$192 **2 a** \$972 **b** \$152 **3 a** \$4800 **b** \$27 200 **c** \$37 440 **d** \$5440 **e** 5% **4 a** \$1 380 000 **b** \$25 200

Page 4 **1 a** $0.\dot{6}\%$ **b** 2% **c** 4% **d** $2.\dot{6}\%$ **2 a** $0.541\dot{6}\%$ **b** $0.8\dot{3}\%$ **3 a** 2.25% **b** 1.5% **4 a** 72 **b** 16 **c** 16 **d** 6 **5 a** 20 **b** 3% **6 a** 14% **b** 9.6% **c** 15% **d** 12.775%

Page 5 **1 a i** \$10 **ii** \$111 **iii** \$249 **b** The principal has yet to be compounded with the interest earned in the first year. **2 a** \$1198.08 **b** \$652.05 **c** \$2674.22 **d** \$1414.06 **e** \$4316.13

Page 6 **1 a** \$15 861.08 **b** \$36 405.78 **c** \$199 476.01 **d** \$68 436.52 **e** \$177 133.43 **2 a** \$3438.29 **b** \$3149.72 **c** \$7724.67 **d** \$45 384.93 **e** \$41 720.78 **f** \$5720.65

Page 7 **1 a** \$7129.86 **b** \$7424.70 **c** \$7811.98 **d** \$5504.50 **e** \$5224.69 **f** \$4371.09 **2 a** \$23 309.70 **b** \$25 866.66 **c** \$12 966.59 **d** \$31 388.93 **e** \$41 847.56

Page 8 **1 a** \$12 750.40 **b** \$22 076 **c** \$20 529 **2** \$31 046.26

Page 9 **1 a** \$12 459.01 **b** \$3478.44 **c** \$1107.42 **d** \$12 282.50 **2 a** \$46 080 **b** \$43 920 **3 a** 14 961 **b** \$29 524.50

Page 10 **1 a** \$4200 **b** \$4830.62 **2** 7.7% **3 a** \$5400 **b** \$5407.33 **c** compound interest by \$7.33 **4** \$25 000 **5** 9

Page 11 **1** C **2** D **3** A **4** C **5** B **6** B **7** D **8** A **9** C **10** B **11** D **12** A **13** B **14** B **15** B

Page 12 **1 a** \$11 712.80 **b** \$3712.80 **c** 11.6025% **2 a** \$5120 **b** \$1687.44 **3 a** \$2025 **b** \$11 475 **c** \$3213 **d** \$14 688 **e** \$306 **4** \$712.50 **5** 5 years **6** \$3477.60 **7** \$1910.30 **8** \$988 000

Chapter 2 – Algebraic techniques

Page 13 **1 a** $7a$ **b** $4p$ **c** $12a$ **d** $8x$ **e** $4m$ **f** $5q$ **g** a **h** $11ab$ **i** $-3t$ **j** $8x^2$ **k** $-2n$ **l** $-7k$ **2 a** $12a$ **b** $9t$ **c** $3x$ **d** $-10k$ **e** $-m$ **f** xy **g** a **h** $-7p$ **i** $2x^2$ **j** 0 **k** $-7m$ **l** $-10y$ **3 a** $5a+5b$ **b** $8x+2y$ **c** $2a^2-2a$ **d** $-5c-2d$ **e** $3x-6y$ **f** $10a-7b$ **g** $3m+6$ **h** $12-7m$ **i** $-9a+2b$ **j** $7xy-2x-4y$ **k** $6x+3$ **l** $11t-9u$ **4 a** $7-6a$ **b** $-x^2$ **c** $12n-5$ **d** 0 **e** $3x$ **f** $-y$ **g** $11a^2-a-2$ **h** $-2m$ **i** $-6y$ **j** $9k-9n$ **k** $-5ab+2a+b$ **l** $-10a+b$ **m** $14-5x$ **n** $-p$ **o** $4m+6n$ **p** $5x^3-2x^2+3x$

Page 14 **1 a** x^8 **b** n^7 **c** y^6 **d** $6m^9$ **e** $18a^9$ **f** a^{10} **g** $4x^{10}$ **h** $16x^{10}$ **i** $15m^3$ **2 a** x^6 **b** y^5 **c** a^5 **d** $3m^{10}$ **e** $6n^8$ **f** $6a^6$ **g** $6y$ **h** a^6 **i** a^6b^2 **3 a** x^6 **b** a^{20} **c** x^{12} **d** $16m^6$ **e** $4m^6$ **f** $8a^{12}$ **g** $a^{12}b^8$ **h** $27a^3b^{12}$ **i** $5x^{14}y^7$ **4 a** 1 **b** 1 **c** 1 **d** 5 **e** 1 **f** 2 **g** 0 **h** 9 **i** 11 **5 a** a^6 **b** $15x^8$ **c** $2a^{25}$ **d** $125x^6$ **e** $2x^3$ **f** $6a^5b^5$ **g** m^5n^5 **h** $5x^3$ **i** $20x^6y^9$ **j** xy **k** 5 **l** $36x^4$

Page 15 **1 a** $12a$ **b** $15xy$ **c** $6m^2$ **d** $-12ab$ **e** $12xy$ **f** $15ab$ **g** $-55t^2$ **h** $12q^3$ **i** x^3y^2 **j** $24abc$ **k** $24x^{11}$ **l** $8x^3y^2$ **2 a** x^{10} **b** $3x^7$ **c** $2u^6$ **d** $5q^2$ **e** $12a^2b$ **f** $18x^2y^2$ **g** $24t^3$ **h** $-12x^2y^2$ **i** $-24pqr$ **j** $60abc$ **k** $-a^3$ **l** a^3b^3 **3 a** $6a^3b^2$ **b** $24a^2b^2$ **c** $35p^3q^3$ **d** x^6y^8 **e** $a^{10}b^7$ **f** $6m^9n^8$ **g** $5p^4q^3$ **h** $20a^2b^7$ **i** $14x^2y^5$ **j** $27a^3b^8$ **k** $20a^8b^6c^{10}$ **l** $15x^5y^5z^6$ **4 a** $6a^2$ **b** $28a^{-1}$ **c** 18 **d** $8x$ **e** $35t^{-5}$ **f** $24k^{-1}$ **g** $36n^{-1}$ **h** $24a^{-9}$ **i** $-e^{-2}$ **j** $48q^7$

Page 16 **1 a** $5a$ **b** 3 **c** 8 **d** 6 **e** $-4k$ **f** $3m$ **g** 1 **h** $4m$ **i** $3x$ **j** -2 **k** $-6ac$ **l** y **m** a^8 **n** $4b^{15}$ **o** x^4y **p** a^4b^5 **q** $9p^5q$ **r** $3ab^2c^3$ **2 a** $\frac{2x^2}{3}$ **b** $\frac{5}{7a^3}$ **c** $\frac{9t^3}{10}$ **d** $\frac{5b}{3}$ **e** $\frac{7x}{8}$ **f** $\frac{3n}{5m}$ **g** $\frac{3a^2}{4}$ **h** $\frac{2}{3x^3}$ **i** $\frac{3ab^2}{2}$ **j** $\frac{8}{9e}$ **k** $\frac{4n^6}{3m^2}$ **l** $\frac{2a^3}{3b}$ **m** $\frac{a^2}{2}$ **n** $\frac{t^2}{3}$ **o** $\frac{3}{x^5}$ **p** $\frac{2a}{b^2}$ **q** $\frac{mn}{4}$ **r** $\frac{2}{y}$ **s** $\frac{1}{3x}$ **t** $\frac{1}{2x^6}$ **u** $2a$ **v** $\frac{1}{5n}$ **w** $3a^2b$ **x** $\frac{1}{5y}$

Page 17 **1 a** $11x$ **b** $4k$ **c** $2x^2$ **d** 8 **e** p **f** 1 **g** $12x^2+3x$ **h** $6a^{10}$ **i** $15x^3y$ **j** x^7 **k** $-3a$ **l** $-8m$ **m** $3a^2b$ **n** 0 **o** $-2a$ **p** x^8 **q** $12ab$ **r** $-2x$ **2 a** $16x^2$ **b** 3 **c** 2 **d** $13a^5$ **e** $2p^2$ **f** $6a^2b$ **3 a** $2x^5$ **b** $2a$ **c** $13ab$ **d** $-4m^6$ **e** $3p$ **f** $8+3n$

Page 18 **1 a** 8 **b** 45 **c** 24 **d** 15 **e** 15 **f** 135 **g** 2 **h** -1 **i** 24 **j** 81 **k** 36 **l** 75 **2 a** 60 **b** 12 **c** 17 **d** -4 **e** 22 **f** -50 **3 a** 255 **b** 80.5 **4 a** -4 **b** -1 **5 a** 455 **b** 43.3 **6 a** 25 **b** 23.5 (1 d.p.)

Page 19 **1 a** $5x+10$ **b** $7x-21$ **c** $8x+20$ **d** $15x-9y$ **e** $12t-6$ **f** x^2+7x **g** a^2-a **h** $6x^2-15x$ **i** $12n^2+8n$ **j** $16a+8b-8c$ **k** $10a^2+8ab+6a$ **l** $-6x-8$ **m** $-10x+15$ **n** $-4x+8x^2$ **o** $-7a^2-28a$ **p** $-x+y$ **q** $-m-n$ **r** $-3p+1$ **s** $2x^3-10x$ **t** $3a^3b+15a^2$ **2 a** $14x+15$ **b** $a-3$ **c** $15-x$ **d** $13x-4y$ **e** $12x+38$ **f** $12a-15$ **g** 22 **h** $17m-16$ **i** $2a+27$ **j** $x^2+2x+12$ **k** $-a$ **l** $x^2+2xy+3y^2$

Page 20 **1 a** x^2+5x+6 **b** $x^2+5x-14$ **c** $x^2+4x-21$ **d** $x^2+2x-15$ **e** $2x^2+11x+15$ **f** $3x^2-8x+4$ **g** $2x^2-7x-15$ **h** $3x^2+x-10$ **i** $6x^2+7x+2$ **j** $6x^2-7x+2$ **2 a** x^2+5x+6 **b** $x^2+2x-15$ **c** $2x^2-7x-15$ **d** $6x^2-7x+2$ **e** $2x^2+11x+15$ **f** $x^2+5x-14$ **g** $6x^2+7x+2$ **h** $x^2+4x-21$ **i** $3x^2-8x+4$ **j** $3x^2+x-10$ **3 a** x^2, $5x$, $3x$, 15, $x^2+8x+15$ **b** x^2, $7x$, $4x$, 28, $x^2+11x+28$

Page 21 **1 a** x^2+3x+2 **b** x^2+5x+6 **c** $a^2+8a+15$ **d** m^2+7m+6 **e** $p^2+10p+16$ **f** $y^2+10y+21$ **g** $a^2+11a+28$ **h** $d^2+12d+27$ **i** $2a^2+13a+15$ **j** $6a^2+20a+6$ **k** $8a^2+24a+18$ **l** $6x^2+17x+5$ **2 a** a^2+a-6 **b** x^2-x-6 **c** $y^2+2y-24$ **d** $y^2+2y-15$ **e** $a^2+4a-21$ **f** $x^2+4x-12$ **g** $2y^2-3y-2$ **h** $3x^2-7x-6$ **i** $6x^2+x-1$ **j** $2x^2+13x-7$ **k** $3x^2+19x-40$ **l** $x^2-7x+12$ **3 a** $a^2+7a+12$ **b** $a^2+11a+30$ **c** $-2a^2+5a+3$ **d** $x^2+13x+36$ **e** $-n^2-2n+35$ **f** $-x^2+13x-42$ **g** $3x^2+8x+4$ **h** $-3n^2+14n+5$ **i** $2a^2+8a-42$ **j** x^2-y^2 **k** $4m^2-n^2$ **l** a^2-b^2 **m** $4x^2-9y^2$ **n** $a^2-2ab+b^2$ **o** $4x^2-9$ **p** $6x^2+7x-20$

Page 22 **1 a** $\frac{4m}{5}$ **b** x **c** $\frac{5t}{7}$ **d** y **e** $\frac{6a+3b}{11}$ **f** $\frac{33k}{8}$ **g** x **h** $2p$ **i** $\frac{9m}{19}$ **2 a** $\frac{3y}{7}$ **b** $\frac{2x}{11}$ **c** $\frac{a}{3}$ **d** $\frac{7a}{17}$ **e** $\frac{2m}{23}$ **f** $\frac{m}{6}$ **g** $\frac{4a}{7}$ **h** $\frac{x}{2}$ **i** $\frac{5a-2b}{11}$ **3 a** $\frac{5x}{6}$ **b** $\frac{a}{20}$ **c** $\frac{8m}{15}$ **d** $\frac{13x}{20}$ **e** $\frac{3a}{10}$ **f** $\frac{3y}{32}$ **g** $\frac{31y}{15}$ **h** $\frac{3p}{8}$ **i** $\frac{-x}{8}$ **j** $\frac{23y}{12}$ **k** $\frac{10m}{21}$ **l** $\frac{3a-4b}{8}$ **4 a** $\frac{x}{2}$ **b** $\frac{x}{2}$ **c** $\frac{t}{3}$

Answers

PAGE 23 **1 a** $\frac{ab}{35}$ **b** $\frac{mn}{12}$ **c** $\frac{xy}{24}$ **d** $\frac{20xy}{27}$ **e** $\frac{a^2}{33}$ **f** $\frac{6x^2}{35}$ **g** $\frac{a}{8}$ **h** $\frac{2n}{21}$ **i** $\frac{25x}{49}$ **2 a** $\frac{ab}{6}$ **b** $\frac{x^2}{2}$ **c** t^2 **d** $\frac{2cd}{15}$ **e** $\frac{2ab}{3}$ **f** $\frac{2xy}{3}$ **g** $\frac{4x^2}{5}$ **h** $\frac{3mn}{4}$ **i** $\frac{3t^2}{2}$ **3 a** $\frac{3a}{2b}$ **b** $\frac{27x}{14y}$ **c** $\frac{5m}{6n}$ **d** 3 **e** $\frac{6}{5}$ **f** 4 **g** $\frac{2}{m}$ **h** $\frac{x}{3}$ **i** $2p$ **j** $\frac{3b}{4a}$ **k** $\frac{20b}{3}$ **l** $\frac{9}{10}$

PAGE 24 **1 a** $\frac{17a}{5b}$ **b** $\frac{a}{7x}$ **c** $\frac{4}{a}$ **d** $\frac{1}{t}$ **e** $\frac{2}{x^2}$ **f** $\frac{2a}{x}$ **g** $\frac{39}{4x}$ **h** $\frac{a}{2b}$ **i** $\frac{29m}{20n}$ **2 a** $\frac{15}{ty}$ **b** $\frac{20}{ab}$ **c** $\frac{9}{5ab}$ **d** $\frac{16}{3t^2}$ **e** $\frac{x^2}{y^2}$ **f** $\frac{8a}{3mn}$ **g** $\frac{10b^2}{27c^2}$ **h** $\frac{1}{16}$ **i** $\frac{3}{4}$ **j** $2a^2$ **k** 3 **l** 1 **m** x **n** b **o** $\frac{12}{5y^2}$ **3 a** $\frac{12}{p}$ **b** $\frac{8}{5}$ **c** $\frac{5}{2}$ **d** $\frac{1}{5}$ **e** 1 **f** $\frac{x^2y^2}{z^2}$ **g** 21 **h** $\frac{9n}{8}$ **i** $\frac{10n}{m}$

PAGE 25 **1 a** $5(x+2)$ **b** $3(x+2)$ **c** $8(y+2)$ **d** $m(m+1)$ **e** $2x(x+2)$ **f** $3x(y+2)$ **g** $3a(2a-1)$ **h** $3(m+5)$ **i** $x(9+y)$ **j** $4(x+4)$ **k** $5b(b+2a)$ **l** $3(m+7)$ **m** $3m(2-n)$ **n** $5(x+3)$ **o** $y(a-1)$ **p** $7m(n-2p)$ **q** $xy(xy-z)$ **r** $8m^2n(n-2)$ **2 a** $-3(x+2)$ **b** $-4(a+2)$ **c** $-5(y+3)$ **d** $-m(m-1)$ **e** $-x(x-5)$ **f** $-l(l-2m)$ **g** $-x(1-4x)$ **h** $-m(4-m)$ **i** $-x(3+2x)$ **j** $-6a(1+3a)$ **k** $-7(y-3)$ **l** $-8x(1-2y)$ **m** $-3(a+3)$ **n** $-5xy(1-3xy)$ **o** $-ay(ay-1)$ **3 a** $a(b+c+d)$ **b** $p(x+y+z)$ **c** $ab(2a+3ab-5c)$ **d** $5m(m^2+2m+3)$ **e** $2(a+2b+3c)$ **f** $3x(4x+5y+6z)$ **g** $xy(xy+y+x)$ **h** $3a^2b(3-4b)$ **i** $5(a^2-b^2-2c^2)$ **j** $6mp(1+2m-3mp)$ **k** $3a(b-2c-3d)$ **l** $12x^2y^2(1-3xy)$ **4 a** $2a^2b^3(4-5ab^2)$ **b** $2x(8y+3x^2)$ **c** $3pq^2(3p^2+4q^3)$ **d** $3abc^2(2ac-3)$ **e** $3x^2y^4(4x-5y^2)$ **f** $2ab^2c(ab-4)$ **g** $5p^2q^5(2-5p)$ **h** $14x^4y(2y^6+3)$ **i** $2a^4b^2c^6(1-6abc)$ **j** $3tu^3(3t-2u)$ **k** $5xy(3xy-2x^2+4y^2)$ **l** $8pq^2(3q^2+2pq+1)$

PAGE 26 **1 a** $(x+3)(x+4)$ **b** $(x-2)(x-3)$ **c** $(x+1)(x+2)$ **d** $(x+2)^2$ **e** $(y-3)(y-4)$ **f** $(m+2)(m+6)$ **g** $(a+3)^2$ **h** $(x+4)(x+7)$ **i** $(n+3)(n-1)$ **j** $(x+2)(x+7)$ **2 a** $(x-3)(x-5)$ **b** $(y-6)(y+2)$ **c** $(x+6)(x-1)$ **d** $(x+9)(x+10)$ **e** $(x+6)(x-2)$ **f** $(m-8)(m+7)$ **g** $(x-4)(x+1)$ **h** $(y-7)(y+1)$ **3 a** $x(x-8)$ **b** $(m+5)(m+1)$ **c** $(t-3)(t+2)$ **d** $(y-4)(y-5)$ **e** $(a-9)(a+2)$ **f** $(x+4)^2$ **g** $x(x-12)$ **h** $(y-3)(y-8)$

PAGE 27 **1** C **2** D **3** B **4** A **5** A **6** D **7** B **8** D **9** D **10** B **11** D **12** D **13** D **14** A **15** A

PAGE 28 **1 a** $4x^6y^4$ **b** $\frac{1}{2x}$ **c** $9n^8$ **d** $63x^5$ **2** $8x-17$ **3 a** $x^2-2x-24$ **b** $6x^2+29x+35$ **4 a** $3(a+2b-4)$ **b** $6ab^3c(a-2bc)$ **c** $(x+3)(x+4)$ **d** $(p+9)(p-4)$ **5 a** $\frac{x}{5}$ **b** $\frac{xy}{5}$ **c** $\frac{4a}{9}$ **d** $\frac{5x}{7}$

CHAPTER 3 – Equations

PAGE 29 **1 a** $x=4$ **b** $x=11$ **c** $x=-3$ **d** $x=6$ **e** $x=16$ **f** $a=36$ **g** $m=-10$ **h** $n=6$ **i** $x=9$ **j** $a=-7$ **k** $p=8$ **l** $t=9$ **m** $x=16$ **n** $a=9$ **o** $x=-10$ **p** $m=80$ **2 a** $x=5$ **b** $a=2$ **c** $m=7$ **d** $n=6$ **e** $p=-3$ **f** $k=1$ **g** $p=4$ **h** $x=25$ **i** $a=-4$ **j** $x=12$ **k** $a=14$ **l** $t=25$ **m** $x=9$ **n** $a=15$ **o** $b=-14$ **p** $x=20$ **q** $a=12$ **r** $n=-9$

PAGE 30 **1 a** $x=7$ **b** $x=9$ **c** $x=-5$ **d** $x=3$ **e** $x=6$ **f** $x=1$ **g** $a=0$ **h** $p=2$ **i** $e=1$ **j** $k=-3$ **k** $m=1.2$ **l** $k=-1$ **m** $x=2$ **n** $n=4$ **o** $y=2$ **p** $n=8$ **q** $q=2$ **r** $m=17$ **2 a** $x=11$ **b** $q=-2$ **c** $a=6$

PAGE 31 **1 a** $x=2$ **b** $x=7$ **c** $x=-2$ **d** $x=6$ **e** $x=2$ **f** $x=-4$ **g** $x=26$ **h** $x=-13$ **i** $x=-5$ **j** $x=6$ **k** $x=-3$ **l** $x=1$ **2 a** $a=1$ **b** $x=23$ **c** $m=10$ **d** $y=-5$ **e** $a=11$ **f** $k=3$ **g** $m=4$ **h** $a=-30$ **i** $m=-4\frac{1}{4}$ **j** $x=-1$ **k** $x=2$ **l** $k=-8$

PAGE 32 **1 a** $x=15$ **b** $a=16$ **c** $n=15$ **d** $m=78$ **e** $x=7$ **f** $a=17$ **g** $x=2$ **h** $t=-5$ **i** $x=6$ **j** $x=1$ **k** $k=4$ **l** $p=-3\frac{1}{2}$ **m** $n=10$ **n** $x=-3$ **o** $k=2$ **p** $c=-8$ **2 a** $x=8$ **b** $a=32$ **c** $m=-19$ **d** $c=1$ **e** $b=9$ **f** $h=-4$

PAGE 33 **1 a** 4 **b** 20 **c** 7 **d** –15 **e** 90 **f** 12 **g** 27 **h** 26 **i** 9 **2 a** 30, 32, 34 **b** 36 years **c** 14 years, 42 years

PAGE 34 **1 a** $S=463.25$ **b** $S=366.32$ **2 a** $L=42$ **b** $B=45$ **3 a** $h=14$ **b** $h=4$ **4 a** $u=\pm1$ **b** $a=13.125$ **5 a** $r=57.3$ **b** $r=8.7$ **6 a** $u=18$ **b** $t=6$ **7 a** $P=6929.1$ **b** $P=6753.4$

PAGE 35 **1 a** $x<3$ **b** $x\geq-7$ **c** $-1\leq x\leq5$ **d** $x\leq-4$ or $x\geq2$ **e** $x\leq0$ **f** $x>-4$

2 a 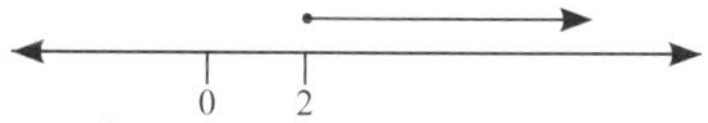

b 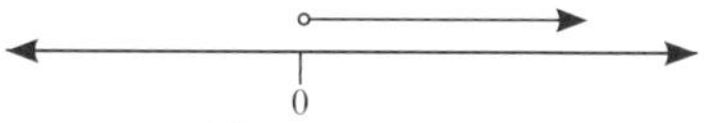

c 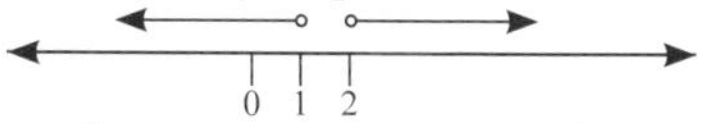

d 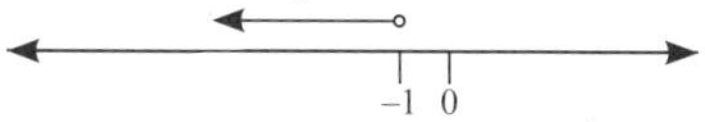

e 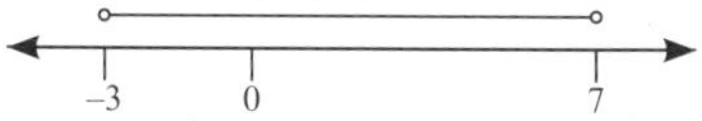

f 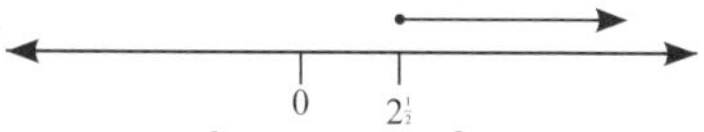

g 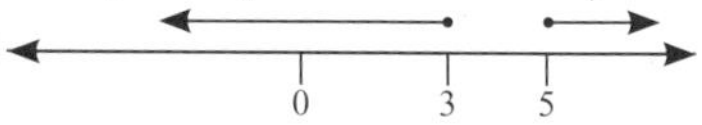

h 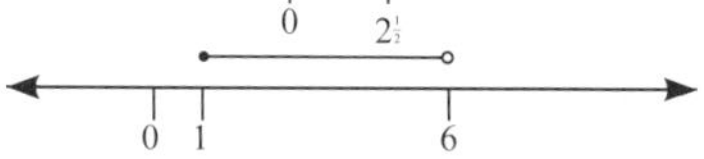

i 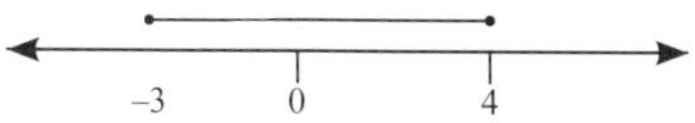

3 a $x<4$ 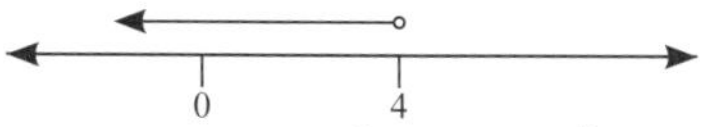

b $a>8$ 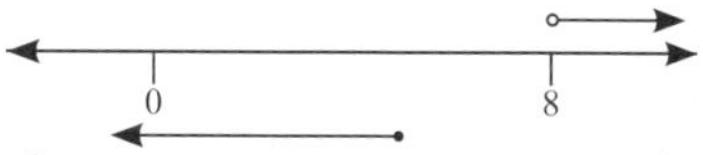

c $m>4$ 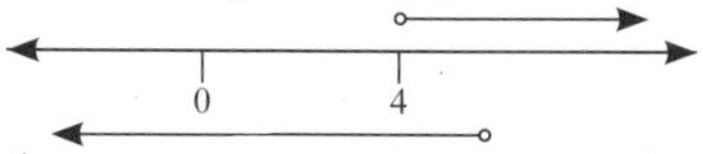

d $y\leq4$ 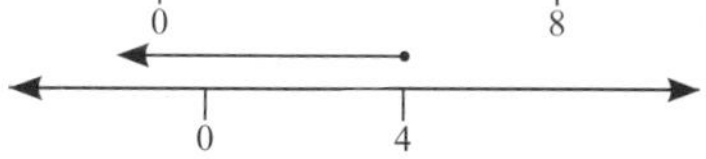

e $y<7$ 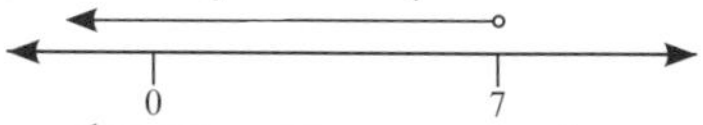

f $x>5$ 

g $m\leq9$ 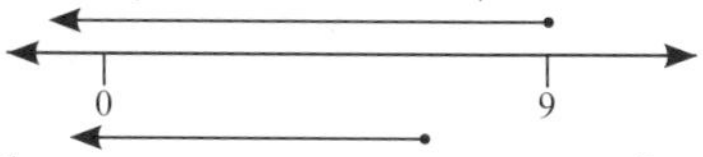

h $m\leq11$ 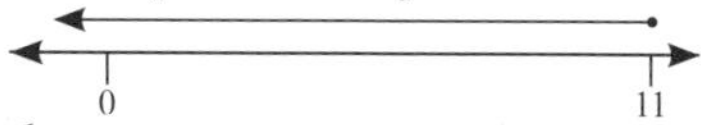

i $x\leq5$ 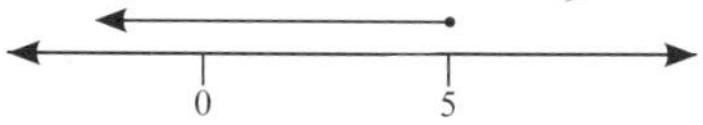

j $m\leq14$ 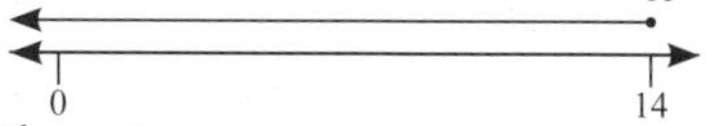

k $m\geq10$ 

l $y<-9$ 

m $a \geq -4$ 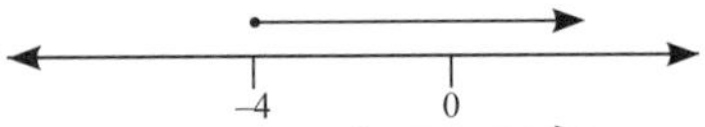 **n** $x \leq -4$ 

o $y > 2$ 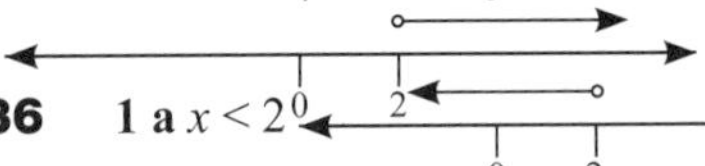

Page 36 **1 a** $x < 2$ 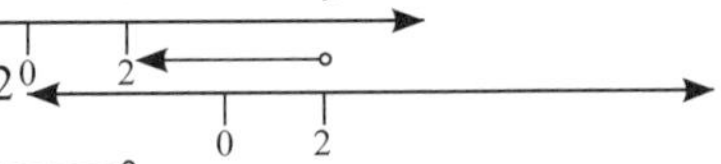 **b** $m > 2$ 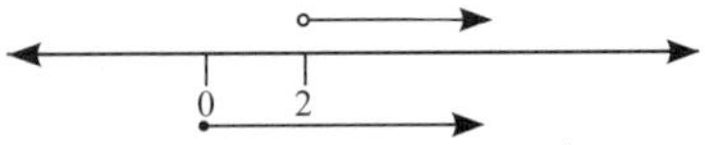

c $t < 4$ 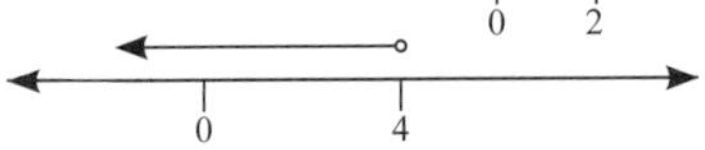 **d** $p \geq -3$ 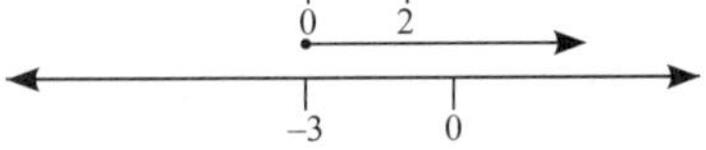

e $y \leq 3$ 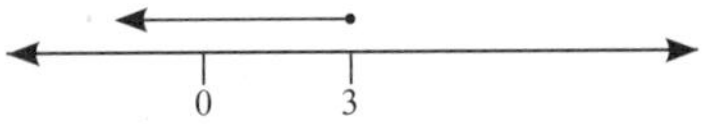 **f** $x \geq 2$ 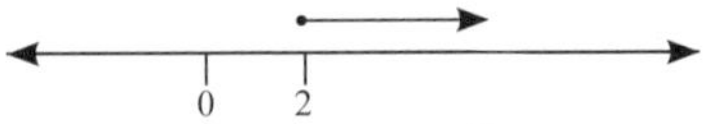

g $x < 4$ 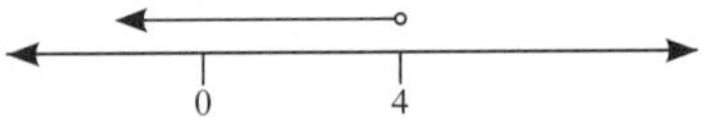 **h** $p \geq 2$ 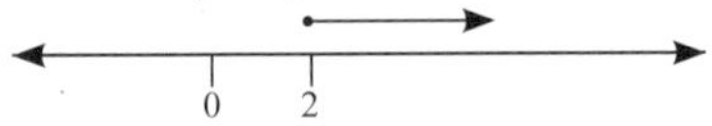

i $m < 2$ 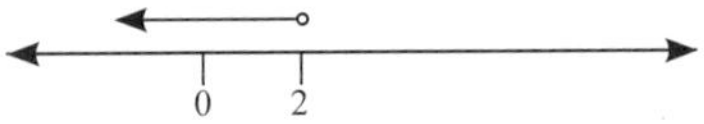 **j** $x \geq 3$ 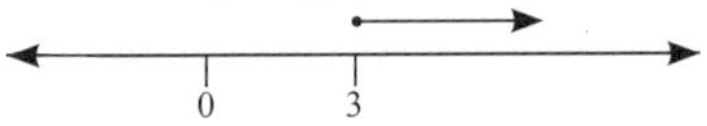

k $x < 6$ 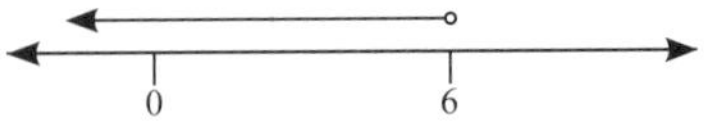 **l** $x \leq 8$ 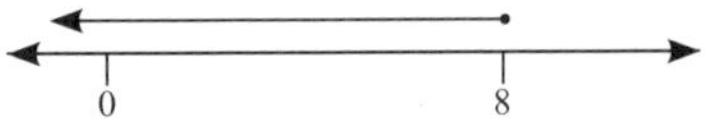

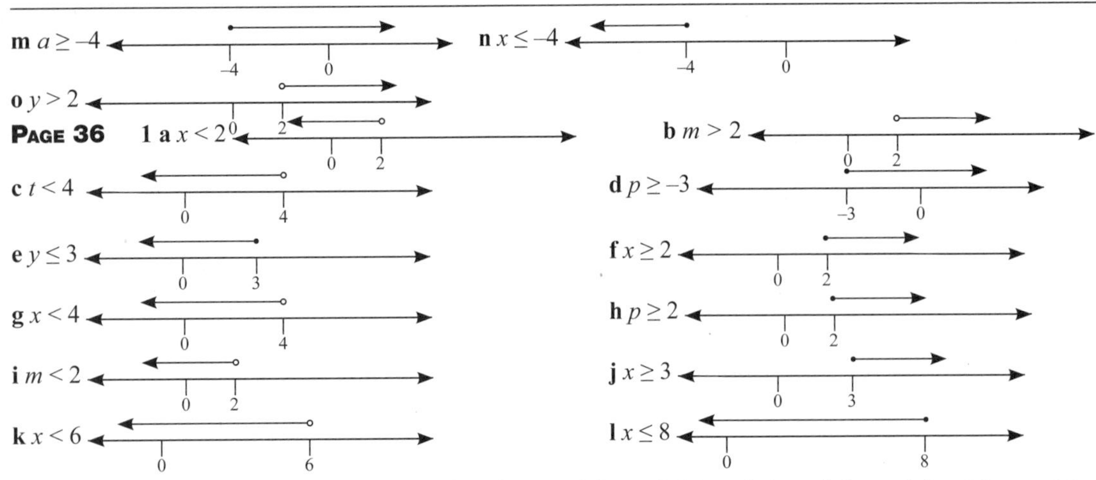

2 a $x < 4$ **b** $x < 4$ **c** $x \leq 2$ **d** $x \geq -1$ **e** $x \leq 4$ **f** $t \geq 2$ **g** $y < 2$ **h** $x \geq 2$ **i** $x \leq -1$ **j** $x \geq 5$ **k** $x < 5$ **l** $x \leq 2$ **m** $x < 6$ **n** $y > 5$ **o** $m \leq 1\frac{1}{5}$

Page 37 **1 a** True **b** False **c** True **d** False **e** True **f** False **g** True **h** False **2 a** $x > 4$ **b** $x < -4$ **c** $x < 4$ **d** $x > -4$ **e** $a \geq -3$ **f** $p < 2$ **g** $k \geq 2$ **h** $q > 14$ **i** $x > -7$ **j** $m \leq 2$ **k** $x \geq 18$ **l** $x < -4$ **3 a** $x \geq 1$ **b** $a < -2$ **c** $m > -6$ **d** $x \leq 2$ **e** $x > -10$ **f** $x \geq -7$ **g** $x \leq 8$ **h** $x \leq 1$ **i** $a > 70$ **j** $x \leq 9$ **k** $x \geq 17$ **l** $x < -13$

Page 38 **1 a** $x = 9, y = 1$ **b** $p = 5, q = -2$ **c** $x = 3, y = 4$ **d** $x = 9, y = 3$ **e** $x = 0, y = 6$ **f** $x = 3, y = 1$ **2 a** $m = -3, n = 4\frac{2}{3}$ **b** $x = 3, y = 2$ **c** $x = \frac{1}{2}, y = -2$ **d** $x = 3, y = 7$

Page 39 **1 a** $a = 7, b = 4$ **b** $x = 6, y = 5$ **c** $m = 9, n = 5$ **d** $p = 4, q = 3$ **e** $x = 11, y = 9$ **f** $k = 5, d = -2$ **2 a** $x = 7, y = 2$ **b** $p = 8, q = 7$ **c** $a = 12, b = -2$ **d** $a = 7, b = 4$ **e** $m = 15, n = 6$ **f** $x = 9, y = 2$

Page 40 **1 a** $x = 5, y = 3$ **b** $x = 8, y = 3$ **c** $x = 1, y = 2$ **2 a** $a = 1, b = -1$ **b** $x = 3, y = 7$ **c** $x = -5, y = 7$ **d** $m = 4, n = -1$ **e** $a = -2, b = -5$ **f** $a = 8, b = 2$ **g** $x = 11, y = -3$ **h** $y = 10, z = 8$ **i** $m = -1, n = -4$

Page 41 **1 a** $x = \pm 3$ **b** $x = \pm 4$ **c** $x = \pm 5$ **d** $x = \pm 1$ **e** $x = \pm 2$ **f** $x = \pm 8$ **g** $x = \pm 6$ **h** $x = \pm 7$ **2 a** $x = 2$ **b** $x = 10$ **c** $x = -1$ **d** $x = -3$ **3 a** $x = \pm 10$ **b** $x = \pm 9$ **c** $x = \pm 13$ **d** $x = \pm 30$ **e** $x = \pm 6$ **f** $x = \pm 3$ **g** $x = \pm 5$ **h** $x = \pm 12$ **i** $x = 2\frac{1}{2}$ or $-2\frac{1}{2}$ **j** $x = \frac{4}{3}$ or $-\frac{4}{3}$ **k** $x = \frac{5}{4}$ or $-\frac{5}{4}$ **l** $x = \frac{3}{2}$ or $-\frac{3}{2}$ **m** $x = \frac{1}{3}$ or $-\frac{1}{3}$ **n** $x = 1$ or -1 **o** $x = 3$ or -3 **p** $x = 3$ or -3 **q** $x = \frac{3}{2}$ or $-\frac{3}{2}$ **r** $x = \frac{6}{5}$ or $-\frac{6}{5}$ **s** $x = 2$ or -2 **t** $x = -3$ or -7 **4 a** $x = \pm 4.80$ **b** $m = \pm 7.28$ **c** $y = \pm 2.41$ **d** $k = \pm 4.36$

Page 42 **1 a** $x = 1$ or 2 **b** $x = 2$ or -3 **c** $x = 1$ or 3 **d** $x = 0$ or -5 **e** $x = 0$ or 4 **f** $x = 3$ or 7 **g** $x = 3$ or 5 **h** $x = -1$ or 3 **i** $x = -2$ or 4 **j** $x = -3$ or 3 **k** $x = -2$ or 2 **l** $x = -5$ or 5 **m** $x = -1$ or 6 **n** $x = -3$ or -2 **o** $x = 0$ or -8 **2 a** $x = 0$ or $\frac{1}{2}$ **b** $x = -6$ or $\frac{1}{2}$ **c** $x = -1$ or $\frac{2}{3}$ **d** $x = 2$ or $\frac{1}{3}$ **e** $x = 0$ or $\frac{1}{2}$ **f** $x = 0$ or 2 **g** $x = -3$ or $\frac{1}{3}$ **h** $x = 0$ or $2\frac{1}{2}$ **i** $x = 0$ or 1 **j** $x = 0$ or $-\frac{1}{3}$ **k** $x = 3$ **l** $x = 0$ or 3 **3 a** $x = 4$ or 5 **b** $x = 8$ or -8 **c** $x = 0$ or 3 **d** $x = 0$ or 2 **e** $x = 7$ or 9 **f** $x = -1$ or 5 **g** $x = -4$ or $\frac{1}{2}$ **h** $x = -1\frac{1}{2}$ or $1\frac{1}{2}$ **i** $x = -\frac{5}{4}$ or $\frac{4}{5}$

Page 43 **1 a** $x = 0$ or 5 **b** $x = 0$ or 4 **c** $x = 0$ or 2 **d** $x = 0$ or -7 **e** $x = 0$ or -5 **f** $x = 0$ or -9 **g** $x = 0$ or 4 **h** $x = 0$ or 9 **i** $x = 0$ or 12 **j** $x = 0$ or 2 **k** $x = 0$ or -8 **l** $x = 0$ or 10 **m** $x = 0$ or -7 **n** $x = 0$ or $\frac{1}{5}$ **o** $x = 0$ or -3 **2 a** $x = 0$ or 4 **b** $x = 0$ or -5 **c** $x = 0$ or 1 **d** $x = 0$ or 2 **e** $x = 0$ or 1 **f** $x = 0$ or 1 **g** $x = 0$ or $-\frac{1}{3}$ **h** $x = 0$ or $\frac{7}{3}$ **i** $x = 0$ or $\frac{3}{5}$ **j** $x = 0$ or 3 **k** $x = 0$ or 3 **l** $x = 0$ or $\frac{1}{2}$

Page 44 **1 a** $x = -3$ or -2 **b** $x = 7$ or -5 **c** $x = 6$ or -1 **d** $x = -3$ or -4 **e** $x = 2$ or 3 **f** $x = -8$ or 6 **g** $x = 4$ **h** $x = 3$ or -5 **i** $x = -4$ or -5 **j** $x = 3$ or 5 **k** $x = 2$ or -6 **l** $x = 5$ or -2 **m** $x = -5$ or -6 **n** $x = 2$ or 7 **o** $x = 4$ or -7 **p** $x = 11$ or -9 **q** $x = -2$ or -4 **r** $x = 1$ or -7 **s** $x = 1$ or 5 **t** $x = -4$ **u** $x = 10$ or -6 **2 a** $x = 6$ or -3 **b** $x = 5$ or 8 **c** $x = -9$ or 4 **d** $x = 6$ or 9 **e** $x = 6$ or -4 **f** $x = 3$ or -8

Page 45 **1** A **2** C **3** A **4** C **5** D **6** C **7** C **8** D **9** B **10** C

Page 46 **1 a** $y = 5$ **b** $x = 1$ **c** $x = 4$ **d** $x = -3$ **2 a** $x > 5$ 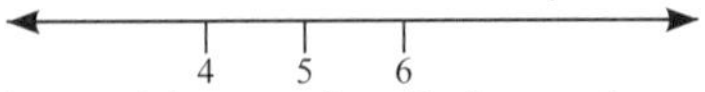

b $x \geq -1$ 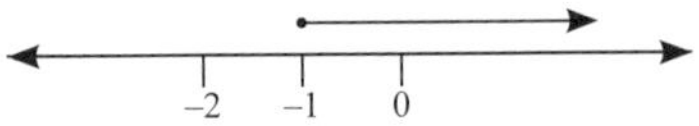 **3 a** $x = \pm 12$ **b** $x = \pm 4$ **c** $x = 3$ or 9 **d** $x = -4$ or -9 **e** $x = 9$ or -10 **f** $x = 4$ or -1

4 $x = 4, y = -10$

Chapter 4 – Further algebra and indices

Page 47 **1 a** $x^2 + 7x + 10$ **b** $x^2 + 11x + 28$ **c** $x^2 + 9x + 18$ **d** $x^2 + 6x - 16$ **e** $x^2 - 6x - 16$ **f** $x^2 - 10x + 16$ **g** $x^2 - x - 12$ **h** $x^2 + 6x - 7$ **i** $x^2 - 19x + 90$ **j** $x^2 - 8x + 15$ **k** $x^2 + 8x + 15$ **l** $x^2 - x - 42$ **m** $x^2 + x - 42$ **n** $x^2 - 13x + 42$ **o** $x^2 + 13x + 42$ **2 a** $(x + 2)(x + 6)$ **b** $(x + 2)(x + 7)$ **c** $(x + 3)(x + 7)$ **d** $(x + 5)(x + 6)$ **e** $(x + 4)(x + 1)$ **f** $(x + 7)(x + 8)$ **g** $(x - 3)(x - 6)$ **h** $(x - 7)(x + 4)$ **i** $(x - 7)(x - 4)$ **j** $(x + 7)(x - 4)$ **k** $(x + 10)(x - 3)$ **l** $(x - 7)(x + 5)$ **m** $(x - 6)(x + 2)$ **n** $(x + 7)(x - 2)$ **o** $(x - 4)(x + 3)$

Answers

p $(x+9)(x-8)$ **q** $(x+5)(x+1)$ **r** $(x-2)(x-3)$ **3 a** $2x^2+9x+10$ **b** $8x^2-2x-1$ **c** $14x^2+11x-15$ **d** $15a^2-37a+20$ **e** $3m^2+11m-42$ **f** $56p^2-37p-45$

PAGE 48 **1 a** $(x+2)(y+z)$ **b** $(b+3)(a+7)$ **c** $(2x+3)(m+5)$ **d** $(7+x^2)(y^2+8)$ **e** $(p-3)(p-2)$ **f** $(a+7)(t-5)$ **g** $(x-1)(y-2)$ **h** $(m-n)(a-b)$ **i** $(x+y)(6+z)$ **j** $(m-n)(x-y)$ **k** $(3x-5)(2a-1)$ **l** $(3a-2)(p-q)$ **2 a** $(a+b)(x+y)$ **b** $(2+y)(a+b)$ **c** $(x+7)(a+b)$ **d** $(x^2+z^2)(1-y)$ **e** $(x+1)(x^2+1)$ **f** $(b+1)(a+1)$ **3 a** $(a+d)(b+c)$ **b** $(a-b)(a+7)$ **c** $(a-1)(a^2+5)$ **d** $(m+n)(a-b)$ **e** $(pq-1)(pq+a)$ **f** $(x-3y)(3x+8)$ **4 a** $(x-1)(x^2+3)$ **b** $(y^2+1)(y+1)$ **c** $(9+4a)(a-b)$ **d** $(q-p)(pq+7)$ **e** $(a-2)(m-5)$ **f** $(3x+2)(y+z)$

PAGE 49 **1 a** x^2-4 **b** x^2-9 **c** y^2-1 **d** m^2-25 **e** n^2-49 **f** p^2-16 **g** $64-x^2$ **h** y^2-36 **i** a^2-b^2 **j** x^2-y^2 **k** m^2-n^2 **l** l^2-m^2 **2 a** $9a^2-1$ **b** $4x^2-9$ **c** $16a^2-25$ **d** $49m^2-n^2$ **e** $16q^2-9$ **f** $25x^2-49$ **g** $16a^2-9b^2$ **h** $4x^2-y^2$ **i** $25x^2-16y^2$ **j** x^2-81y^2 **k** $4a^2-49b^2$ **l** $25m^2-n^2$ **m** $81a^2-121b^2$ **n** $9a^2-64b^2$ **3 a** $25x^2-1$ **b** $49a^2-4$ **c** $64x^2-49$ **d** $4x^2-9y^2$ **e** $16x^2-81y^2$ **f** $36x^2-49y^2$ **g** a^2-144 **h** $4x^2-81$ **i** $9x^2-100$ **j** $4m^2-n^2$ **k** $25-4q^2$ **l** $25x^2-121$ **m** $64a^2-121b^2$ **n** $9a^2-49b^2$

PAGE 50 **1 a** x^2+6x+9 **b** y^2+4y+4 **c** $m^2+14m+49$ **d** $x^2-8x+16$ **e** $x^2-18x+81$ **f** x^2-6x+9 **g** $y^2+22y+121$ **h** $x^2-10x+25$ **i** m^2-4m+4 **j** $x^2+2xy+y^2$ **k** $a^2-2ab+b^2$ **l** $m^2+2mn+n^2$ **2 a** $4x^2+12x+9$ **b** $4m^2+4m+1$ **c** $9y^2-6y+1$ **d** $16a^2+8a+1$ **e** $9x^2-24x+16$ **f** $4x^2-12xy+9y^2$ **g** $4a^2+4a+1$ **h** $25m^2-10m+1$ **i** $36y^2+12y+1$ **j** $9n^2+12n+4$ **k** $4x^2+20xy+25y^2$ **l** $a^2+6ab+9b^2$ **m** $4x^2+4xy+y^2$ **n** $x^2-6xy+9y^2$ **3 a** x^2+9x+6 **b** $4a^2-8a+13$ **c** $2y^2-4y-5$ **d** $2ab+2b^2$ **e** $2a^2+2b^2$ **f** $2a^2-8b^2+2ab$ **g** $10x^2-15y^2+2xy$ **h** $2x^2+6x+5$

PAGE 51 **1 a** $(x+2)(x-2)$ **b** $(x+3)(x-3)$ **c** $(x+4)(x-4)$ **d** $(x+1)(x-1)$ **e** $(x+5)(x-5)$ **f** $(x+6)(x-6)$ **g** $(a+b)(a-b)$ **h** $(x+y)(x-y)$ **i** $(m+n)(m-n)$ **j** $(a+7)(a-7)$ **k** $(y+8)(y-8)$ **l** $(t-9)(t+9)$ **m** $(p+2q)(p-2q)$ **n** $(x-3y)(x+3y)$ **o** $(m+5n)(m-5n)$ **p** $(5a-b)(5a+b)$ **q** $(7x+y)(7x-y)$ **r** $(8p-q)(8p+q)$ **s** $(2x+3y)(2x-3y)$ **t** $(3m+4n)(3m-4n)$ **u** $(4x+5y)(4x-5y)$ **2 a** $(x+11)(x-11)$ **b** $(5y+4)(5y-4)$ **c** $(1+2y)(1-2y)$ **d** $(10x+7y)(10x-7y)$ **e** $(y+2z)(y-2z)$ **f** $(1+5m)(1-5m)$ **g** $(7m+10n)(7m-10n)$ **h** $(4a+7)(4a-7)$ **i** $(3x+5y)(3x-5y)$ **j** $(3x+4y)(3x-4y)$ **k** $(a+bc)(a-bc)$ **l** $(ab+c)(ab-c)$ **m** $(6x+7y)(6x-7y)$ **n** $(p+8q)(p-8q)$ **o** $(5+8a)(5-8a)$ **3 a** $(12+5a)(12-5a)$ **b** $(a+x)(a-x)$ **c** $(4x+3y)(4x-3y)$ **d** $(2x+5)(2x-5)$ **e** $(9a+11b)(9a-11b)$ **f** $(2x+1)(2x-1)$ **g** $(9+z)(9-z)$ **h** $(2+ab)(2-ab)$ **i** $(3y+10)(3y-10)$ **j** $(2a+7)(2a-7)$ **k** $(6y+x)(6y-x)$ **l** $(4x+9y)(4x-9y)$ **m** $(1+10x)(1-10x)$ **n** $(m+13)(m-13)$ **o** $(5x+11y)(5x-11y)$ **4 a** $(x+2)(x-2)(x^2+4)$ **b** $(1-x)(1+x)(1+x^2)$ **c** $(x+5)(x-1)$ **d** $(y+6)(y-4)$ **e** $(x+1)(x-7)$ **f** $4(x+4)$

PAGE 52 **1 a** $2(a+2)(a+3)$ **b** $3(x+4)(x-1)$ **c** $4(x+4)(x+5)$ **d** $2(x-1)(x-3)$ **e** $3(m-4)(m-5)$ **f** $3(t+9)(t-1)$ **g** $2(x+2)(x+9)$ **h** $4(a-2)(a-6)$ **i** $5(y-1)(y-2)$ **j** $6(n-1)(n-6)$ **2 a** $3(x-3)(x-6)$ **b** $2(y-4)(y-6)$ **c** $4(a-5)(a-6)$ **d** $5(m+7)(m-2)$ **e** $3(n+7)(n-3)$ **f** $6(p+7)(p-4)$ **g** $4(y+7)(y-5)$ **h** $2(n-7)(n+6)$ **3 a** $a(m+5)(m-4)$ **b** $2(t+2)(t+5)$ **c** $2(y-3)(y-6)$ **d** $3(x-3)(x-7)$ **e** $p(n-3)(n-9)$ **f** $2(x-3)(x-10)$ **g** $b(a+7)(a-1)$ **h** $2(y+1)(y+7)$

PAGE 53 **1 a** $3(x+3)(x-3)$ **b** $5(a+2)(a-2)$ **c** $3(x-2)(x-3)$ **d** $14a(1-3a)$ **e** $(a^2+4b^2)(a+2b)(a-2b)$ **f** $(4a+9b)(4a-9b)$ **g** $3(x-3)(x-4)$ **h** $12t(1-4t)$ **i** $(a+5b)(a-5b+4)$ **j** $(2m-3n+5p)(2m-3n-5p)$ **k** $(1+7t)(1-7t)$ **l** $(a+2)(3a-4b)$ **2 a** $4y(2-3y)$ **b** $x(x+1)(x-1)$ **c** $4a(a-2)$ **d** $4(x+3)(x-1)$ **e** $9(x-1)$ **f** $5(t+2)(t+5)$ **g** $(8+abc)(8-abc)$ **h** $(a+1)(b+c)$ **i** $(ab+c)(ab-c)$ **j** $(x+6)(x-4)$ **k** $3(x+1)(x+2)$ **l** $(x-3)(x-13)$ **m** $(mn+1)(mn-1)$ **n** $4a(a-x)$ **o** $(a-1)(m+n)$

PAGE 54 **1 a** $7(x-1)$ **b** $(x+3)(x-3)$ **c** $(m+5)(m-5)$ **d** $x(x-2y)$ **e** $-5(m+n)$ **f** $a(y+b)$ **g** $4a(a-2)$ **h** $(x+y)(2+m)$ **i** $(x+11)(x-11)$ **j** $a^2(a-3b)$ **k** $n(n-9)$ **l** $(3x+4y)(3x-4y)$ **m** $3(x-2)$ **n** $-a(a+2+y)$ **2 a** $6y(3-2y)$ **b** $4a(a-x)$ **c** $(a+1)(b+c)$ **d** $(mn+1)(mn-1)$ **e** $(ab+c)(ab-c)$ **f** $(x-y+z)(x-y-z)$ **g** $x(x+1)(x-1)$ **h** $(m^2+1)(m+1)$ **i** $(y-7)(x+m)$ **j** $(m+n)(a-1)$ **3 a** $(x+6)(x-4)$ **b** $(x-9)(x+3)$ **c** $(t-4)(t+2)$ **d** $(x-3)(x-7)$ **e** $(a-2)(a-3)$ **f** $(x-2)(x+1)$ **g** $(m+5)^2$ **h** $(y-4)(y-5)$ **4 a** $4(x+3)(x-1)$ **b** $2(x-2)(x-3)$ **c** $3(x+2)(x+1)$ **d** $2(x+1)(x+2)$ **e** $9(x-2)(x+1)$ **f** $3(x-5)(x+2)$

PAGE 55 **1 a** $2a$ **b** x **c** $2m$ **d** $\frac{5a}{6}$ **e** $\frac{11a}{15}$ **f** $\frac{29a}{35}$ **g** $\frac{11}{2x}$ **h** $\frac{19}{9x}$ **i** $\frac{5m}{3n}$ **2 a** a **b** $\frac{x}{3}$ **c** $\frac{m}{3}$ **d** $\frac{x}{6}$ **e** $\frac{a}{10}$ **f** $\frac{11q}{15}$ **g** $\frac{1}{2x}$ **h** $\frac{-5}{9x}$ **i** $\frac{4m}{3n}$ **3 a** $\frac{xy}{20}$ **b** $\frac{a^2}{12}$ **c** $\frac{am}{bn}$ **d** $\frac{b}{2}$ **e** 1 **f** 2 **g** $3q$ **h** $\frac{8m^2}{3}$ **i** $2\frac{2}{3}$ **4 a** 3 **b** $1\frac{1}{3}$ **c** $\frac{2}{7}$ **d** 2 **e** $\frac{1}{4}$ **f** $\frac{1}{3}$ **g** 6 **h** a^2 **i** $16\frac{4}{5}$

PAGE 56 **1 a** $\frac{1}{2^5}$ **b** $\frac{1}{7^2}$ **c** $\frac{1}{3^4}$ **d** $\frac{1}{5^6}$ **e** $\frac{1}{8^3}$ **f** $\frac{1}{10^8}$ **g** $\frac{1}{x^4}$ **h** $\frac{1}{a^2}$ **i** $\frac{9}{m^4}$ **j** $\frac{1}{(-7)^3}$ **k** 2^4 **l** $(\frac{6}{5})^2$ **2 a** $\frac{1}{9}$ **b** $\frac{1}{8}$ **c** $\frac{1}{64}$ **d** $\frac{1}{125}$ **e** $\frac{1}{100\,000}$ **f** $\frac{1}{27}$ **g** $2\frac{1}{4}$ **h** $\frac{8}{27}$ **i** 16 **j** 16 **k** 27 **l** $1\frac{11}{25}$ **3 a** 3^{-2} **b** 3^{-5} **c** a^{-1} **d** y^{-2} **e** $4x^{-3}$ **f** $8x^{-5}$ **g** $7y^{-1}$ **h** $6a^{-4}$ **i** $3^{-1}x^{-4}$ **j** $5^{-3}a$ **k** $5^{-1}m^{-2}$ **l** $3m^{-3}n$ **4 a** $\frac{1}{25}$ **b** $\frac{1}{216}$ **c** $\frac{1}{64}$ **d** $\frac{1}{18}$ **e** $\frac{7}{8}$ **f** $\frac{1}{8}$ **g** $\frac{1}{125}$ **h** $\frac{2}{25}$ **i** $\frac{1}{200}$ **j** $\frac{1}{24}$ **k** 25 **l** $\frac{1}{27}$ **5 a** $x=-3$ **b** $x=3$ **c** $x=-1$ **d** $x=-4$ **e** $x=11$ **f** $x=4$ **g** $x=4$ **h** $x=-9$ **i** $x=3$ **j** $x=2$ **k** $x=3$ **l** $x=-5$ **6 a** 3^2 **b** 2^3 **c** 5^{-3} **d** 8^{-14} **e** 7^7 **f** 6^{-16} **g** 7^{-14} **h** 4^{-21} **i** 8^{-2} **j** 4^{-9} **k** x^{12} **l** a

PAGE 57 **1 a** x^{17} **b** a^5 **c** m^3 **d** $2n^{12}$ **e** $16p^6q^{10}$ **f** $16k^7$ **g** $2h^{-12}$ **h** $30y^{-7}$ **i** $5q^{11}$ **j** $4xy^5$ **k** $2a^2b^{11}$ **l** $2p^{-2}q^2$ **2 a** x^2 **b** $\frac{a^2}{b^4}$ **c** $\frac{p}{2q}$ **d** $\frac{1}{n^6}$ **e** $2z^5$ **f** q^9 **g** x^{21} **h** $\frac{x^2}{y^6}$ **i** 1 **3 a** $\frac{y^{12}}{x^8}$ **b** $\frac{k}{3}$ **c** $\frac{2}{a^4}$ **d** $\frac{49}{4x^4}$ **e** $\frac{m^9}{k^6}$ **f** $\frac{n^2}{2}$ **g** $\frac{x^{10}}{a^{15}}$ **h** $\frac{1}{2p^3}$ **i** $\frac{1}{4x^2}$ **4 a i** 4 **ii** 9 **iii** 25 **iv** 100 **b i** 4 **ii** 9 **iii** 25 **iv** 100 **c** $\frac{1}{2}$ **d i** 9 **ii** 4 **iii** 11 **iv** 6 **v** 12 **vi** 8 **vii** 7

PAGE 58 **1** D **2** C **3** D **4** B **5** B **6** C **7** C **8** A **9** B **10** A

PAGE 59 **1 a** $6x^2-x-15$ **b** $20a^2-39a+7$ **c** $9a^2-4$ **d** x^2+6x+9 **2 a** $(m+6)(m-6)$ **b** $(a-b)(x+y)$ **c** $(m-10)(m+8)$ **d** $(a-3)(a-6)$ **e** $2(n+3)(n+5)$ **f** $4(1-x)(1+x)$ **3 a** $\frac{29x}{15y}$ **b** $\frac{4b^2}{15a^2}$ **4 a** $\frac{5t^5}{12u^2}$ **b** $\frac{14}{3x^{11}}$ **c** $\frac{3x^2}{4}$

Answers

Chapter 5 – Right-angled triangles and trigonometry

Page 60 **1 a** 25 m **b** 15.3 m **c** 85 cm **2 a** 15.6 m **b** 228 mm **c** 14.3 m **3 a** 11.4 **b** 9.4 **c** 14.5 **d** 22.0 **e** 23.0 **f** 68.5

Page 61 **1 a** a = adj, b = opp, c = hyp **b** a = adj, b = hyp, c = opp **c** a = opp, b = adj, c = hyp **2 a** BC **b** EF **c** HI
3 a $\sin\theta = \frac{y}{z}$, $\cos\theta = \frac{x}{z}$, $\tan\theta = \frac{y}{x}$ **b** $\sin\theta = \frac{p}{r}$, $\cos\theta = \frac{q}{r}$, $\tan\theta = \frac{p}{q}$ **c** $\sin\theta = \frac{b}{c}$, $\cos\theta = \frac{a}{c}$, $\tan\theta = \frac{b}{a}$ **4 a** $\sin\theta = \frac{7}{25}$, $\cos\theta = \frac{24}{25}$, $\tan\theta = \frac{7}{24}$
b $\sin\theta = \frac{3}{5}$, $\cos\theta = \frac{4}{5}$, $\tan\theta = \frac{3}{4}$ **c** $\sin\theta = \frac{15}{17}$, $\cos\theta = \frac{8}{17}$, $\tan\theta = \frac{15}{8}$ **5 a** tan **b** cos **c** sin

Page 62 **1 a** 27° **b** 46° **c** 30° **d** 78° **e** 78° **f** 83° **g** 22° **h** 55° **i** 65° **2 a** 83°25' **b** 89°34' **c** 63°28' **d** 27°16' **e** 41°45' **f** 30°46' **g** 24°46' **h** 57°21' **i** 54°28' **3 a** 0.848 **b** 0.839 **c** 0.788 **d** 0.139 **e** 0.866 **f** 1.376 **4 a** 0.882 **b** 2.53 **c** 0.770 **d** 6.34 **e** 0.958 **f** 1.26 **g** 40.8 **h** 51.5 **i** 15.3 **5 a** 0.100 **b** 0.049 **c** 65.670 **d** 0.149 **e** 0.183 **f** 15.749 **g** 0.627 **h** 0.814 **i** 55.210 **6 a** 35° **b** 56° **c** 74° **d** 34° **e** 70° **f** 46° **g** 54° **h** 20° **i** 56° **7 a** 59°18' **b** 18°23' **c** 55°26' **d** 20°14' **e** 62°50' **f** 39°55' **g** 30°27' **h** 70°33' **i** 32°37'

Page 63 **1 a** 2.8 cm **b** 4.0 cm **c** 11.0 cm **d** 5.9 cm **e** 5.0 cm **f** 1.0 m **2 a** 38.83 mm **b** 20.38 mm **c** 62.93 mm **d** 4.75 cm **e** 4.80 cm **f** 7.35 cm

Page 64 **1 a** 11.40 cm **b** 8.75 cm **c** 15.40 cm **d** 10.64 cm **e** 23.60 cm **f** 12.91 cm **2 a** 85.2 mm **b** 201.8 mm **c** 38.9 mm **d** 278.1 mm **e** 26.2 mm **f** 538.3 mm **3 a** 10.10 cm **b** 7.98 cm **c** 17.36 cm

Page 65 **1 a** 53° **b** 46° **c** 30° **2 a** α = 30°10' **b** β = 65°43' **c** θ = 23°41' **d** θ = 60°31' **e** α = 27°32' **f** β = 67°4'
3 a α = 57°9' **b** β = 36°36' **c** θ = 58°50' **4 a** α = 35°25' **b** β = 39°35' **c** θ = 60°56'

Page 66 **1** 11.4 cm **2** 2.14 m **3** θ = 42°43' **4** 8.40 m **5** 12 m **6** 17.4 cm

Page 67 **1** 564 m **2** 81.04 m **3** 2° **4** 27 m **5** 54° **6** 55°

Page 68 **1** 41 m **2** 1°14' **3** 11°39' **4** 196 m **5** 44° **6** 119 m

Page 69 **1 a** 90° **b** 180° **c** 45° **d** $67\frac{1}{2}$° **e** 45° **f** $67\frac{1}{2}$° **2 a** ENE **b** SSE **c** WNW **d** SSW **e** NNW **f** ESE **3** 7.07 n miles
4 a $\angle PRQ = 90°$, $\angle PQR = 45°$ so ΔPQR is isosceles **b** 113 km **5 a** 90° **b i** 23 km **ii** 30 km

Page 70 **1 a** 125° **b** 067° **c** 235° **d** 290° **e** 140° **f** 210° **g** 125° **h** 248° **2 a** **b** **c** **3** 301°

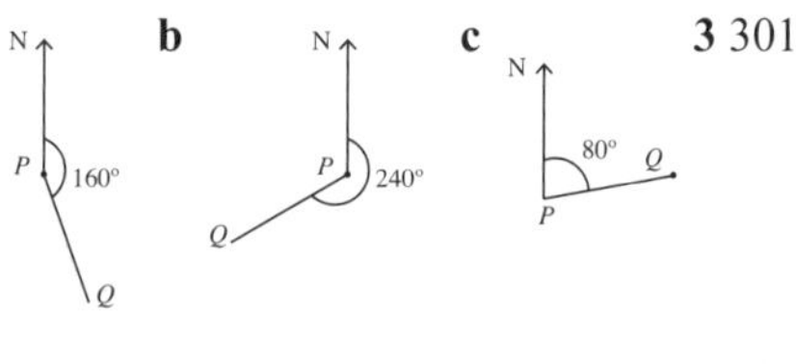

4 a **b** 90° **c** 50° **d** 256 km

Page 71 **1 a** N71°E **b** N39°W **c** S52°E **d** S17°W **e** S50°E **f** S70°W **g** S62°E **h** N33°W **2 a**

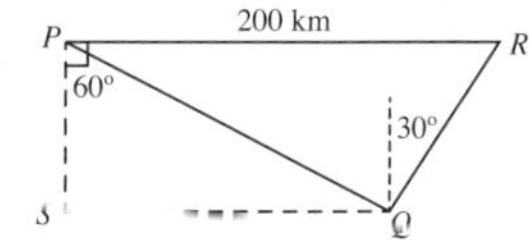

b 90° **c** 100 km **d** 150 km **3 a** 062°T **b** 230°T **c** 320°T **d** 160°T **e** 043°T **f** 293°T **g** 146°T **h** 193°T **4 a** N55°E **b** S60°E **c** N20°W **d** S10°W **e** S27°E **f** S65°W **g** N39°W **h** S84°E

Page 72 **1** B **2** C **3** B **4** C **5** D **6** B **7** B **8** D **9** D **10** A

Page 73 **1** 19.5 m **2 a** 045°T **b** S45°W **c** 99 m **3 a**

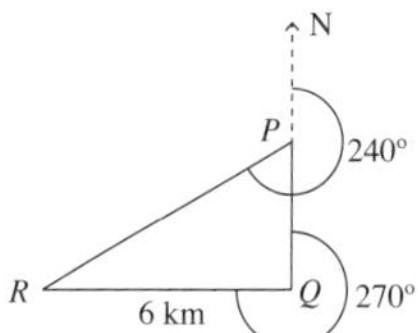

b i 90° **ii** 60° **c** 6.9 km **4 a** 433 m **b** 516 m **c** 266 m **5 a**

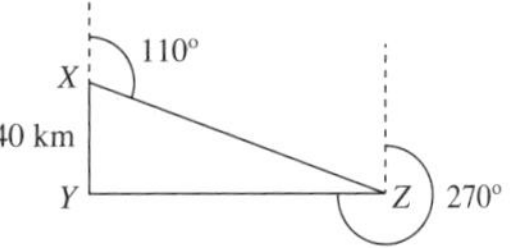

b i 90° **ii** 70° **c** 117 km

Chapter 6 – Linear relationships

Page 74 **1 a i** $\frac{3}{4}$ **ii** $-\frac{3}{4}$ **b i** (2, 4) **ii** $(4, 2\frac{1}{2})$ **c i** 10 units **ii** 5 units **2 a** (–1, 3) **b** $\frac{3}{4}$ **c** 20 units **3 a** (3, –4) **b** (1, 1) **c** $(1\frac{1}{2}, -1\frac{1}{2})$
4 a 4 **b** –1 **c** $-\frac{2}{3}$ **5 a** 13 units **b** $\sqrt{13}$ units **c** $\sqrt{41}$ units

Page 75 **1 a i** 2 **ii** 1 **b i** 1 **ii** –4 **c i** –1 **ii** 2 **2** gradient, y-intercept **3 a** $y = 2x + 3$ **b** $y = \frac{1}{2}x + 4$ **c** $y = -2x + 5$ **d** $y = 3x - 2$ **e** $y = -3x - 1$ **f** $y = -\frac{3}{4}x$ **g** $y = \frac{2}{3}x + \frac{1}{2}$ **h** $y = 1$ **4 a** $m = 4, c = 1$ **b** $m = 2, c = -5$ **c** $m = -3, c = 2$ **d** $m = 1, c = -4$ **e** $m = \frac{3}{2}, c = \frac{1}{2}$ **f** $m = 0, c = -8$ **g** $m = -1, c = -7$ **h** $m = -1, c = 6$ **i** $m = -5, c = 2$ **5 a**

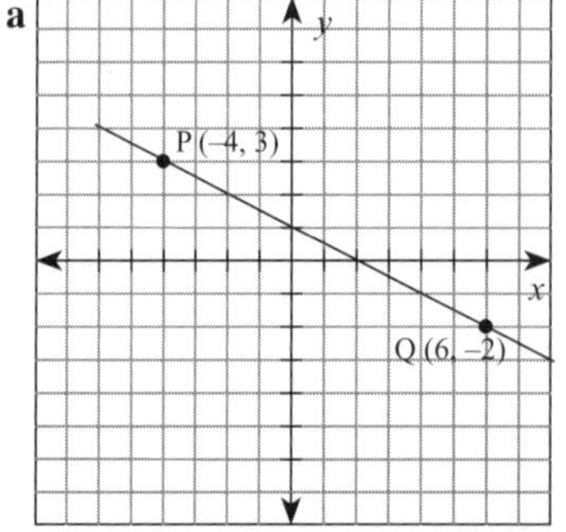

b $-\frac{1}{2}$ **c** 1 **d** $y = -\frac{1}{2}x + 1$

Answers

PAGE 76 1 a

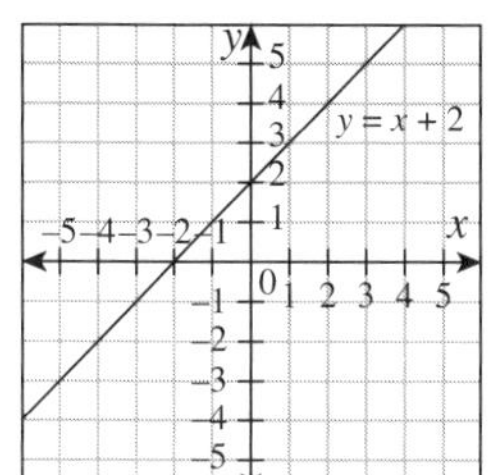

b

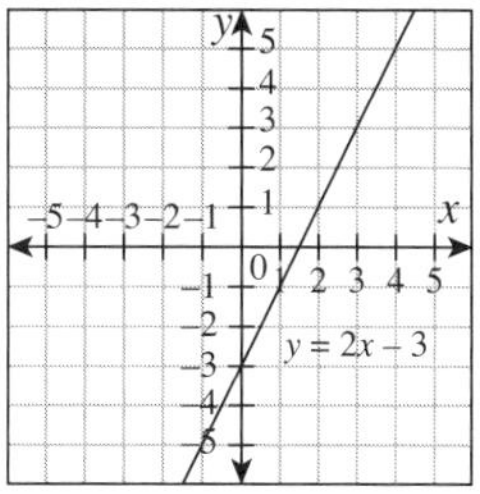

$y = 2x - 3$

c

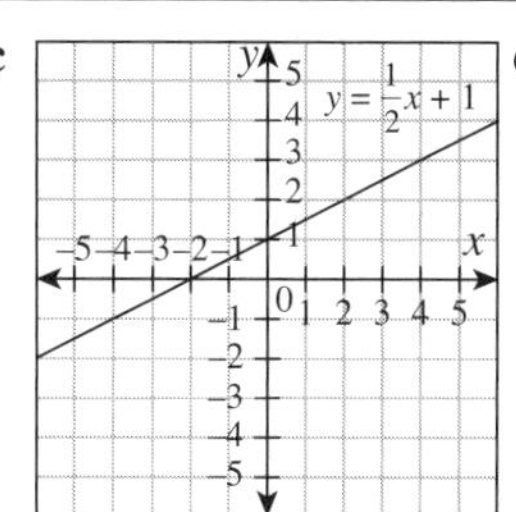

d

$y = -2x + 4$

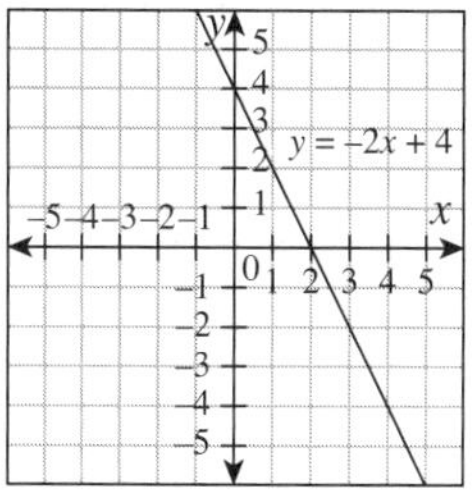

e

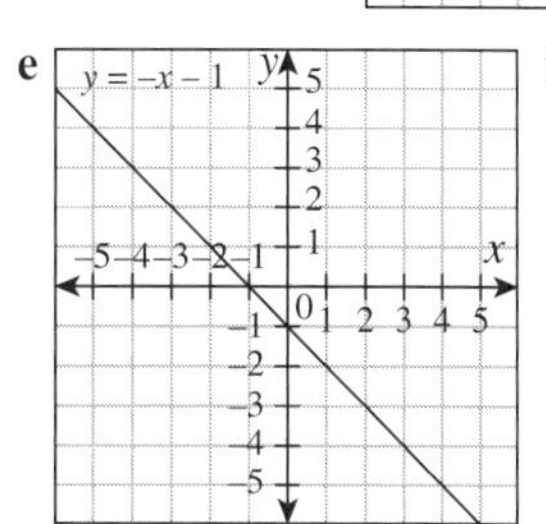

f

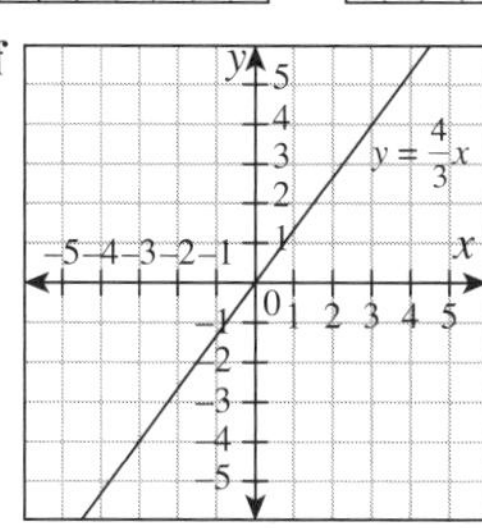

2 a

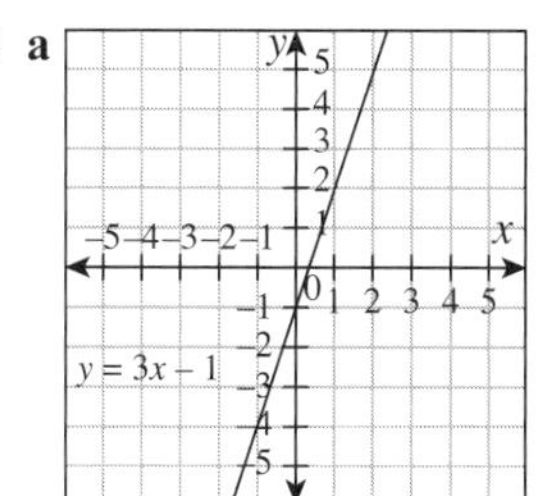

b

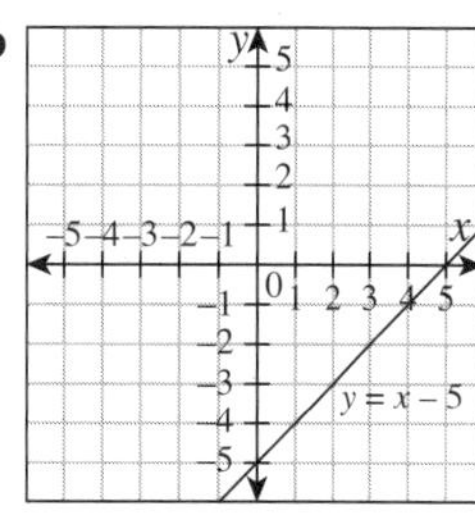

c

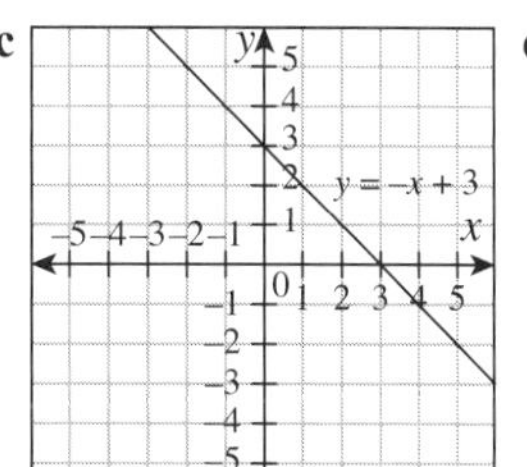

d

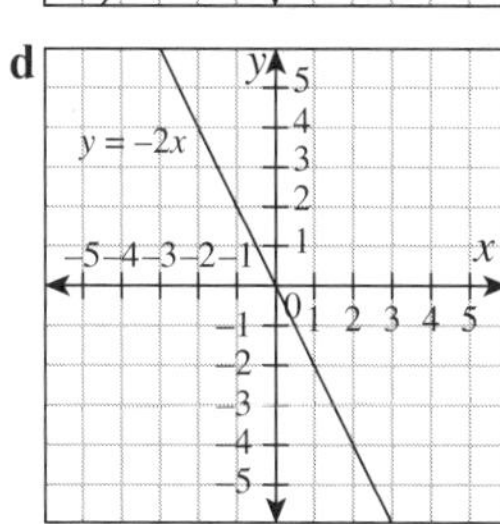

e

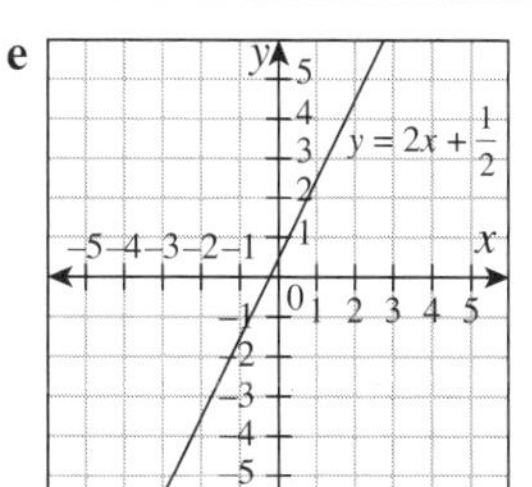

f

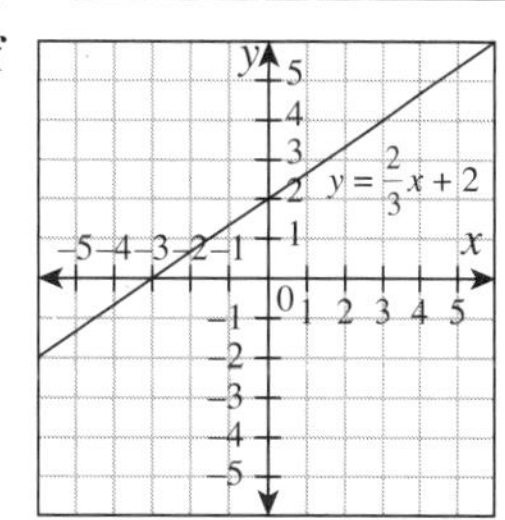

PAGE 77 1 **a** 5 **b** the fixed amount of pocket money per week, \$5 **c** 5 **d** The rate that Andrew's mother pays him per hour when he helps. **2 a** 90 **b** Clair is 90 km from Baxton **c** –15 **d** Melissa rides at a constant speed of 15 km/h **e** $d = -15t + 90$

PAGE 78 1

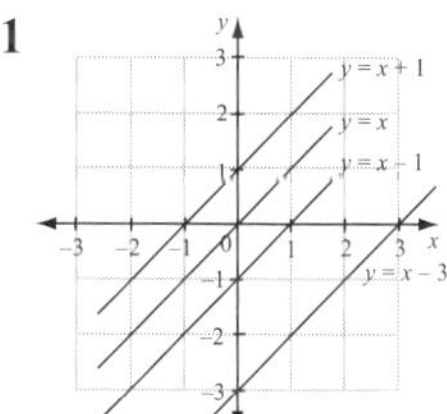

2

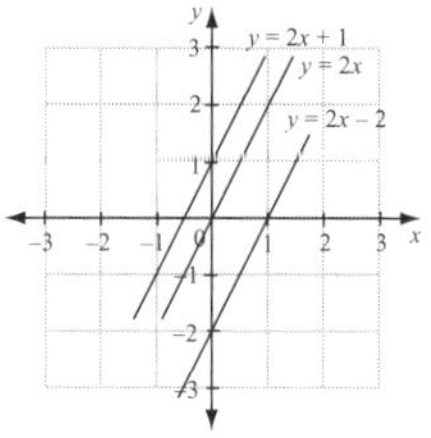

$y = 2x + 1$, $y = 2x$, $y = 2x - 2$

3

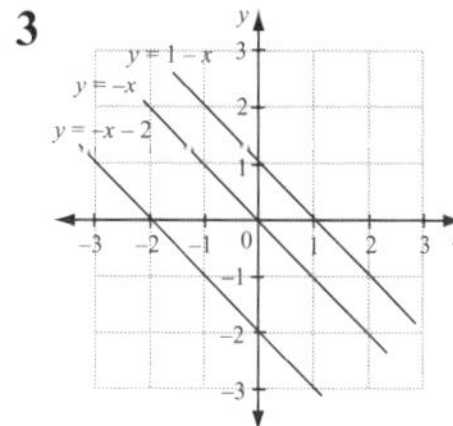

4

$y = \frac{1}{3}x + 2$, $y = \frac{1}{3}x$, $y = \frac{1}{3}x - 1$

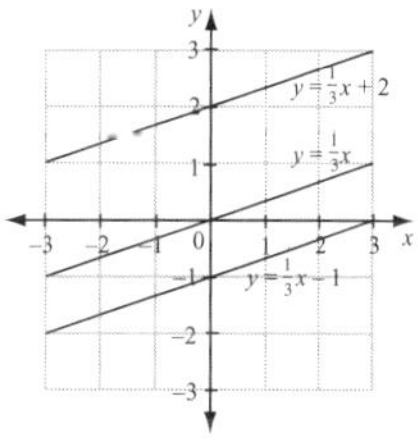

5

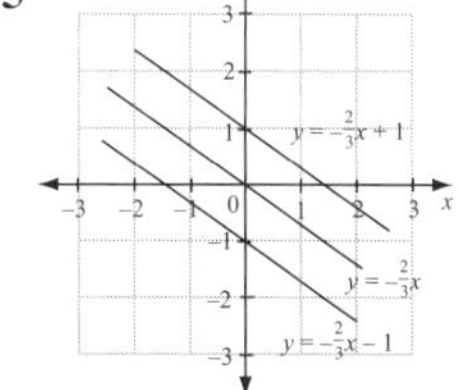

6 parallel

PAGE 79 1 **a** –1 **b** –1 **c** –1 **d** –1 **e** –1 **f** –1 **g** –1 **h** –1 **2** negative one

3

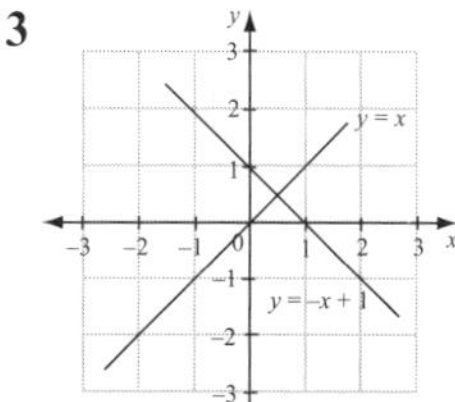

4

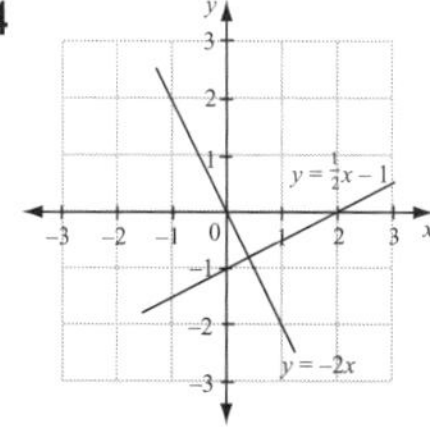

5

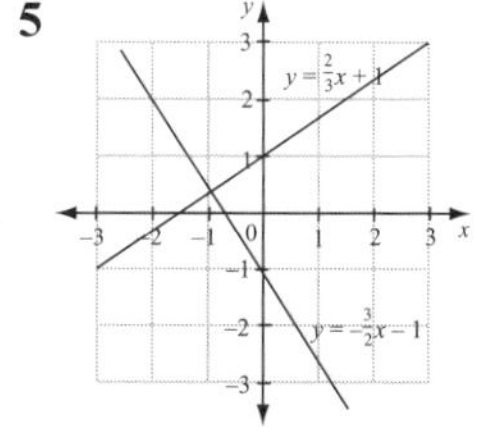

6

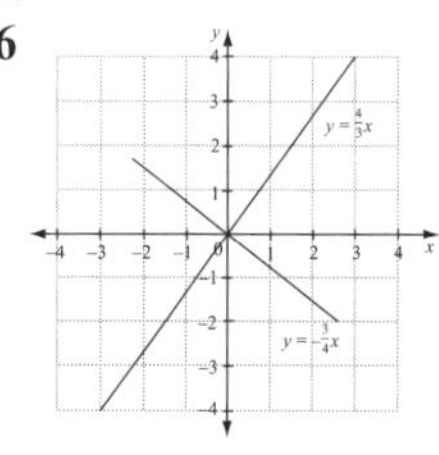

7 perpendicular

PAGE 80 1 **a** the same **b** negative reciprocals **2 a** parallel **b** neither **c** perpendicular **d** perpendicular **e** parallel **f** perpendicular **g** neither **h** perpendicular **3 a** $y = 3x + 2$ **b** $y = -2x + 2$ **c** $y = \frac{1}{2}x + 2$ **d** $y = -\frac{5}{3}x + 2$ **4 a** $y = -\frac{1}{4}x$ **b** $y = 3x$ **c** $y = \frac{1}{2}x$ **d** $y = -\frac{2}{3}x$ **5 a** $\frac{2}{3}$ **b** $\frac{2}{3}$ **c** $-\frac{3}{2}$ **d** 90°

PAGE 81 1 **a** $2x - 5y - 9 = 0$ **b** $3x + 4y - 8 = 0$ **c** $5x - 2y - 7 = 0$ **d** $4x - 8y + 3 = 0$ **e** $2x + y - 9 = 0$ **f** $8x - y + 7 = 0$ **g** $2x - 3y + 6 = 0$ **h** $8x - 9y + 12 = 0$ **i** $x - 6y + 3 = 0$ **2 a** $y = -\frac{2}{3}x + \frac{8}{3}$; $m = -\frac{2}{3}$, $c = \frac{8}{3}$ **b** $y = -\frac{1}{5}x + \frac{7}{5}$; $m = -\frac{1}{5}$, $c = \frac{7}{5}$

c $y = \frac{3}{2}x - \frac{3}{2}$; $m = \frac{3}{2}$, $c = -\frac{3}{2}$ **d** $y = x + 7$; $m = 1$, $c = 7$ **e** $y = -2x + 9$; $m = -2$, $c = 9$ **f** $y = \frac{5}{6}x + \frac{11}{6}$; $m = \frac{5}{6}$, $c = \frac{11}{6}$
g $y = \frac{3}{2}x - 3$; $m = \frac{3}{2}$, $c = -3$ **h** $y = -\frac{4x}{5} - \frac{3}{5}$; $m = -\frac{4}{5}$, $c = -\frac{3}{5}$ **i** $y = 2x + 6$; $m = 2$, $c = 6$ **3 a** $y = 4x + 3$; $4x - y + 3 = 0$
b $y = 2x - 5$; $2x - y - 5 = 0$ **c** $y = 3x + 7$; $3x - y + 7 = 0$ **d** $y = \frac{1}{2}x + 4$; $x - 2y + 8 = 0$ **e** $y = \frac{2}{3}x + 6$; $2x - 3y + 18 = 0$
f $y = -\frac{5}{6}x + 3$; $5x + 6y - 18 = 0$

PAGE 82 **1 a** yes **b** no **c** yes **d** yes **e** yes **f** yes **g** yes **h** no **2 a** no **b** yes **c** no **d** yes **e** no **f** yes **g** no **h** yes
3 a parallel **b** parallel **c** neither **d** perpendicular **e** perpendicular **f** neither **g** parallel **h** perpendicular **4 a** $3x - y - 1 = 0$
b $4x - 5y = 0$ **c** $x - 2y + 8 = 0$ **d** $x + y + 1 = 0$

PAGE 83 **1 a**

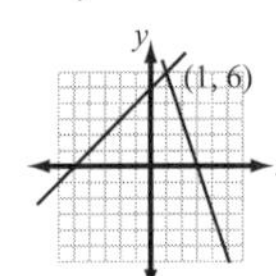

b

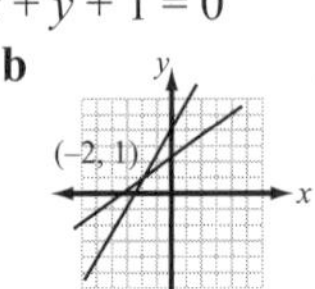

2 a

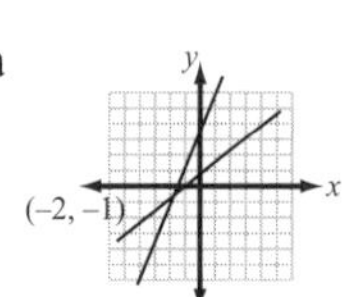

b

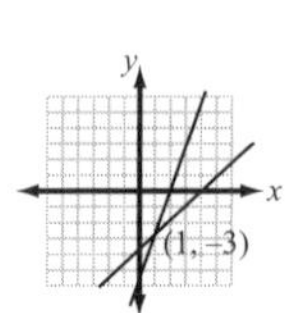

3 a

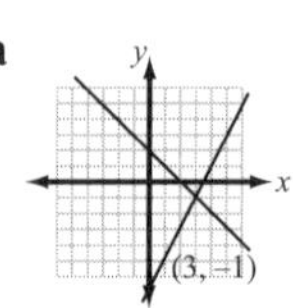

b

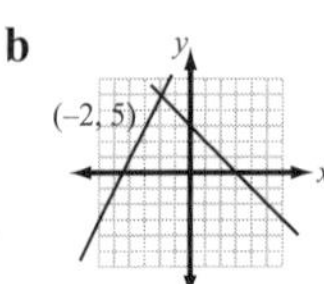

PAGE 84 **1 a** $y = 4x + 14$ **b** $y = 2x - 9$ **c** $y = \frac{1}{2}x + \frac{9}{2}$ **d** $y = \frac{2}{3}x + \frac{22}{3}$ **e** $y = -\frac{1}{3}x + \frac{17}{3}$ **f** $y = -3x + 8$ **2** $x = -7, y = 7$ **3 a** $m = \frac{2}{3}$
b $y = \frac{2}{3}x + \frac{7}{3}$ **4 a** (5, 4) **b** (−1, −3) **c** (1, −2) **d** (−1, 2) **e** (2, −1) **f** (−2, 1) **5 a** 2 **b** −3 **c**

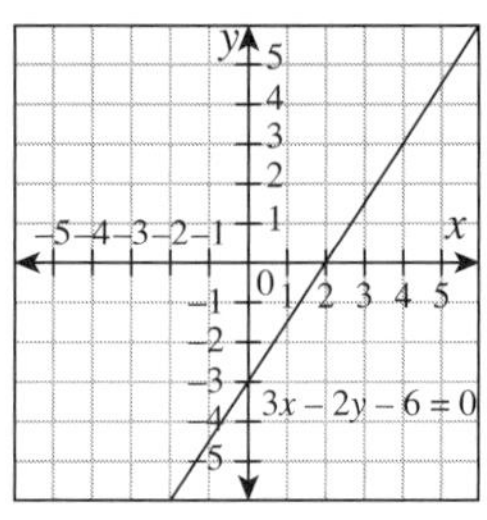

PAGE 85 **1** A **2** D **3** C **4** C **5** C **6** B **7** A **8** B **9** D **10** D

PAGE 86 **1 a** $y = 2x - 3$ **b** 2 **c** −3 **d** yes **2 a** −1 **b** 1 **c** yes, gradients multiply to −1 **d** $(1\frac{1}{2}, 1\frac{1}{2})$ **e** yes, both have gradient 1
3 a $-\frac{4}{3}$ **b** $y = -\frac{4}{3}x + \frac{7}{3}$ **c** $4x + 3y - 7 = 0$ **d** yes **e** $y = -\frac{4}{3}x$ **f** 10 units

CHAPTER 7– Non-linear relationships

PAGE 87 **1**

x	−3	−2	−1	0	1	2	3
$y = x^2$	9	4	1	0	1	4	9
$y = 2x^2$	18	8	2	0	2	8	18
$y = \frac{1}{2}x^2$	$4\frac{1}{2}$	2	$\frac{1}{2}$	0	$\frac{1}{2}$	2	$4\frac{1}{2}$

2

x	−3	−2	−1	0	1	2	3
$y = x^2$	9	4	1	0	1	4	9
$y = x^2 + 1$	10	5	2	1	2	5	10
$y = x^2 - 1$	8	3	0	−1	0	3	8

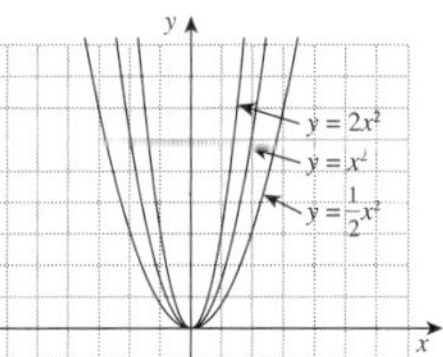

3

x	−3	−2	−1	0	1	2	3
$1 - x^2$	−8	−3	0	1	0	−3	−8

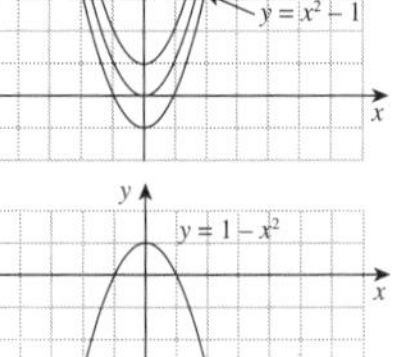

a $x = 0$ **b** (0, 1) **c** $y = 1$ **d** $x = -1, x = 1$ **e** $y = 1$

4

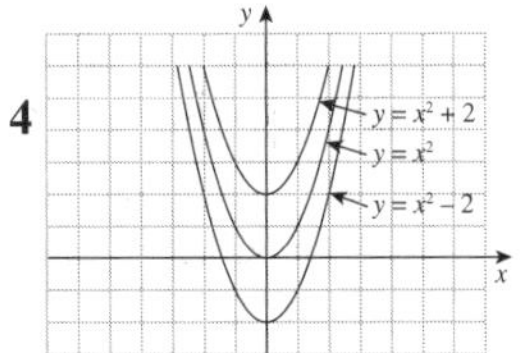

d While sketching $y = x^2 + 2$, move the parabola $y = x^2$ two units vertically upwards and for the parabola $y = x^2 - 2$ move two units vertically downwards.

PAGE 88 **1 a** Yes, $x = 0$ **b** Yes, (0, 0) **c** The curve is turned upside down. If the coefficient of x^2 is positive the parabola is concave up and if the coefficient is negative the curve is concave down. **d** It determines the width of the parabola. The larger the absolute value of a, the narrower the parabola. **2 a** Yes, $x = 0$ **b** No **c** c is the y-intercept. It is the value where the curve cuts the y-axis. It has the effect of moving the curve up, if c is positive, or down, if c is negative.

3 a –2 **b** –1 and 1 **c**

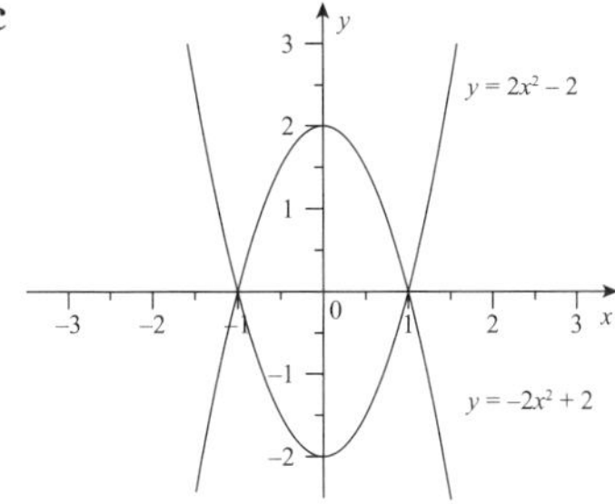

PAGE 89 **1 a** $x = 0$ **b** (0, 3) **c** 5 **d** 2 **e** The parabola is symmetrical, so whenever a point other than the vertex is known a matching point will be found on the other side of the axis of symmetry. So, for this parabola (–1, 5) and (–2, 11) will be found on the parabola. **2 a i** –2 **ii** –3 **iii** No, the curve is concave down and the vertex is below the x-axis. **b i** –8 **ii** $\frac{1}{2}$ **iii** Yes, $x = -4$ and $x = 4$ **3 a** $y = x^2 - 4$ **b** $y = \frac{1}{4}x^2 - 1$ **c** $y = -2x^2 + 5$

PAGE 90 **1 a**

x	–2	–1	0	1	2	3
$y = 2^x$	$\frac{1}{4}$	$\frac{1}{2}$	1	2	4	8

b

x	–3	–2	–1	0	1	2	3
$y = 2^{-x}$	8	4	2	1	$\frac{1}{2}$	$\frac{1}{4}$	$\frac{1}{8}$

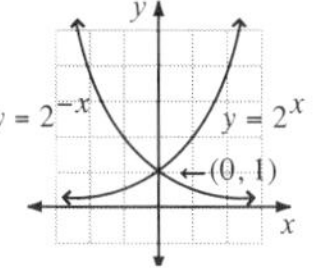

2 a

x	–2	–1	0	1	2	3
$y = 2^x$	$\frac{1}{4}$	$\frac{1}{2}$	1	2	4	8

b

x	–1	0	1	2	3
$y = 3^x$	$\frac{1}{3}$	1	3	9	27

c

x	–2	$-1\frac{1}{2}$	–1	$-\frac{1}{2}$	0	$\frac{1}{2}$	1	2
$y = 5^x$	0.04	0.09	0.2	0.44	1	2.24	5	25

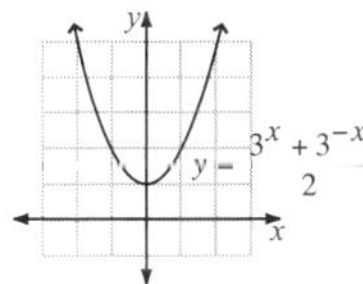

3

x	–3	–2	–1	0	1	2	3
3^x	27	9	$\frac{1}{3}$	1	3	9	27
3^{-x}	$\frac{1}{27}$	$\frac{1}{9}$	3	1	$\frac{1}{3}$	$\frac{1}{9}$	$\frac{1}{27}$
$\frac{3^x + 3^{-x}}{2}$	13.5	4.6	1.7	1	1.7	4.6	13.5

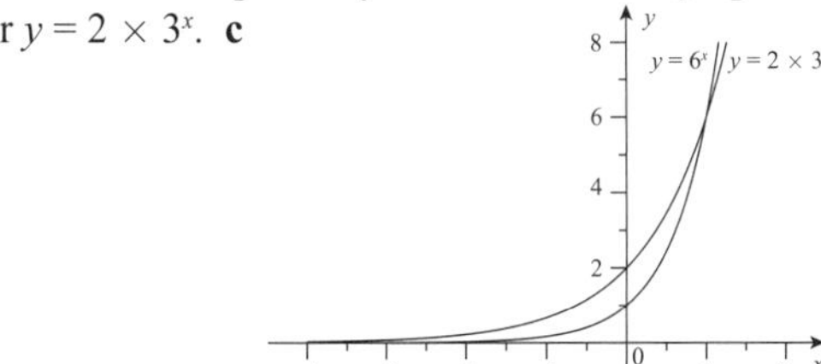

PAGE 91 **1 a** 1 **b** No. 2^x is always positive. **c** $y = 0$ **d i** 8 **ii** 16 **iii** 32 **iv** 1024 **v** 1 048 576 **vi** 1 073 741 824 **e** The values of 2^x increase exponentially as x increases (positively). They get very large very quickly. **2 a** 1 **b** 1 **c** The y-intercept is found when $x = 0$ and $a^0 = 1$ for all values of a. **d** No. If $x > 0$, $a^x > 1$. If $x < 0$, a^x is a positive fraction. **e** All curves of the form $y = a^x$ have the x-axis as an asymptote. **f** All the curves have a similar shape. They all increase exponentially when $x > 0$. **3 a** 2×3^x is not the same as $(2 \times 3)^x$. $(2 \times 3)^x = 6^x$ but in 2×3^x, 3 is first raised to the power of x then that value is multiplied by 2. **b** Both curves have a similar shape. They have the same asymptote and increase exponentially. The y-intercept is different, being 1 for $y = 6^x$ but 2 for $y = 2 \times 3^x$. **c**

PAGE 92 **1 a** (0, 0), 2 units **b** (0, 0), 7 units **c** (0, 0), $\frac{2}{3}$ units **d** (0, 0), 9 units **2 a** $x^2 + y^2 = 9$ **b** $x^2 + y^2 = 49$ **c** $x^2 + y^2 = 4$ **d** $x^2 + y^2 = 100$ **3 a** $x^2 + y^2 = 4$ **b** $x^2 + y^2 = 25$ **c** $x^2 + y^2 = 64$

4 a $r = 4$, centre (0, 0)

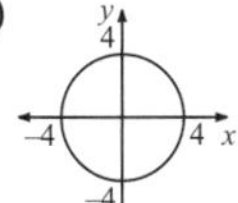

b $r = 1$, centre (0, 0)

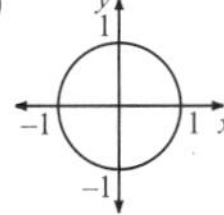

c $r = 3$, centre (0, 0)

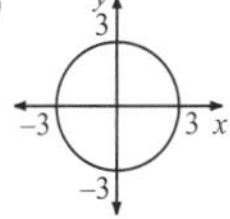

d $r = 6$, centre (0, 0)

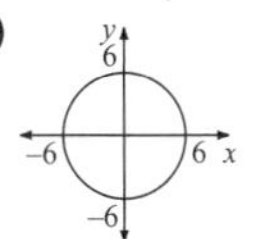

Answers

PAGE 93 **1 a** straight line **b** parabola **c** straight line **d** parabola **e** parabola **f** straight line **g** parabola **h** exponential **i** none of these **j** circle **k** exponential **l** circle **2 a** D **b** H **c** F **d** G **e** I **f** C **g** A **h** B **i** E **j** L **k** J **l** K
3 a **b** **c**

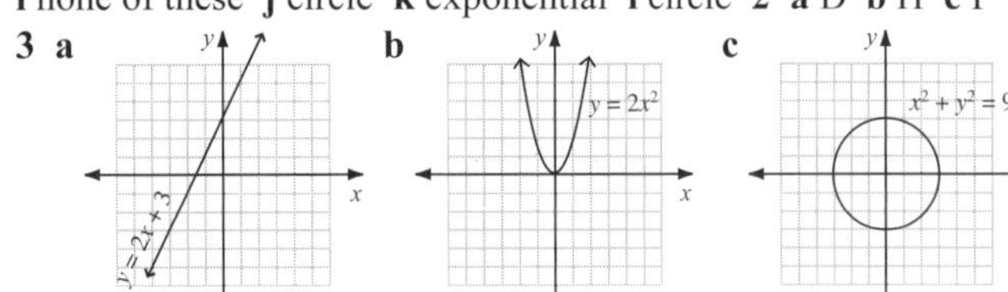

PAGE 94 **1** D **2** C **3** A **4** A **5** A **6** C **7** B **8** A **9** C **10** C
PAGE 95 **1 a** concave down **b** (0, 3) **c** $x = -3$ and 3 **d** The curve is concave up and the vertex at (0, 3) is above the x-axis so the curve will not cut the x-axis. **2 a** $y = 1 - x^2$ **b** $y = 5^x$ **c** $y = \frac{1}{8}x^2$ **3 a** 1 **b** (0, 1) and (1, 3) **c** $y = 2x + 1$ **d** No **4 a** parabola **b** –2 **c** $\frac{1}{2}$ **d** 16

CHAPTER 8 – Area, surface area and volume

PAGE 96 **1 a** $A = \frac{1}{2}bh$ **b** $A = s^2$ **c** $A = lb$ **d** $A = bh$ **e** $A = \frac{1}{2}h(a + b)$ **f** $A = \frac{1}{2}xy$ **g** $A = \frac{1}{2}xy$ **h** $A = \pi r^2$ **2 a** 10.32 cm² **b** 56 cm² **c** 40.74 cm² **d** 27.52 cm² **e** 1385.44 cm² **f** 106 cm² **g** 288.3 cm² **h** 96 cm² **i** 49 cm²
PAGE 97 **1 a** 520 cm² **b** 1947 cm² **c** 24 cm² **2 a** 348 m² **b** 600 m² **c** 1028.3 cm² **d** 235.6 cm² **e** 1764 m² **f** 850 cm²
PAGE 98 **1 a** 325.58 m² **b** 447.6 m² **c** 168.2 cm² **d** 308.5 cm² **e** 50.3 cm² **f** 100.5 m² **g** 88.2 cm² **h** 34.4 cm² **i** 222.5 cm²
PAGE 99 **1 a** 1032 cm² **b** 2649.92 cm² **c** 138.24 m² **2 a** 81.72 m² **b** 2408.98 mm² **3 a** 267.2 cm² **b** 589.8 cm² **c** 863.6 cm²
PAGE 100 **1 a** 336 cm² **b** 720 cm² **c** 1195.8 cm² **d** 4696 cm²
PAGE 101 **1 a** 864 m² **b** 5212 cm² **c** 8999.5 m² **d** 4602.8 cm²
PAGE 102 **1 a i** 78.54 cm² **ii** 439.82 cm² **b i** 415.48 cm² **ii** 2456.73 cm² **c i** 4.52 m² **ii** 27.90 m² **d i** 47.78 mm² **ii** 480.29 mm² **2 a** 26.88π cm² **b** 2.97π m² **c** 96π m² **d** 702π mm²
PAGE 103 **1 a i** 20.4 cm² **ii** 29.4 cm² **iii** 49.8 cm² **b i** 42.5 m² **ii** 35.9 m² **iii** 78.4 m² **c i** 14.1 m² **ii** 70.7 m² **iii** 84.8 m² **d i** 3620 mm² **ii** 14 600 mm² **iii** 18 200 mm² **2** 60.3 m² **3** 3.8 m²
PAGE 104 **1 a** 1600 cm² **b** 510 cm² **c** 38 m² **d** 2200 cm²
PAGE 105 **1 a** 1944 cm³ **b** 26.9 m³ **c** 14 000 cm³ **d** 1892.4 m³ **2 a** 110 cm³ **b** 2.3 m³ **c** 13 000 cm³ **3 a** 270 cm³ **b** 11 074 cm³ **c** 4560 cm³ **4 a** 39 m³ **b** 2520 cm³ **c** 175 m³
PAGE 106 **1 a** 1600 cm³ **b** 410 cm³ **c** 17 m³ **2 a** 9052 cm³ **b** 1472 cm³ **c** 764.7 m³ **d** 132.7 cm³
PAGE 107 **1 a** 750 cm³ **b** 1200 cm³ **c** 9.2 m³ **d** 13 m³ **e** 1.1 m³ **f** 49 m³ **2 a** 19.09 cm³ **b** 1272.79 cm³ **c** 4.07 m³ **d** 2.24 m³ (or 2 237 750.56 cm³) **3 a i** $V = 6283$ cm³ **ii** $V = 12\,566$ cm³ (has the larger volume) **b** No; the surface area of the first cylinder is 600π cm² while for the second cylinder it is 1200π cm². **4 a** 2 times **b** 4 times
PAGE 108 **1 a** 40π cm³ **b** 40π cm³ **c** 40π cm³ **d** 40π cm³ **2 a** 2460 m³ **b** 2150 cm³ **c** 1.53 m³ **d** 21.3 m³ **e** 452 cm³ **f** 15 000 cm³
PAGE 109 **1 a** 1.98 m³ **b** 1980 L **c** 28 cm **2 a** 22.75 m² **b** 120.5 m² **c** 115.1 m² **d** 18 L **3 a** 0.08 m³ **b** 94
PAGE 110 **1** C **2** C **3** B **4** B **5** D **6** B **7** C **8** B **9** A **10** C
PAGES 111–112 **1 a** 260 cm³ **b** 250 cm² **2 a** 1885 cm³ **b** 1.885 L **c** 911 cm² **3 a** 2.4 m **b** 2.4 m² **c** 7.68 m³ **d** 27.84 m² **4 a** 1809 cm² **b** 86 830 cm³ **5** 6.72 m² **6 a** 16.9 cm **b** 183.6 cm² **c** 661 cm³

CHAPTER 9 – Further surface area and volume

PAGE 113 **1 a** 654 cm² **b** 1734 cm² **c** 4990 cm² **2 a** 660 cm² **b** 1240 cm² **c** 690 cm² **3 a** 428 m² **b** 313.6 m² **c** 1120 m² **4 a** 1099.6 cm² **b** 1193.8 cm² **c** 1281.8 cm²
PAGE 114 **1** 89 mm **2** 23.1 cm **3 a** 13 cm **b** 20 cm **4 a** 8 m **b** 17 m **5** 28 cm **6** 9.6 m, 3 m
PAGE 115 **1 a** 504.0 cm² **b** 236.3 cm² **c** 288.7 cm² **2 a** 1144 cm² **b** 445.1 cm² **c** 855.0 cm²
PAGE 116 **1 a** 301.6 cm² **b** 339.3 cm² **c** 364.0 cm² **d** 1617.2 cm² **2 a** 264π cm² **b** 800π cm² **3** 23 cm **4** 8 cm
PAGE 117 **1 a** 196π cm² **b** 324π cm² **c** 3136π cm² **d** 1764π cm² **e** 275.56π cm² **f** 571.21π cm² **2 a** 576π cm² **b** 1024π cm² **3 a** 942.48 cm² **b** 1847.26 cm² **4 a** 1260 cm² **b** 2890 cm² **c** 1230 cm² **5** 5.35 cm
PAGE 118 **1 a** 614.1 cm³ **b** 461.9 cm³ **c** 663.3 cm³ **2 a** 2167 cm³ **b** 471.0 cm³ **c** 3348 cm³ **3 a** 4477.20 cm³ **b** 336.00 cm³ **c** 2483.02 cm³ **4 a** 1170.0 m³ **b** 2470.0 m³ **c** 77 584.8 cm³
PAGE 119 **1 a** 361.6 cm³ **b** 174.1 cm³ **c** 216.2 cm³ **2 a** 306.05 cm³ **b** 2.33 m³ **c** 776.83 cm³ **3 a** 1230 cm³ **b** 392 cm³ **4 a** 1.9 m³ **b** 326.5 cm³ **c** 396.8 cm³
PAGE 120 **1 a** 121.5 cm³ **b** 223.3 cm³ **c** 1838.6 cm³ **d** 1766.9 cm³ **e** 72804.1 cm³ **f** 37.7 m³ **2 a** 1005.3 cm³ **b** 55.9 cm³ **3 a** 73.45 m³ **b** 86.86 m³ **c** 49.18 m³
PAGE 121 **1 a** 3053.6 cm³ **b** 4188.8 cm³ **c** 113 097.3 cm³ **d** 22 449.3 cm³ **e** 15 002.5 cm³ **f** 91 952.3 cm³ **2 a** 4188.8 cm³ **b** 150 532.6 cm³ **c** 17 157.3 cm³ **d** 102 160.4 cm³ **3 a** 1526.8 cm³ **b** 15 529.7 cm³
PAGE 122 **1** D **2** B **3** C **4** B **5** C **6** A **7** B **8** B **9** B **10** D

Answers

PAGE 123 **1 a** 57.2 cm **b** 13896.4 cm^2 **c** 184 506.1 cm^3 **2 a** 1018 cm^2 **b** 3054 cm^3 **3 a** 360 cm^2 **b** 400 cm^3 **4 a** 302 cm^2 **b** 302 cm^3 **5 a** 7304 cm^2 **b** 45 946 cm^3 **6 a** 20 cm **b** 14 cm **c** 1144 cm^2 **d** 2240 cm^3

CHAPTER 10 – Properties of geometrical figures

PAGE 124 **1 a** $x = 138°$ (angles at a point) **b** $x = 40°$ (straight angle) **c** $t = 35$ (vertically opposite) **d** $a = 70$ (straight angle) **e** $x = 55$ (complementary angles) **f** $x = 70$ (straight angle) **g** $x = 120$ (corresponding angles and parallel lines) **h** $x = 70$ (co-interior angles and parallel lines) **i** $x = 40$ (alternate angles and parallel lines) **j** $x = 50$ (angle sum of triangle) **k** $x = 60$ (equilateral triangle) **l** $x = 139$ (exterior angle of a triangle) **m** $x = 80$ (a quadrilateral) **n** $x = 230$ (a quadrilateral) **o** $x = 36$ (a quadrilateral)
PAGE 125 **1 a** $x = 60$ **b** $x = 90$ **c** $x = 40$ **d** $x = 54$ **e** $x = 60$ **f** $x = 45$ **g** $x = 35$ **h** $x = 47$ **i** $x = 40$ **2 a** $x = 2$ **b** $x = 5$ **c** $x = 9$ **d** $m = 17$ **e** $x = 25$ **f** $x = 20, y = 30$ **g** $a = 12$ **h** $x = 50$ **i** $x = 20, y = 22$
PAGE 126 **1 a** 4 **b** 720° **c** 120° **2 a** 540° **b** 1080° **c** 1800° **3 a** 108° **b** 135° **c** 150° **4** 360° **5 a** 36° **b** 144° **6 a** 130 **b** 30 **c** 51° (nearest degree)
PAGE 127 **1** 35° **2** 45° **3** 30°, 60° **4** 30°, 60°, 90° **5** 13 cm **6** 55° **7** 60° **8** 44° **9** 47° **10** 60°
PAGE 128 **1 a** shape, size **b** equal, length **c** ≡ **2 a** *SP* **b** *SR* **c** *RQ* **d** *PQ* **e** *QS* **f** *PR* **g** $\angle SPR$ **h** $\angle SQP$ **i** $\angle PRQ$ **j** $\angle RSP$ **k** $\angle QPR$ **l** $\angle RQS$ **3 a i** 8 m **ii** 15 m **iii** 17 m **b i** 28° **ii** 62° **iii** 90° **4** ΔABC and ΔMLP; ΔGHI and ΔYZA; ΔMNO and ΔJKL; ΔVWX and ΔPRQ; ΔDEF and ΔSUT
PAGE 129 **1 a** shape, size **b** $\angle A, \angle B, \angle C$ **2 a i** $\angle A$ and $\angle C$; $\angle ADB$ and $\angle CDB$; $\angle ABD$ and $\angle CBD$ **ii** $AD = CD$; $AB = CB$; $BD = BD$ **b i** $\angle E = \angle G, \angle EHF = \angle GFH, \angle EFH = \angle GHF$ **ii** $EH = GF, EF = GH, HF = FH$ **c i** $\angle I = \angle K, \angle ILJ = \angle KJL, \angle IJL = \angle KLJ$ **ii** $IJ = KL, IL = KJ, LJ = JL$ **d i** $\angle P = \angle N, \angle PMO = \angle NOM, \angle POM = \angle NMO$ **ii** $MN = OP, MP = ON, MO = OM$ **e i** $\angle QTS = \angle QTR, \angle S = \angle R, \angle SQT = \angle RQT$ **ii** $QS = QR, QT = QT, ST = RT$ **f i** $\angle U = \angle X, \angle UWV = \angle XVW, \angle UVW = \angle XWV$ **ii** $UV = XW, VW = WV, UW = XV$ **3 a** $\angle F = \angle H$; $\angle FEG = \angle HGE$; $\angle FGE = \angle HEG$ **b** $EF = GH$; $FG = HE$; EG is common **4 a** *KLM* **b** *LJK*
PAGE 130 **1 a** three sides **b** two angles and a corresponding side **c** two sides and the included angle **d** the hypotenuse and one side **2 a** RHS **b** SAS **c** AAS **d** SSS **e** AAS **f** RHS **g** SSS **h** SAS **3 a** *OC* **b** $OA = OB$ **c** Yes **d** RHS
PAGE 131 **1 a** $a = 65$ and $b = 25$ (corresponding $\angle$s of congruent Δs) **b** $x = 20, y = 70$ (corresponding $\angle$s of congruent Δs) **c** $y = 93$ (vertically opposite), $x = 12$ cm (corresponding sides of congruent Δs) **2 a** $x = 90$ **b** $x = 90, y = 45$ **c** $x = 75, y = 40, z = 65$ **3 a** $\Delta ADE \equiv \Delta BCE$ RHS, $x = 42, y = 48$ **b** $\Delta ABD \equiv \Delta CDB$ SAS, $y = 15, m = 63$
PAGE 132 **1 a** ||| **b** two angles **c** same ratio **d** one angle, same ratio **e** hypotenuse, side, right-angled **2 a** alternate angles, parallel lines **b** vertically opposite **c** *DEC* **d** equiangular **e** 2 **3 a** common angle **b** corresponding angles, parallel lines **c** *ABC* **d** equiangular **4 a** $\frac{2}{3}$ **b** $\frac{2}{3}$ **c** 2 sides in proportion, included angle **d** 1.5 **e** $\angle A = \angle D, \angle B = \angle E, \angle C = \angle F$, **f** *AC* and *DF*, *BC* and *EF*, *AB* and *DE*
PAGE 133 **1 a** $x = 6\frac{2}{3}, y = 6$ **b** $x = 73°, y = 58°, z = 49°$ **c** $x = 8, y = 20$ **d** $x = 16, y = 12.5$ **2 a** $x = 15, y = 61$ **b** $x = 26, y = 5$ **c** $y = 9$ **d** $x = 4$
PAGE 134 **1** B **2** D **3** D **4** C **5** B **6** A **7** C **8** D **9** B **10** A
PAGE 135 **1 a** $\Delta OCA \equiv \Delta OCB$ **b** RHS **c** $OA = OB, AC = BC, OC = OC$ **d** $\angle OCA = \angle OCB, \angle OAC = \angle OBC, \angle AOC = \angle BOC$ **2 a** opposite sides of the rectangle are equal **b** both 90°, angles of a rectangle **c** opposite sides of a rectangle are equal **d** SAS **e** corresponding sides of congruent triangles **3 a** *PQR* **b** SSS **4 a** SAS **b** 108° **c** 72° **5** 14.4

CHAPTER 11 – Data analysis

PAGE 136 **1 a** 7 **b** 8 **c** $7.5\dot{4}$ **d** 7 **2 a** 2 **b** 5 **c** 5 **d** 7 **3** 55 kg **4 a**

Score (x)	Frequency (f)	f × x	Cumulative frequency
5	2	10	2
6	6	36	8
7	8	56	16
8	9	72	25
9	7	63	32
10	5	50	37
Total	37	287	

b 7.8 **c** 8 **d** 8 **e** 5 **5 a** 43.45 **b** 58 **c** 45.5 **d** 42 **e** negatively skewed

6 a $\bar{x} = 6.9, \sigma_n = 2.2$ **b** $\bar{x} = 37.6, \sigma_n = 4.4$

c $\bar{x} = 13.6, \sigma_n = 4.4$ **d** $\bar{x} = 20.7, \sigma_n = 8.4$

PAGE 137 **1 a** 27.5 **b** 23 **c** 29 **d** 6 **2 a** 100% **b** 50% **c** 25% **d** 75% **e** 25% **f** 50% **3 a** 11 **b** 25 **c** 17 **d** 14.5 **e** 20.5 **f** 6 **4 a i** 16 **ii** 11 **iii** 19 **iv** 8 **b i** 37 **ii** 36 **iii** 39 **iv** 3 **c i** 65 **ii** 52.5 **iii** 72 **iv** 19.5 **d i** 11 **ii** 10 **iii** 13 **iv** 3
PAGE 138 **1 a** 18 **b** 2.6 **2 a** stay the same, the score being added is equal to the mean **b** decrease, the spread of all the scores relative to the mean has decreased **3 a** decrease, the score being added is less than the mean **b** increase, the spread of scores relative to the mean has increased **4 a** 23 **b** 2.6 **5 a** All the scores have increased by 5. **b** The mean has increased by 5 because all the scores have increased by 5. **c** Standard deviation does not change because the spread of marks relative to the mean does not change. **6 a** 36 **b** 5.3 **7 a** The scores have doubled. **b** The mean has doubled because all the scores have doubled. **c** The standard deviation has doubled because the spread of marks relative to the mean has doubled.

Answers

Page 139 **1 a i** 6 **ii** 4.5 **iii** 7 **iv** 2.5 **b i** 18 **ii** 18 **iii** 20 **iv** 2.0 **2 a** positively skewed **b** The box plot will be positively skewed, the median will be closer to the lower quartile than upper and closer to the lower extreme than upper extreme. **3 a** technical **b** negatively skewed **c i** [8, 29, 41.5, 50.5, 56] **ii** [5, 17, 28.5, 42.5, 54] **d**

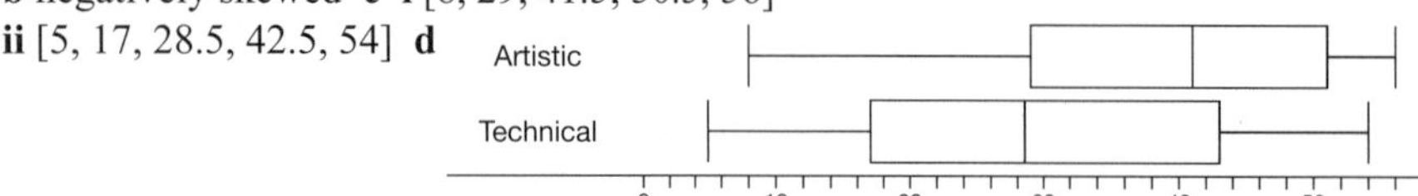

e Stem-and-leaf plot shows all the scores and shape. Shape can also be seen with the box plot as well as measures of spread.

Page 140 **1 a** 9 **b** 5 **c** 2.8 **d** The range gives the total spread of all the scores; the interquartile range gives the spread of the middle 50% of scores. The standard deviation shows which scores are most consistent. **2 a** 32 **b** 6 **c** 7.6 **d** range **e** interquartile range **f** The range is affected the most because it is the difference between the highest and lowest score. The interquartile range is affected the least because it is concerned with the middle 50% of scores. **3 a i** 43 **ii** 46 **b i** 23.5 **ii** 17 **c i** 13.5 **ii** 12.6 **d** The theory marks have the slightly higher range but lower interquartile range and standard deviation. This is reflected in the shape of the distribution where the theory marks have a more even distribution about the mean.

Page 141 **1 a** one **b** two **2 a** association **b** causal relationship **c** association **3 a** independent is blood alcohol level, dependent is time **b** independent is amount of coffee, dependent is amount of sleep **4 a** no relationship **b** strong positive **c** strong negative **d** weak negative **e** weak positive **f** no relationship

Page 142 **1 a** 25 **b** 48% **c** $9\frac{1}{2}$ **d** 32% **e** 85% **f** 1 **2 a**

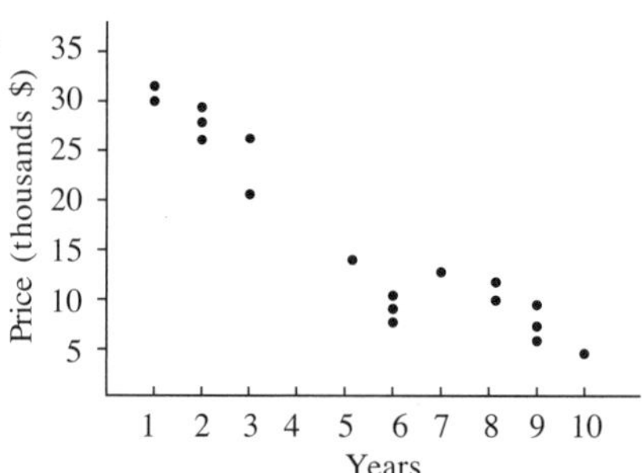

b Around $18 000 (using the pattern from the scatter plot)

Page 143 **1 a**

b As the hours of study increase, the marks also increase. The relationship is not linear.

2 a

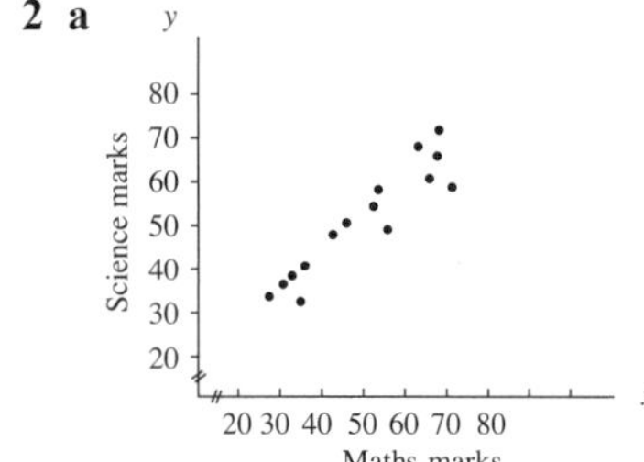

b 65. Most students have science marks that are close to their maths marks.

Page 144 **1 a and b**
c $y = x + 10$

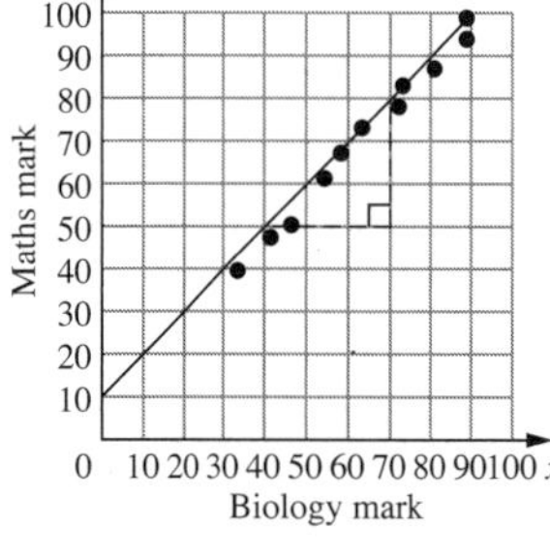

2 a and b
c i 100 **ii** 25
d $y = x + 5$

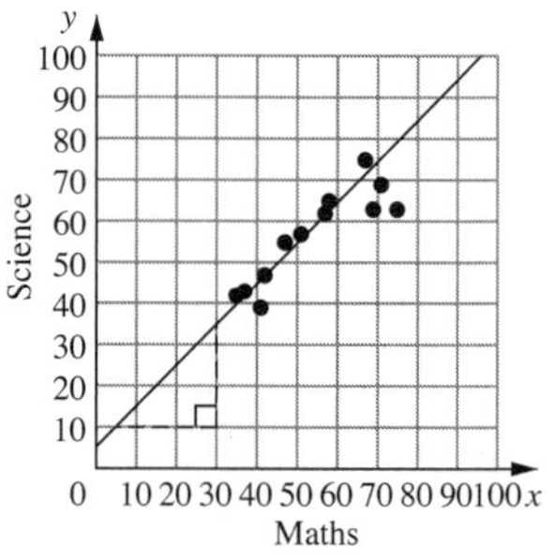

Note: Slightly different lines of fit, with slightly different equations, are acceptable.

Answers

PAGE 145 **1 a** $h = 10n$ **b i** 85 cm **ii** 130 cm **c i** $5\frac{1}{2}$ weeks **ii** 16 weeks **2 a** $h = 7n + 20$ **b i** 79.5 cm **ii** 111 cm **c i** 5 weeks **ii** 20 weeks **d** There is not a lot of difference for small values of n but as n gets larger, the difference between the predictions increases. **3 a** $V = 25\,000 - 2000a$ **b** \$15 000 **c** 10 years **d** After $12\frac{1}{2}$ years the equation will give a value that is negative.

PAGE 146 **1 a** census **b** sample **c** random sample **d** systematic sample **e** stratified sample **2** No, it is not a random sample. Every office worker does not have an equal chance of being selected. For example, all those selected might have been from the same floor or the same office. **3 a** systematic **b** every item **c** More items could be selected, e.g. every tenth one, or items could be randomly selected to check. **4 a** 9 **b** 40

PAGE 147 **1 a** The scale on the vertical axis is not even. It goes up by 5 at first and then by 2. **b** The scale on the vertical axis does not begin at 0 (or a gap shown to exist). **c** The dots are not the same size. **2 a** No. Babies don't drink alcohol **b** 'On average, Australians each drink ...' **c** You need to consider the source of the information and whether statements can be verified. **d** It immediately suggested that the statement could not be true. **3 a** not important **b** How much information did the viewers have? Was it a biased report? How many people responded to the survey? Was it a good sample of the population? **4 a** to establish the credibility of the product **b** Yes, if the chemists were biased, e.g. they were all employed by the company that produced the product.

PAGE 148 **Suggested answers: 1** No, we don't have any information about the sample so we cannot tell whether it is representative of all students. **2** the size of the study, the age and sex of the participants, where it was conducted, who paid for it **3** They were trying to introduce doubt into voters' minds by implying that the politician might change her mind. **4** The results might be biased because of the content of the program shown. Viewers of that program might not be representative of the general public. **5** The previous driving record of the drivers and the number of previous claims made **6** the need to provide facilities for young families such as child care and schools

PAGE 149 **1** B **2** B **3** A **4** B **5** B **6** A **7** D **8** C **9** C **10** C

PAGE 150 **1 a** 51 **b** 72 **c** 23 **d** the individual scores; the mode, mean and standard deviation **e** the spread of scores, the five-number summary and the range and interquartile range; the shape of the distribution

2 a

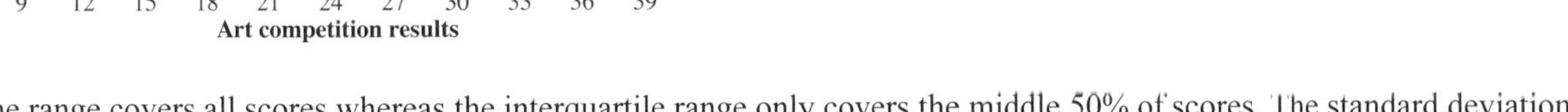

b 25 **c** 5 **d** 4.9

e The range covers all scores whereas the interquartile range only covers the middle 50% of scores. The standard deviation measures the spread from the mean. **3** Different lines with different equations are possible. **a**

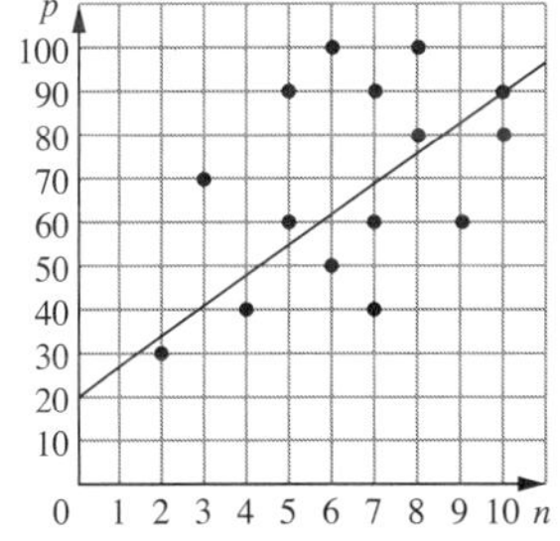

b $p = 7n + 20$ **c** \$139 **d** 40 toys

4 Probably not, as this sample is too small to be significant. There is no mention also of the 'recommended' daily fruit amount in terms of quantity nor how many students are close to this amount.

CHAPTER 12 – Probability

PAGE 151 **1 a** 1 **b** 4 **c** unlikely **d** impossible **e** likely **f** certain **2 a** $\frac{1}{6}$ **b** $\frac{1}{2}$ **c** $\frac{2}{3}$ **3 a** $\frac{3}{10}$ **b** $\frac{1}{5}$ **c** $\frac{1}{2}$ **d** 0 **e** $\frac{7}{10}$ **f** $\frac{4}{5}$ **4 a** $\frac{1}{52}$ **b** $\frac{1}{13}$ **c** $\frac{1}{2}$ **d** $\frac{1}{4}$ **e** $\frac{1}{26}$ **f** $\frac{3}{4}$ **5 a** $\frac{2}{5}$ **b** $\frac{1}{4}$ **c** $\frac{3}{4}$ **d** $\frac{13}{20}$ **e** $\frac{13}{20}$ **f** $\frac{1}{4}$ **6 a** $\frac{1}{26}$ **b** $\frac{1}{13}$ **c** $\frac{25}{26}$ **d** $\frac{3}{26}$ **e** $\frac{5}{26}$ **f** $\frac{21}{26}$

PAGE 152 **1 a** $\frac{1}{8}$ **b** $\frac{3}{8}$ **c** $\frac{7}{8}$ **2 a** 12 **b** $\frac{2}{3}$ **c** $\frac{1}{3}$ **d** $\frac{1}{2}$ **3 a** $\frac{3}{10}$ **b** $\frac{1}{10}$ **c** $\frac{3}{5}$ **4 a** $\frac{1}{8}$ **b** $\frac{3}{8}$ **c** $\frac{3}{8}$ **d** $\frac{1}{2}$ **e** $\frac{1}{2}$ **f** $\frac{1}{8}$

PAGE 153 **1 a**

–	1	2	3	4	5	6
1	0	1	2	3	4	5
2	1	0	1	2	3	4
3	2	1	0	1	2	3
4	3	2	1	0	1	2
5	4	3	2	1	0	1
6	5	4	3	2	1	0

b $\frac{1}{6}$ **c** 0 **d** $\frac{5}{6}$ **2** rolling one dice

Answers

3 a 123, 132, 213, 231, 312, 321 **b** $\frac{1}{6}$ **c** $\frac{1}{3}$ **d** $\frac{1}{3}$ **e** $\frac{2}{3}$ **f** $\frac{1}{2}$ **4 a**

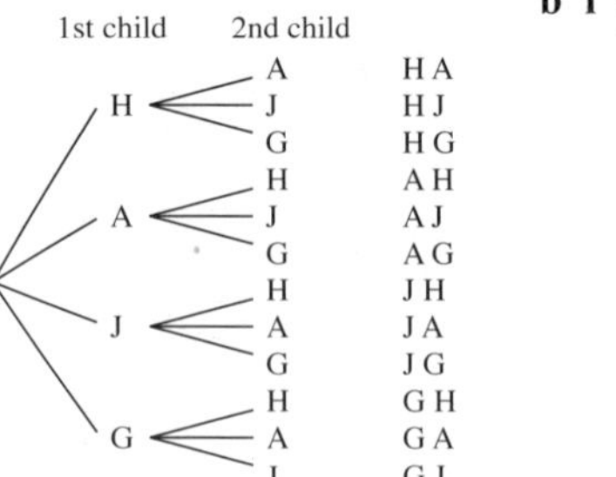

b i $\frac{1}{2}$ **ii** $\frac{1}{3}$

PAGE 154 **1 a** 90 **b** 58 **c** 36 **d** $\frac{7}{30}$ **e** $\frac{73}{90}$ **2 a**

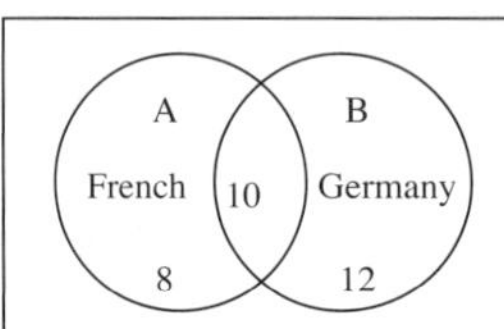

b 10 **c** $\frac{4}{15}$

3 a 14 **b** 1 **c** 22 **d** 18 **e** 3 **f** 7 **g** 23 **h** 25 **4 a**

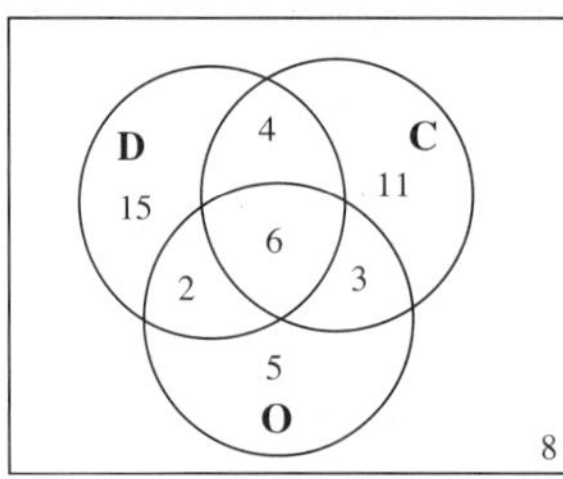

b $\frac{5}{12}$ **c** $\frac{10}{27}$ **d** $\frac{2}{27}$

PAGE 155 **1 a** men 110; women 100; drivers 128; passengers 82; total 210 **b** 44 **c** 110 **d** $\frac{36}{55}$ **e** $\frac{9}{16}$ **f** $\frac{11}{25}$ **g** $\frac{22}{41}$ **h** $\frac{12}{35}$ **i** $\frac{22}{105}$

2 a

	Plays football	Does not play football	Total
Plays cricket	19	28	47
Does not play cricket	31	22	53
Total	50	50	100

b $\frac{1}{2}$ **c** $\frac{19}{50}$ **d** $\frac{19}{47}$ **e** $\frac{11}{50}$ **f** $\frac{19}{100}$ **g** $\frac{22}{53}$ **h** $\frac{14}{25}$

3

	A	Not A	Total
B	37	41	78
Not B	24	18	42
	61	59	120

A 24 37 B 41 18

4

	X	Not X	Total
Y	19	28	47
Not Y	36	17	53
	55	45	100

PAGE 156 **1 a** union, everything in A together with everything in B **b** intersection, everything that is in both A and B **c** complement, everything not in A **2 a** {2, 4, 6, 8, 10} **b** {3, 6, 9} **c** {2, 3, 5, 7} **d** {1, 3, 5, 7, 9} **e** {1, 2, 4, 5, 7, 8, 10} **f** {1, 4, 6, 8, 9, 10} **g** {2, 3, 4, 6, 8, 9, 10} **h** {2, 3, 5, 6, 7, 9} **i** {2, 3, 4, 5, 6, 7, 8, 10} **j** {6} **k** {3} **l** {2} **m** {1, 3, 5, 6, 7, 9} **n** {6, 9} **o** {2, 3, 4, 5, 6, 7, 8, 9, 10} **p** {2, 6} **q** {2, 3, 5, 6, 7} **r** {3} **3 a**

A B C

b

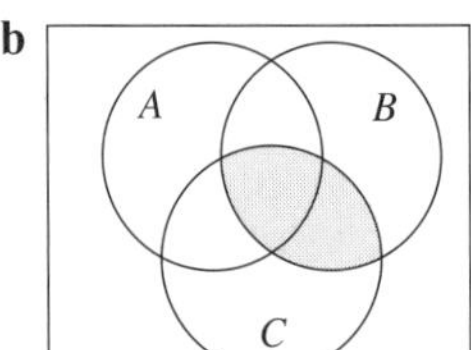

c

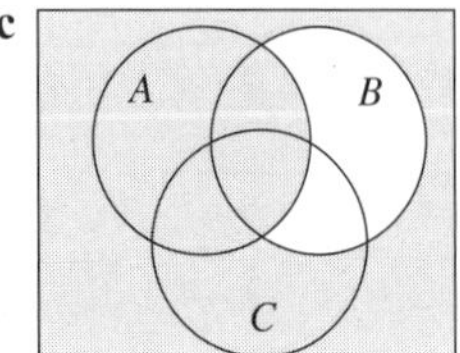

d

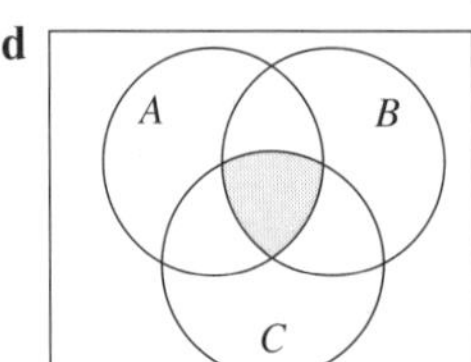

e

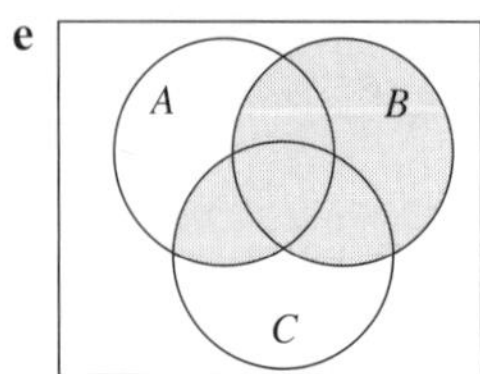

f

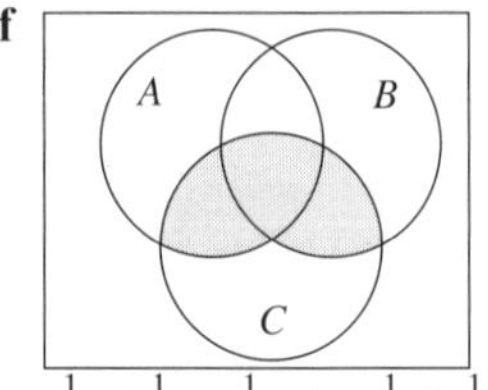

PAGE 157 **1** not affect **2 a** independent **b** independent **c** dependent **d** independent **e** dependent **3 a** $\frac{1}{2}$ **b** $\frac{1}{6}$ **c** $\frac{1}{12}$ **4 a** $\frac{1}{2}$ **b** $\frac{1}{2}$ **c** $\frac{1}{2}$ **d** $\frac{1}{8}$ **5 a** $\frac{1}{2}$ **b** $\frac{1}{2}$ **c** $\frac{1}{4}$ **d** $\frac{9}{100}$ **e** $\frac{1}{25}$ **f** $\frac{3}{10}$ **g** $\frac{16}{25}$ **h** $\frac{9}{25}$ **6 a** $\frac{1}{8}$ **b** $\frac{1}{27}$ **c** $\frac{1}{216}$ **d** $\frac{8}{27}$ **e** $\frac{19}{27}$ **f** $\frac{91}{216}$

PAGE 158 **1** affects **2 a** independent **b** independent **c** dependent **d** dependent **e** independent **3 a** $\frac{1}{200}$ **b** $\frac{5}{999}$ **c** $\frac{4}{999}$ **d** $\frac{1}{49\,950}$ **4 a** $\frac{1}{2}$ **b** $\frac{4}{9}$ **c** $\frac{2}{9}$ **d** $\frac{1}{15}$ **e** $\frac{1}{45}$ **f** $\frac{1}{3}$ **g** $\frac{28}{45}$ **h** $\frac{17}{45}$ **5 a** $\frac{1}{11}$ **b** $\frac{1}{55}$ **c** 0 **d** $\frac{14}{55}$ **e** $\frac{41}{55}$ **f** $\frac{5}{11}$

Answers

PAGE 159 **1 a** $\frac{1}{12}$ **b** $\frac{7}{12}$ **c** $\frac{1}{3}$ **d** $\frac{1}{4}$ **2 a** $\frac{1}{12}$ **b** $\frac{1}{2}$ **c** $\frac{1}{2}$ **d** $\frac{5}{12}$ **e** $\frac{7}{36}$ **f** $\frac{1}{6}$ **3** $\frac{1}{2}$ **4 a** 6 **b** $\frac{1}{2}$ **c** $\frac{1}{3}$ **5 a** $\frac{1}{8}$ **b** $\frac{3}{8}$ **c** $\frac{1}{2}$ **d** $\frac{1}{2}$

PAGE 160 **1 a** *AB AC AD BA BC BD CA CB CD DA DB DC* **b** $\frac{1}{2}$ **2 a** $\frac{1}{3}$ **b** $\frac{1}{27}$ **c** $\frac{4}{27}$ **d** $\frac{4}{9}$ **3 a** $\frac{1}{8}$ **b** $\frac{1}{8}$ **c** $\frac{1}{4}$ **d** $\frac{3}{8}$ **4** 0.08 **5** $\frac{1}{256}$ **6** 0.343 **7 a** $\frac{3}{20}$ **b** $\frac{51}{380}$ **c** $\frac{3}{190}$ **d** $\frac{68}{95}$ **e** $\frac{27}{95}$ **f** $\frac{51}{190}$

PAGE 161 **1 a** $\frac{1}{6}$ **b** $\frac{1}{3}$ **c** $\frac{1}{2}$ **d** 1 **2 a** $\frac{1}{3}$ **b** $\frac{1}{12}$ **c** $\frac{1}{6}$ **3 a** $\frac{2}{3}$ **b** $\frac{1}{3}$ **c** $\frac{2}{3}$ **d** $\frac{1}{2}$ **e** $\frac{1}{2}$ **f** $\frac{1}{2}$ **4 a** $\frac{25}{144}$ **b** $\frac{1}{16}$ **c** $\frac{7}{16}$ **d** $\frac{5}{33}$ **e** $\frac{1}{22}$ **f** $\frac{5}{11}$ **g** $\frac{20}{99}$ **h** 0 **i** $\frac{3}{11}$ **5** $\frac{1}{18}$

PAGE 162 **1 a** no **b** The events are not equally likely. **2 a** no **b** The events are independent. **3 a** yes **b** There are 26 letters and x is 1 of those letters and each letter is equally likely to be drawn. **4 a** no **b** Each letter of the alphabet will not appear the same number of times on the page so the events are not equally likely. **5 a** no **b** The events are independent and the coin is fair. **6 a** 27.1% **b** Bill found the probability of rain on every day.

PAGE 163 **1** C **2** B **3** D **4** C **5** D **6** A **7** C **8** C **9** C **10** B

PAGE 164 **1 a** $\frac{1}{26}$ **b** $\frac{1}{104}$ **2 a** $\frac{7}{30}$ **b** $\frac{7}{30}$ **c** $\frac{1}{15}$ **d** $\frac{7}{15}$ **e** $\frac{7}{15}$ **f** $\frac{14}{15}$ **3 a** $\frac{1}{8}$ **b** $\frac{7}{8}$ **c** $\frac{1}{4}$ **d** $\frac{3}{4}$ **4 a** $\frac{1}{216}$ **b** $\frac{1}{27}$ **c** At each toss the probability of getting a 6 is twice what it was before but when these are multiplied together it is 8 times greater.

Exam Paper 1

PAGE 165 **Part A** **1** D **2** A **3** D **4** B **5** A **6** C **7** D **8** D **9** C **10** D **11** D **12** A **13** B **14** A **15** B **16** D **17** C **18** D **19** C **20** C **21** C **22** D **23** D **24** D **25** A **26** B **27** A **28** D **29** B **30** A **31** D **32** B **33** D **34** C **35** A **36** D **37** D **38** C **39** A **40** A **41** D **42** D
43 B **44** C **45** C **46** A **47** B **48** C **49** D **50** D

PAGE 169 **Part B** **1 a** \$1260 **b** \$8337 **c** \$77 **2 a** $\frac{1}{22}$ **b** $\frac{1}{16}$ **3 a** $\frac{5x}{6}$ **b** $\frac{-a}{15}$ **4 a** 10 240 **b** 40 000 **5 a** 67.2 cm **b** 92.8 cm **c** 12 813 cm^2 **6 a** $2x^3y^2(y^2 - 3x)$ **b** $(x - 3)(x - 1)$ **c** $(a + 9)(a - 2)$ **d** $(x + 6)(x - 6)$ **e** $(m + 5)(m - n)$ **7 a** S64°W **b** 90° **c** 29° **d** 215° **e** 64 km **8 a** $x = 1$ **b** $x = \pm 5$ **9** $x = 5, y = -2$ **10 a** $x \leq 2$ **b**

11 a –2 **b** –2 **c** $\frac{1}{2}$ **d**

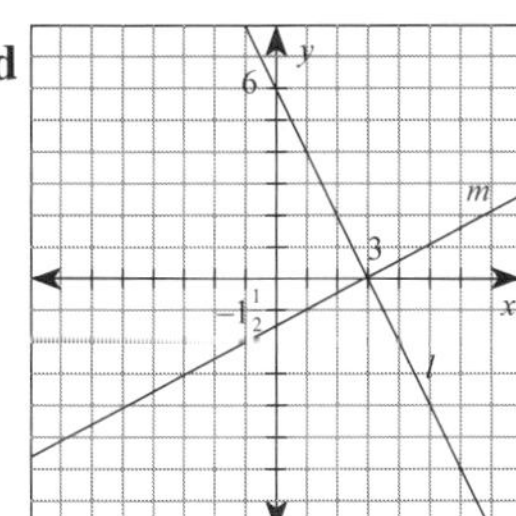

e $y = \frac{1}{2}x - \frac{3}{2}$ **f** 11.25 units^2 **12 a** $a^2 + 9a + 20$ **b** $6x^2 + 7x - 5$

13 a 0.06 m^2 **b** 160 L **14 a** $AB = AD$ (side of a square) and $AP = RD$ (given) **b** SAS **c** 90° **d** 45° **15 a** 12 million **b** 1975 **c** 7.2 million **d** about 21 million **e** exponential **16 a** 6 **b** 3 **c** **d** both negatively skewed

Exam Paper 2

PAGE 174 **Part A** **1** B **2** C **3** B **4** B **5** A **6** B **7** D **8** D **9** B **10** A **11** A **12** B **13** B **14** A **15** C **16** C **17** C **18** C **19** A **20** B **21** C **22** D **23** B **24** A **25** C **26** D **27** B **28** D **29** C **30** A **31** D **32** B **33** C **34** A **35** C **36** C **37** B **38** D **39** B **40** C **41** A **42** A **43** D **44** D **45** C **46** C **47** D **48** C **49** B **50** D

PAGE 178 **Part B** **1 a** ΔPST and ΔPQR **b** equiangular **c** 21 **2 a** 15 cm **b** trapezuim **c** 162 cm^2 **d** 1296 cm^3 **e** 804 cm^2
3 a $4, \frac{1}{2}, -2, -3\frac{1}{2}, -4, -3\frac{1}{2}, -2, \frac{1}{2}, 4$ **b and c**

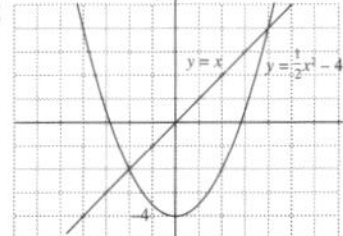

d 2 **e** –2 and 4 **4 a** $\frac{19x}{24}$ **b** $\frac{x^2}{10}$ **5 a** $2x^2 + 8x - 2$ **b** $12x^3y^5 + 15x^5y^4$ **6 a** $2a^2b^3(3ab - 1)$ **b** $(x + 10)(x - 2)$ **c** $(p + q)(p + r)$ **d** $6(x + 1)(x - 1)$ **7 a** \$490.91 **b** \$10.91 **8 a** $a = -19$ **b** $x = 180$ **c** $x \geq -2$ **d** $x > 1$ **e** $x = \pm 9$ **f** $x = 2$ or 4 **9 a** $a = 4, b = 3$ **b** $x = 3, y = 1$ **10 a** SSS **b** corresponding angles of congruent triangles **c** SAS **d** 90° **e** The diagonals of a kite meet at right-angles
11 a i positive **ii** strong **b** 7 **c** 9 **d** 7, average of marks of others who scored 7 in arithmetic **12 a** 3 in arithmetic **b** median **c** both plots are fairly symmetrical. Both the range and interquartile range are greater for arithmetic **13 a** 9 cm **b** 729 cm^3 **14 a** 0.64 **b** 0.36 **15 a** 16.3 m **b** 51.3 m

Notes

Notes

Notes